Physiology

Physiology

An Illustrated Review with Questions and Explanations

Second Edition

Charles H. Tadlock, M.D.
Chief of Anesthesiology
Mendocino Coast District Hospital
Medical Director
Mendocino Pain Center
Fort Bragg, California

with
Leslie C. Andes, M.D.
Attending Staff
Department of Anesthesiology
Marin General Hospital
Greenbrae, California

Douglas R. Rank
Jason Goldsmith
Pauline Terebuh
Jeffrey Ustin
Stanford University School of Medicine
Stanford, California

Foreword by
Roy H. Maffly, M.D.
Dean of Medical Education
Stanford University School of Medicine
Stanford, California

Little, Brown and Company
Boston New York Toronto London

Second Edition

Previous edition titled *Physiology: A Review with Questions and Explanations* copyright © 1987 by Benjamin Hsu, Charles H. Tadlock, Saied Assef, and Jack M. Percelay

Library of Congress Cataloging-in-Publication Data

Tadlock, Charles H.
Physiology : an illustrated review with questions and explanations
Charles H. Tadlock ; with Leslie C. Andes . . . [et al.]. — 2nd ed.
p. cm.
Includes bibliographical references and index.
ISBN 0-316-82764-9
1. Human physiology — Examinations, questions, etc. I. Andes, Leslie C.
QP40.T33 1995 95-10424
612'.0076 — dc20 CIP

Printed in the United States of America

MV-NY

The figures on pages123 and 128 are reprinted by permission of the *New England Journal of Medicine* (296; 26, 1977). The figure on page 271 is reprinted from A. C. Guyton. *Textbook of Medical Physiology* (8th ed.). Philadelphia: W.B. Saunders Company, 1991.

Editorial: Evan R. Schnittman, Kristin Odmark
Production Editor: Anne Holm
Copyeditor: Libby Dabrowski
Indexer: Dorothy Hoffman
Production Supervisor: Louis C. Bruno, Jr.
Cover Designer: Michael A. Granger
Cover Art: Todd Buck

Contents

Foreword

In the foreword for the first edition of *Physiology: An Illustrated Review with Questions and Explanations,* I emphasized that as an internist and nephrologist, I have an abiding love and appreciation for the discipline of physiology. The principles of physiology have provided the critical foundation on which I have based the management of my patients, and I have been privileged to have had the joy of teaching the subject of physiology to my students for many years. I expressed pride that the new book had been written by students who had been turned on by the subject I had taught them.

Such an appreciation shines through in this second edition, for which I again congratulate (the now) Dr. Charles Tadlock and his coauthors. Like the remarkably successful first edition, this updated edition is aimed directly at medical students seeking a firm understanding of physiologic principles. Because it is written either directly by, or with the aid of, medical students, the book carries a "guarantee" of relevance. The writing style is lucid, the material chosen for coverage is that which is important, and the authors have strived to distinguish what is significant from what may sometimes appear to be a confusing array of facts. The result, as was the case for the first edition, is an excellent blend of quality of content and practicality in utilization. I commend the authors for their superb accomplishment.

Roy H. Maffly, M.D.

Preface

This text is a concise review of human physiology intended for medical and allied health students seeking a review book that is manageable in size and covers all pertinent material. Trivial, redundant, or confusing experimental data were eliminated in favor of discussions that elucidate the underlying physiology. While brevity and comprehensiveness were always paramount, a major effort was made to make *Physiology: An Illustrated Review with Questions and Explanations,* Second Edition, as user-friendly as possible. In this, I believe the second edition continues the excellent tradition begun with Little, Brown's *Biochemistry: An Illustrated Review with Questions and Explanations,* Fifth Edition, by Paul Friedman, and by the authors of the first edition of this physiology text.

Every chapter was revised to maintain the high standards set by the first edition and to cover in detail the content and new format of the United States Medical Licensure Examination (Step 1). Questions and answers were completely rewritten to reflect the new changes and to make the learning process as efficient as possible.

This edition introduces a new group of chapter authors. Four of the five new authors are Stanford medical students and their perspective has been crucial in keeping the tight focus of the first edition. My thanks to Jason Goldsmith, Pauline Terebuh, Douglas Rank, and Jeffrey Ustin. Leslie Andes, M.D., took responsibility for revising the neurophysiology chapter and did a great job.

Numerous friends and faculty members have reviewed chapters and provided invaluable suggestions for the second edition. Roy Maffly, M.D., has again been kind enough to write the Foreword and give aid and assistance to me in my work on the renal chapter. Eugene Debs Robin, M.D., deserves accolades for the patience he always exhibited in explaining pulmonary and acid-base physiology. Ron Pearl, M.D., reviewed the revised cardiovascular physiology chapter.

Numerous students at Stanford as well as readers of the first edition have contributed useful constructive criticism. Thank you all. Kristin Odmark of Little, Brown provided support, patience, and the occasional necessary spur that made the second edition possible.

C. H. T.

Notice

1

General and Cellular Physiology

Douglas R. Rank

Energy Transformation

Guyton describes physiology as the attempt "to explain the physical and chemical factors that are responsible for the origin, development, and progression of life." At the heart of Guyton's description is the process of **energy transformation,** the conversion of energy from one form to another. For example, energy stored in the form of chemical bonds in food may be transformed into the energy of phosphodiester bonds within adenosine triphosphate (ATP) molecules and subsequently may be transformed by muscle cells into kinetic energy and heat.

Energy is required for all life processes: reproduction, growth, development, movement, communication, sensation, thought, and maintenance of **homeostasis,** the state of maintaining one's internal environment within the narrow limits required for life. This can be conceptualized if one remembers the second law of thermodynamics: Without energy input, disorder (**entropy**) tends to increase. Therefore, energy is required simply to maintain the complex arrangement of molecules, cellular membranes, cells, and tissues that comprise the human body.

Cellular Membranes

Cellular membranes are essential components of all living cells. They are the barrier between the cellular processes of life and the external environment as well as the barrier that separates compartments within eukaryotic cells from one another. As barriers, membranes allow different compartments to contain unique compositions of molecules, thereby conferring specific functions vital for life.

Fluid Mosaic Model

Membranes are composed primarily of a **phospholipid bilayer** in which the hydrophilic heads are on the surface and the hydrophobic tails are facing within (Fig. 1-1A). This structure naturally forms in water solutions because of the polar nature of the water molecule.

Cholesterol and **proteins** are embedded within the phospholipid bilayer (Fig. 1-1B) such that the hydrophobic portions of globular and helical proteins are in contact with the hydrophobic tails of the phospholipids, and the charged portions of proteins are on the surface of the bilayer or line the inside of a membrane pore. Cellular membranes are said to be "fluid" since the phospholipid molecules freely move within the layer of the membrane in which they are found. Motion within a single layer of the

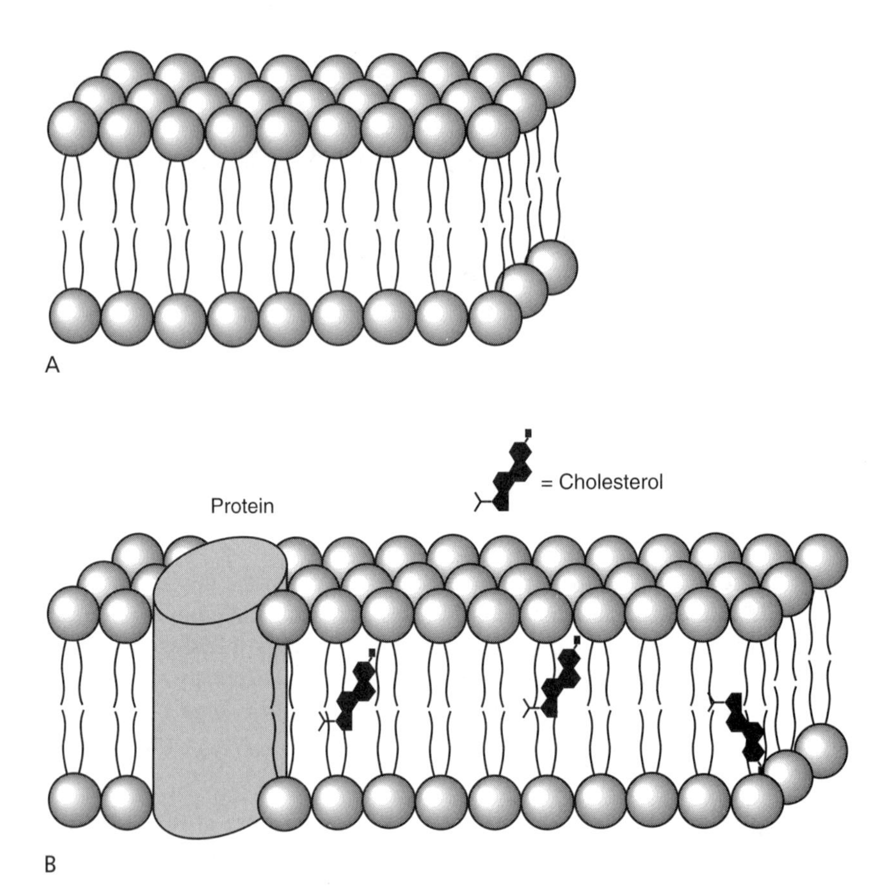

Fig.1-1. A. Phospholipid bilayer showing the lipid molecules' hydrophilic heads on the surface and internal hydrophobic tails. B. Cholesterol and proteins embedded within a phospholipid bilayer.

bilayer is relatively uninhibited; however, flipping of phospholipids between layers of the bilayer requires a great deal of molecular kinetic energy and therefore occurs at a much slower rate. Cholesterol and nonsaturated fatty acids serve to increase membrane fluidity by disrupting the internal matrix of the phospholipid bilayer.

Membrane Functions

Membranes function to separate cellular compartments from one another or from the external environment. Membranes allow cellular functions to occur and be regulated in distinct compartments within a eukaryotic cell. This quality is termed **specialization.** For instance, the nucleus of a eukaryotic cell contains all of the cellular DNA, the proteins responsible for the condensation of the DNA, the enzymes necessary for replication, and transcription of DNA as well as many other factors. Another example of a specialized membrane-bound compartment is the **lysosome,** a membrane-bound vesicle that contains hydrolytic enzymes. These enzymes are only active in an acidic environment. Fortunately, this acidic environment exists within the lysosome, but not within the cytoplasm of the cell; therefore, leakage of small amounts of these enzymes does not cause widespread cellular destruction.

Furthermore, membranes interact with each other in complex ways. For example, bacteria may be engulfed by certain eukaryotic cells such as neutrophils and macrophages, a process termed **phagocytosis** (or more generally termed endocytosis, a process that is discussed later in this chapter). When bacteria contact the outer membrane of these cells, a small portion of the cellular membrane buds off internally to become a membrane-bound vesicle called a **phagosome.** The phagosome containing the bacteria then fuses with a lysosome within the cell, creating a phagolysosome by a poorly understood process. Within the phagolysosome, the lysosomal enzymes are able to hydrolyze the bacterial constituents, thereby killing the bacteria.

Transport

The transport of materials between body or cellular compartments can be divided into **passive** and **active** processes. Active processes require energy expenditure on the part of the organism, whereas passive processes do not. These are summarized in Fig. 1-2.

Passive Transport

Diffusion is the net movement of particles from an area of high concentration to an area of low concentration due to the random molecular movement of the particles.

Simple diffusion of a neutral particle through a barrier such as the cell membrane is described by **Fick's law:**

Fig.1-2. A. Simple diffusion directly through the bilayer and via protein channels. B. Facilitated diffusion. C. Active transport—the NA$^+$-K$^+$ ATPase. D. Cotransport via a symport protein. ADP = adenosine diphosphate; Gl = glucose in the presence of insulin. (From A. C. Guyton. *Textbook of Medical Physiology* [8th ed.]. Philadelphia: Saunders, 1991. Pp. 39, 47, 48.)

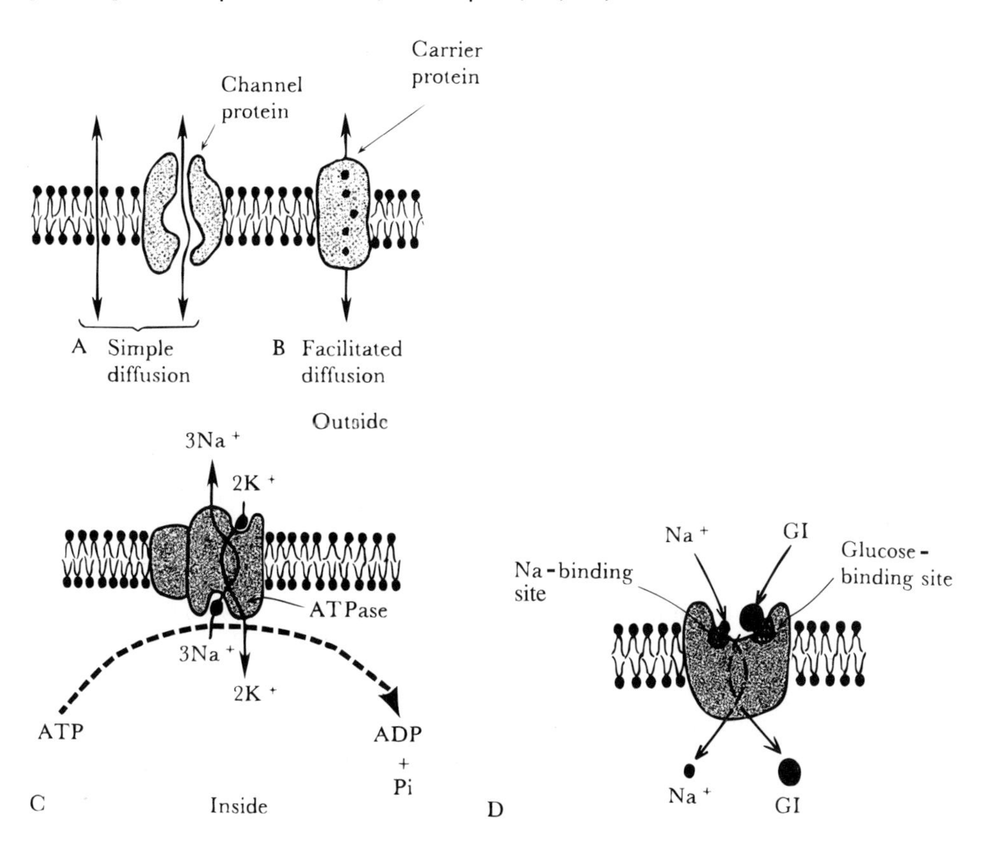

$$\text{Flux} = \frac{pA}{d}(C_1 - C_2), \text{ or more generally, } \quad \text{flux} = \frac{1}{(\text{resistance})}(\text{driving force})$$

Flux is the amount of material moved, p is the permeability of the barrier to the material, A is the cross-sectional area of the barrier, C_1 and C_2 are the concentrations of the material in the respective two regions, and d is the thickness of the barrier.

The high lipid solubility of such small **neutral molecules** as O_2, N_2, N_2O, the alcohols, and urea enables them to pass directly through the interstices of the lipid bilayer (Fig. 1-2A). Some **weak acids and bases** are able to cross cell membranes in their nonionic form. A different environment (i.e., pH) on the other side of the membrane may ionize the molecule, trapping it and resulting in a net movement of particles. **Ions,** because of their charge, are much less able to pass through the small hydrophobic pores of the lipid bilayer but can pass through the membrane via channels created by protein molecules (Fig. 1-2A). These protein channels are selectively permeable to specific compounds based on channel characteristics such as shape, diameter, and charge. Diffusion of charged particles is affected by the electrical gradient in addition to the concentration gradient across the barrier. This relationship is described by the Nernst equation, which is discussed later in this chapter.

Facilitated or carrier-mediated diffusion differs from simple diffusion in that the transported molecule actually binds to the carrier. The carrier then undergoes a conformational change to allow the molecule to pass through to the other side of the cell membrane (Fig. 1-2B). This intermediate complex is analogous to the enzyme-substrate complex of Michaelis-Menten kinetics. Similarly, carrier-mediated diffusion can be **saturated** (i.e., reach the limit of transport per unit time, $\mathbf{V_{max}}$), after which addition of substrate cannot affect rate of transport because all carriers are in use. In contrast, simple diffusion does not saturate and continues to increase in rate with increasing substrate concentration.

Osmosis can be viewed as the passive diffusion of water, or more generally any solvent, down its concentration gradient across a selectively permeable membrane. Thus, one can regard osmosis as the tendency of water to move from a dilute solution to a more concentrated solution. Water will diffuse from a 1-mM salt solution into a 10-mM solution when the two solutions are separated by a membrane permeable to water but not permeable to the salt.

An osmotic pressure gradient is the result of differing water concentrations in two adjacent regions. **Osmotic pressure** is determined by the number of osmotically active particles in solution and is effectively independent of particle type. A solution of 1 M glucose, therefore, has an osmolarity of 1, but a solution of 1 M NaCl has an osmolarity of approximately 2 because in solution NaCl exists as separate Na^+ and Cl^- ions.

Active Transport

Active transport describes any process that uses energy to move substances across cellular membranes. Energy may be required to move a substance against its electrochemical gradient or to move large substances into or out of cells. Such transport usually depends on **coupling** ATP hydrolysis to movement across the cell membrane. As an enzymatic process, active transport, like facilitated diffusion, can be saturated.

Primary active transport describes the process of directly using the hydrolysis of ATP to move an ion against its electrochemical gradient. One example of primary active transport is the sodium-potassium pump—the **NA^+-K^+ ATPase.** It pumps three Na^+ out of the cell and two K^+ into the cell for each ATP consumed (Fig. 1-2C). Note that both ions are pumped against their concentration gradients.

Secondary active transport, sometimes called cotransport, uses the energy of the Na^+ concentration gradient created by the Na^+-K^+ ATPase to drive the transport of other substances against their concentration gradients. The sodium moves into the cell and provides the energy to move another substance against its concentration gradient, either into the cell via a **symport protein** or out of the cell via an **antiport protein.** Symport proteins are used in the mucosa of the small intestine to couple Na^+ influx to glucose and amino acid uptake (Fig. 1-2D). The Na^+–Ca^{2+} exchange protein of cardiac muscle is an important antiport protein.

Endocytosis and exocytosis are transport processes that also require energy. Endocytosis refers to the process of engulfing material by invaginating the outer membrane of a cell until it buds off within the cytoplasm of the cell as a vesicle containing external material. When large substances (i.e., bacteria) are taken within the cell, the process is termed phagocytosis, whereas pinocytosis refers to this process when smaller substances (i.e., large molecules) are engulfed (Fig. 1-3A). **Receptor-mediated endocytosis** refers to this process, when the material engulfed has a specific receptor on the surface of the cell (Fig. 1-3B). Lysosomal enzymes have such receptors on the surface of certain cells. Some leakage of lysosomal enzymes into the extracellular spaces occurs constantly. These enzymes can be recovered by cells that have a transmembrane protein, mannose-6-phosphate receptor protein, on their surface. The lysosomal enzymes are bound to the outer membrane of the cell by the receptor and internalized within endosomes by the process of receptor-mediated endocytosis. The endosome then delivers the lysosomal enzymes to the lysosomes.

Exocytosis refers to this process in reverse. An example encountered later in the chapter is the release of many types of neurotransmitter from an axon terminal (Fig. 1-3C).

Fig.1-3. A. Phagocytosis showing expansion of the plasma membrane to engulf a bacterium, contrasted with pinocytosis of protein molecules. Both processes involve fusion of the plasma membrane. B. Receptor-mediated endocytosis resembles pinocytosis but involves a specific protein receptor on the surface of the cell. C. Exocytosis.

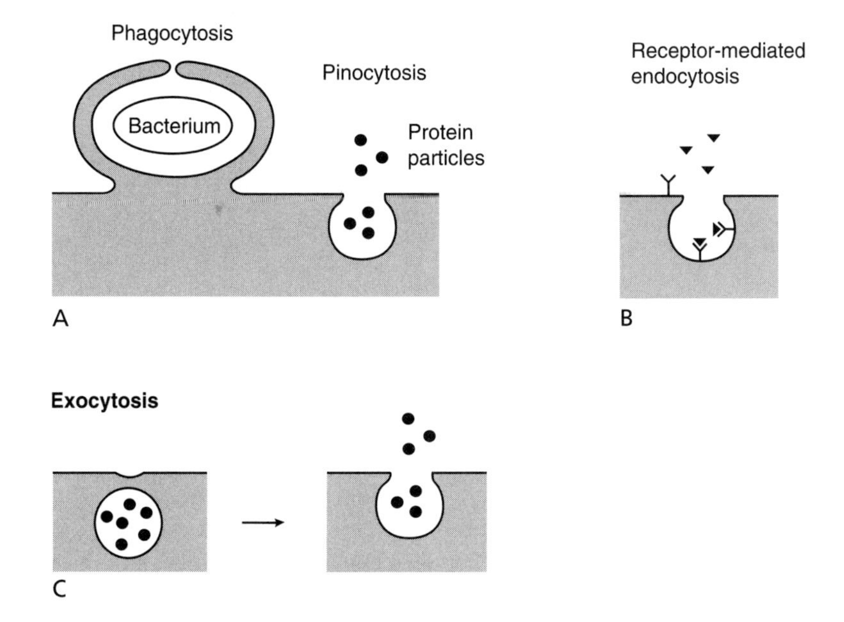

Transepithelial Transport

Transepithelial transport is the movement of substances across a polarized sheet of cells, such as the renal or intestinal epithelium. The basic mechanism involves active transport through the cell membrane on one side of the sheet, and diffusion (facilitated or simple) through the other side. In the cell illustrated in Fig. 1-4, Na^+ is passively absorbed along its electrochemical gradient into the cell across the luminal membrane, increasing the intracellular Na^+ concentration. The Na^+ is then actively pumped against its electrochemical gradient into the interstitial space by Na^+ ion pumps in the basolateral membrane. This cation flux results in a corresponding movement of anions to balance charge and of water to maintain osmolarity. The sodium has thus been reabsorbed from the glomerular filtrate into the extracellular space, but it has passed through a layer of cells to do so. Tight junctions interconnecting the epithelial cells along their luminal aspects prevents sodium from diffusing back into the lumen. The role of transport between cells is as yet undetermined.

Receptors

Receptors are proteins located within the cell or in the outer membrane of the cell. The binding of a ligand to its receptor conveys a signal to the machinery of the cell to alter some aspect of its activity. The binding of a ligand to its receptor is specific, may be saturated, and may be in competition with the binding of other molecules. In addition, receptors may be up-regulated, down-regulated, or both, by their own activity. Regulation may be achieved by regulating the number of receptors per cell or by regulating the affinity (the degree to which a substance stays bound to the receptor) of each receptor for its chemical messenger. For example, prolonged exposure to insulin (which causes its target cells to take up circulating glucose) results in a decreased ability of the target cells to take up glucose from the blood.

Specificity

This property reflects the ability of a receptor to react with only one type or a limited number of structurally related types of molecules. Cell types express receptors, or

Fig. 1-4. Transepithelial transport. Passive transport at the lumen combined with active transport at the base of the cell and along the cell borders to move substances through a sheet of cells. (From A. C. Guyton. *Textbook of Medical Physiology* [8th ed.]. Philadelphia: Saunders, 1991. P. 49.)

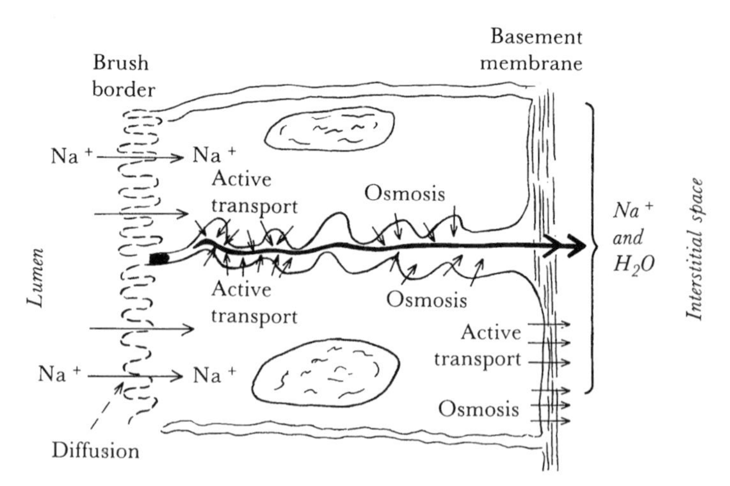

combinations of receptors, that are specific for that type of cell. In this way, chemical messengers (i.e., hormones or drugs in the blood) may contact many cell types and affect only the desired cell types.

It is important to realize that a given chemical messenger may affect different cell types that contain the appropriate receptor differently. For example, the neurotransmitter, norepinephrine, causes smooth muscle cells of blood vessels to contract; however, through an interaction with the same receptor type, norepinephrine causes cells in the pancreas to secrete less insulin.

Receptor Types

In general, there are receptors for **lipid-soluble messengers** and **lipid-insoluble messengers.** The lipid-soluble messengers include steroid receptors and thyroid hormone receptors found in the cytoplasm and the nucleus cells. In general, these are homodimeric proteins that bind to their messenger and contain "zinc finger" regions that bind directly with certain DNA sequences affecting the synthesis of new messenger RNA molecules.

The lipid-insoluble messengers are proteins (or glycoproteins) in the outer membrane of the cell. These receptors are often multimeric proteins comprised of several peptides. These receptors may be (1) ion channels (i.e., ligand-gated ion channels); (2) protein kinases, which then phosphorylate or dephosphorylate specific molecules; or (3) proteins that activate or deactivate **G proteins.** G proteins in turn act on other **"effector proteins,"** such as ion channels, phospholipase C (forms inositol triphosphate and diacylglycerol), adenylyl cyclase (forms cyclic adenosine monophosphate), or guanylyl cyclase (forms cyclic guanosine monophosphate). These are discussed further in the next section.

Signal Transduction

Signal transduction is the process by which a cell converts an input signal into a response. The input may be an action potential that ultimately results in the release of neurotransmitters. Other input signals include the binding of a hormone to a receptor, an antigen to an antibody, a binding molecule to an adhesion molecule, and a steroid to its intracellular receptor. Typical end-stage cellular responses include an increase in metabolic activity, an increase or decrease in production, and release of another signaling substance, cell growth and division, and cell death.

The intermediate steps between the signal and the response comprise signal transduction. These may include a complex sequence of events, such as changes in intracellular ion concentration, modulation of enzyme activity, protein phosphorylation-dephosphorylation, modulation of DNA replication, RNA transcription, and protein translation, as well as posttranslational protein modification.

The process begins with a ligand binding a receptor. The ligand is referred to as the **first messenger,** and the initial signal produced is referred to as a **second messenger.** Examples of second messengers (discussed below) are cyclic adenosine monophosphate (cAMP), inositol triphosphate (IP_3), diacylglycerol (DAG), and CA^{2+}.

G-Protein/Adenylate Cyclase

In this signal transduction pathway, the first messenger interacts with its receptor in the plasma membrane. Next, a G protein binds to the receptor-messenger complex.

Finally, the G-protein complex activates (G_s = stimulatory) or deactivates (G_i = inhibitory) the adenylyl cyclase enzyme, resulting in the production of more or less cAMP, the second messenger. Through this control of cAMP levels, the activity of cAMP-dependent protein kinases is modulated (Fig. 1-5).

An important aspect of this process is that various steps of the transduction allow for amplification of the signal. For example, one adenylyl cyclase can produce many cAMP molecules, and each active protein kinase can phosphorylate many other enzymes and so forth.

Phospholipase C

In a similar sequence to the one outlined above, a G-protein complex may activate (G_s) or deactivate (G_i) the enzyme, phospholipase C. This enzyme (when active) cleaves phosphatidylinositol biphosphate (PIP_2) into DAG and IP_3. Each of these then serves as a second messenger in the cell (Fig. 1-6).

DAG activates a protein kinase associated with the inner surface of the outer membrane of the cell, **protein kinase C.** Protein kinase C plays a vital role in the phosphorylation of proteins associated with the plasma membrane.

IP_3, on the other hand, enters the cytosol and acts on the membrane of the endoplasmic reticulum (i.e., the sarcoplasmic reticulum in muscle), increasing its permeability to Ca^{2+}. Ca^{2+} then diffuses down its concentration gradient into the cytoplasm.

Fig. 1-5. Cyclic AMP second-messenger system. Combination of the first messenger—hormone, neurotransmitter, or paracrine agent—with its specific receptor permits the receptor to bind to the membrane G protein termed G_s. Not shown in the figure is G_i (the inhibitory G protein). (Redrawn from A. J. Vander , J. H. Sherman, and D. S. Luciano. *Human Physiology—The Mechanisms of Body Function* [6th ed.]. New York: McGraw Hill, 1994. P. 167. Reproduced with permission of McGraw-Hill, Inc.)

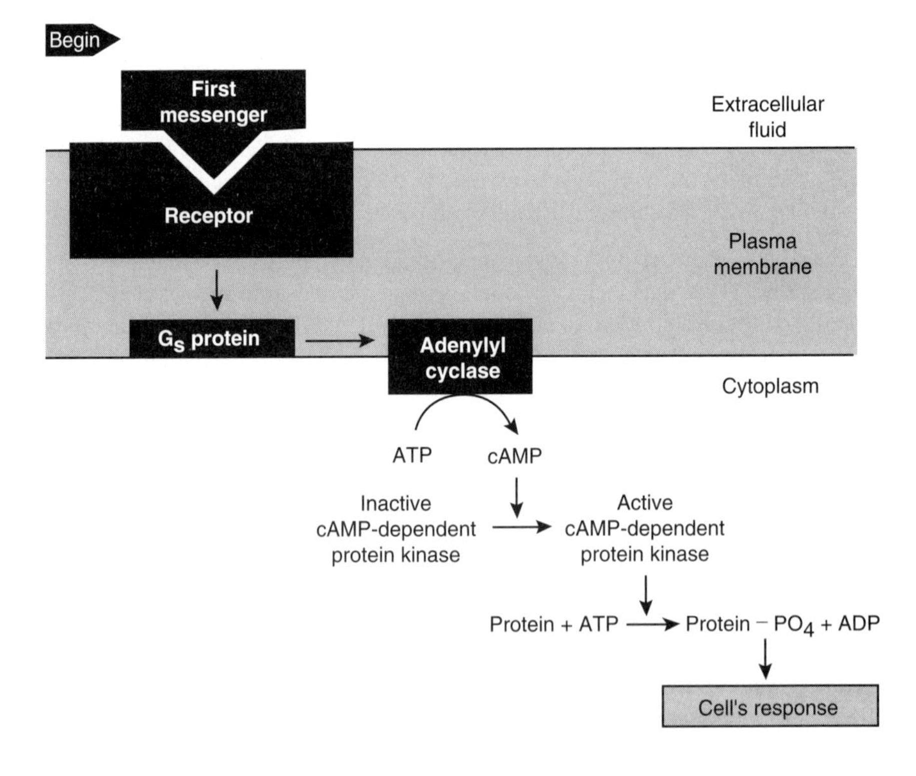

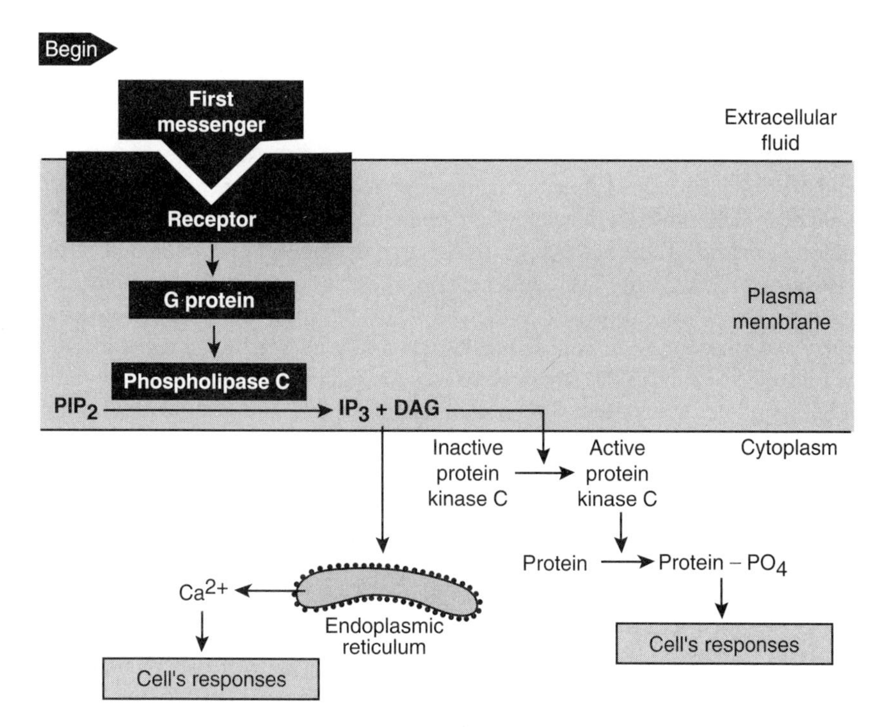

Fig. 1-6. IP_3 and DAG second-messenger system. Combination of the first messenger—hormone, neurotransmitter, or paracrine agent—with its specific receptor permits the receptor to bind to the membrane G protein termed G_s. Not shown in the figure is G_i (the inhibitory G protein). (Redrawn from A. J. Vander , J. H. Sherman, and D. S. Luciano. *Human Physiology—The Mechanisms of Body Function* [6th ed.]. New York: McGraw Hill, 1994. P. 171. Reproduced with permission of McGraw-Hill, Inc.)

This results in an increase in the intracellular Ca^{2+} concentration and further acts to regulate specific cell processes.

Membrane Potential

The **selective permeability** of cell membranes allows for the differences in the composition of extracellular and intracellular fluid, which, in turn, are responsible for the generation of a membrane potential.

Resting Potential

The equilibrium or **resting potential** is the point at which the forces of the concentration gradient and electrical gradient balance. This potential can be calculated by the **Nernst equation:**

$$E_x = \frac{RT}{zF} \ln \frac{[X]_i}{[X]_o}$$

in which E_x = equilibrium potential (in mV) of X, R = gas constant, T = absolute temperature, z = valence of X (including sign of charge), F = Faraday's constant, $[X]_i$ = concentration of X inside the cell, and $[X]_o$ = concentration of X outside the cell. For a T of 37 °C and a valency of +1,

$$E_x = -61.5 \log \frac{[X]_i}{[X]_o}$$

Consider a membrane that is selectively permeable to potassium. Because the intracellular K^+ concentration is greater than the extracellular K^+ concentration, K^+ diffuses out of the cell down its concentration gradient. Continued diffusion results in a separation of charge. There is a net positive charge just outside the membrane (from the diffusion of K^+ out of the cell) and a net negative charge just inside the membrane (from the loss of K^+ ions from within the cell). This partition of charge produces an electrical potential opposing further flux of positively charged K^+. Frequently, the force of the electrical potential equally balances the force of the concentration gradient. There is no net movement of K^+ and the membrane remains at the equilibrium potential for K^+.

In reality, the membrane is permeable to other ions (such as Na^+) in addition to K^+. The membrane contains a **K^+-Na^+ leak channel,** which favors the flux of K^+ (the K^+ permeability at rest is at least 35 times greater than Na^+ permeability); therefore, the K^+ equilibrium potential (E_{K+}) is a good approximation of the cell membrane's resting potential (E_m).

Membrane potential and ion concentrations vary among species and cell types. By convention, cellular potentials are described as the cell interior relative to the cell exterior. Representative ion concentrations for human neurons are shown in Table 1-1. Here the K^+ equilibrium potential is –90 mV (inside negative with respect to the outside). A potential of +60 mV (inside positive with respect to the outside) is required to balance the tendency of Na^+ to flow down its concentration gradient and enter the cell. The normal membrane resting potential is approximately –70 mV. This reflects the much greater contribution of K^+ because of the greater permeability of K^+. Because neither E_{K+} nor E_{Na+} are at the membrane potential, one would expect the cell to gradually gain Na^+ and to lose K^+. In both cases, the electrical potential is insufficient to prevent the tendency of the ions to move down their respective concentration gradients; however, active transport by the **Na^+-K^+ pump** maintains the concentration gradients. The Na^+-K^+ ATPase also contributes to the separation of charge because of its **electrogenic** properties. As seen in Fig. 1-2C, each cycle results in a net movement of one positive charge out of the cell (three Na^+ out, two K^+ in). There are no active chloride pumps; thus, chloride is subject only to the passive forces of electrical and chemical gradients, and the chloride equilibrium potential equals the membrane potential.

Depolarization and Hyperpolarization

If the resting membrane potential is altered by passing a current through the membrane, the electrical gradients that affect ions are altered, and there is a change in ion

Table 1-1. Ion concentrations in representative mammalian neurons

Ion	Concentration (mM)			
	Inside	Outside	Relative permeability	Equilibrium potential (mV)
Na^+	15.0	150.0	1	+60
K^+	150.0	5.5	50–100	–90
Cl^-	9.0	125.0	25–50	–70
Ca^{2+}	0.0001	1.0	—	—

flux to restore the membrane to its resting potential. If, for example, the cell is depolarized (i.e., the membrane potential is made less negative), the electrical gradient that keeps K^+ inside the cell is decreased. The force of the concentration gradient now exceeds the electrical force, and K^+ diffuses out of the cell until the forces balance, at which time the membrane is restored to its resting potential. When the membrane potential is made more negative or hyperpolarized, ions move in the opposite direction.

At low levels of stimulation, this response is proportional to stimulus intensity and decays exponentially with distance and time. These short-lived, local, graded potentials are also called **local or generator receptor potentials** (Fig. 1-7). This ion flow reflects passive properties of the membrane not unique to excitable cells and can be predicted from the Nernst equation and the electrical characteristics of the membrane.

Membrane Excitability

Membrane excitability refers to a rapid depolarization of a **nerve or muscle cell** on reaching a **threshold potential,** the point at which Na^+ influx overcomes K^+ efflux. Membrane excitability results from a voltage-dependent increase in Na^+ permeability.

Gated Ion Channels

Unlike the ion channels discussed in the previous section and illustrated in Fig. 1-2, many ion channels can be opened or closed by **"gates,"** conformational changes in the protein molecule that either allow or prevent passage through the channel. Gates may be modulated by voltage (**voltage-gated**) or by binding of a specific ligand (**ligand-gated**). A given channel may have more than one gate (i.e., an activation gate that opens the channel and an inactivation gate that closes the channel).

Gated ion channels form the basis of the action potential, which is discussed below.

Fig. 1-7. Local potentials and threshold. Changes in membrane potential following application of subthreshold and then threshold stimuli. At the threshold (–55 mV) an action potential is produced. Subthreshold depolarizing stimuli and hyperpolarizing stimuli produce potentials that decay over time (shown here) and distance (not shown). (From W. F. Ganong. *Review of Medical Physiology* [16th ed.]. East Norwalk, CT: Appleton & Lange, 1993. P. 48.)

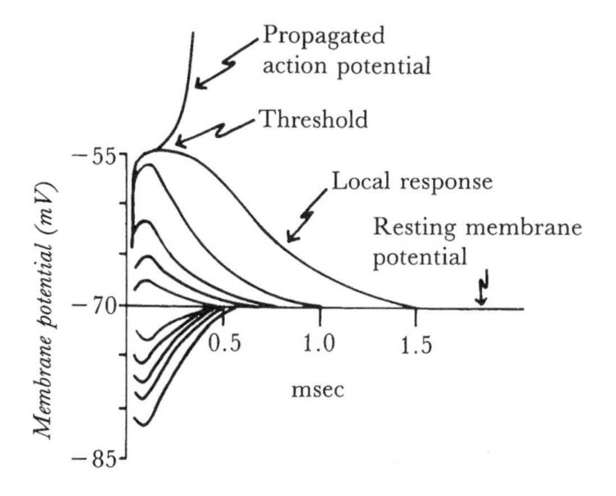

Action Potential

When an excitable membrane is depolarized above a threshold of approximately 15 mV (from –70 mV to –55 mV), an action potential results. This is a self-propagating response mediated by voltage-gated Na^+ and K^+ channels. A typical action potential is depicted in Fig. 1-8, and consists of:

1. A period of **rapid depolarization** characterized by
   a. Opening of (fast-acting) voltage-gated Na^+ channels, increasing Na^+ permeability 50- to 5000-fold
   b. An electrochemical gradient favoring Na^+ influx into the cell
2. An **overshoot period** characterized by
   a. An electrical potential that no longer favors Na^+ influx into cells
   b. An opening of (slow-acting) voltage-gated K^+ channels

Fig. 1-8. The action potential. The components of the action potential are displayed above. Na^+ and K^+ conductances are displayed below, illustrating the molecular events that underlie the action potential. These curves are derived from the experimental system of Hodgkin and Huxley, which differs from that described in the text (thus, the resting membrane potential of approximately –90 mV). (From A. C. Guyton. *Textbook of Medical Physiology* [8th ed.]. Philadelphia: Saunders, 1991. P. 58.)

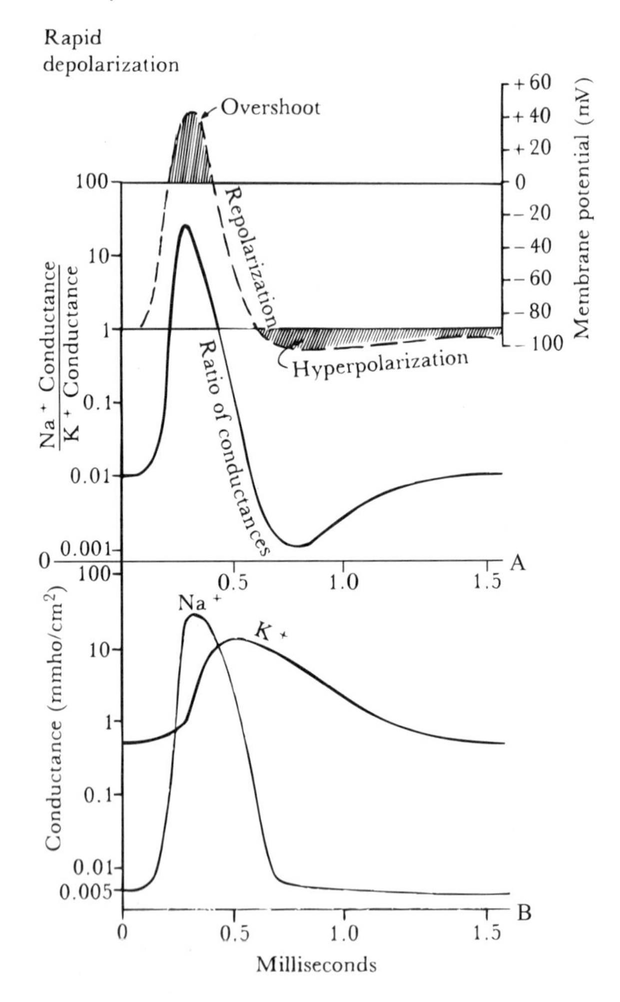

3. A period of **rapid repolarization** characterized by
   a. An increased K^+ permeability
   b. Closure of voltage-gated Na^+ channels
4. A period of **slower repolarization** characterized by
   a. Continued closure of voltage-gated Na^+ channels with a return to low Na^+ permeability
   b. Closure of some voltage-gated K^+ channels
5. A prolonged period of **hyperpolarization** characterized by
   a. A higher than normal resting K^+ permeability
   b. Eventual closure of the voltage-gated K^+ channels

The Refractory Period

Once an excitable cell has been stimulated above threshold, the cell becomes refractory to further stimulation. This refractory period consists of an initial **absolute refractory period** during which no stimulus, regardless of its magnitude, will excite the cell, and a later **relative refractory period** during which stronger than normal stimuli are required to cause excitation. The absolute refractory period results from the inactivation of the voltage-gated Na^+ channels, which cannot be opened again until the cell is repolarized. The relative refractory period represents the period of hyperpolarization mediated by the slow closure of the voltage-gated K^+ channels.

The number of ions involved in the production of the action potential is minute relative to the total ion concentration. There is no measurable change in ion concentrations after a single action potential. Nonetheless, repeated stimulation will lead to significant changes in concentration. The Na^+-K^+ ATPase regenerates and maintains the concentration gradients that characterize the membrane polarization. This is the site at which the cell expends energy in the form of ATP to provide for signaling.

All-or-None Phenomenon

Action potentials travel along nerve axons and are the impulses that carry information in the nervous system. Action potentials are an all-or-none phenomenon. Therefore, depolarization above the threshold may result in more frequently repeated action potentials but will not cause a greater action potential (Fig. 1-9).

Fig. 1-9. Threshold and action potentials. Once a threshold stimulus is reached (here, 65 mV), an action potential results. Increasing receptor potential increases the frequency of the action potential, but it does not change shape. (From A. C. Guyton. *Textbook of Medical Physiology* [8th ed.]. Philadelphia: Saunders, 1991. P. 497.)

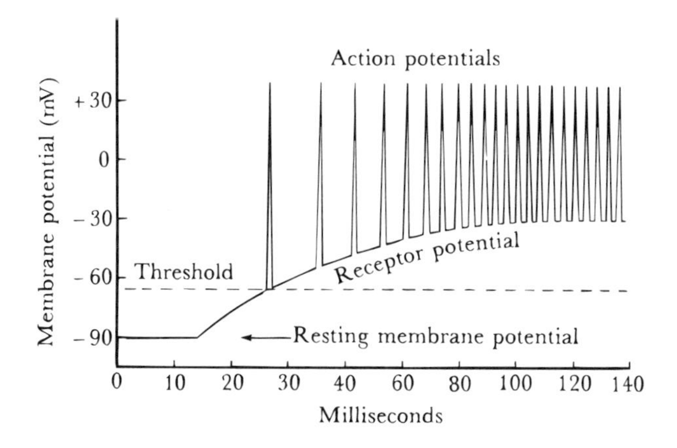

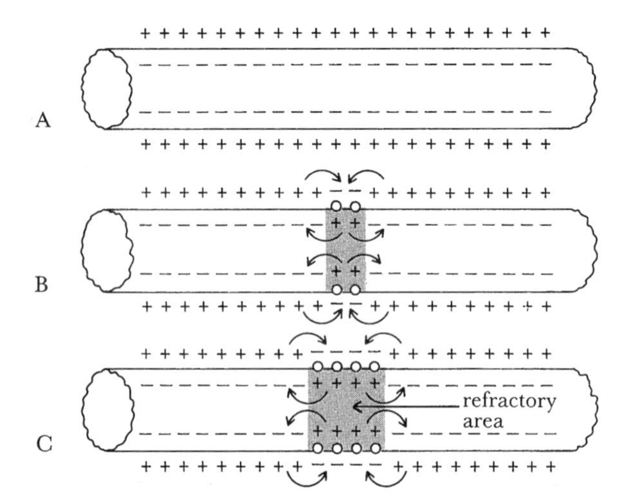

Fig. 1-10. Propagation of action potentials. See text for explanation. Although the figure illustrates propagation of action potentials with reversal of polarity during the overshoot period, local potentials also have a similar circular current flow. (From A. C. Guyton. *Textbook of Medical Physiology* [8th ed.]. Philadelphia: Saunders, 1991. P. 60.)

Propagation

Action potentials propagate as a result of local currents produced at one point of the membrane depolarizing adjacent parts above threshold to induce an action potential. This is demonstrated schematically in Fig. 1-10. In A, the nerve fiber is at rest, with the inside negatively charged. During the action potential this polarity is reversed, producing a local circuit of current flow. Positive charges move toward the area of negativity. In B, the potential is reduced (made less negative) by this electrical current. Note that extracellular movement is in a direction opposite to intracellular movement, resulting in circular current of flow. These local currents are the basis by which local potentials spread along the membrane. When the threshold is reached, an action potential is generated, and the cycle continues.

Myelination accelerates the rate of axonal transmission and conserves energy by limiting the depolarization process to the nodes of Ranvier. Myelin is a very effective insulator and thus limits transmembrane ion flow to the nodes of Ranvier, where voltage-gated channels are found. Myelin also decreases capacitance so that fewer ions are required to charge the membrane. The net result is that more ions can participate in the circular ion flow. Action potentials are thus transmitted from node to node in a process called **saltatory conduction** (Fig. 1-11). By causing the depolarization process to jump long intervals, conduction velocity increases. Conduction velocity also increases with axonal diameter due to decreased resistance to ion flow within the axon. This is true for both unmyelinated and myelinated neurons. Myelination also decreases the metabolic energy required to restore the resting potential because total ion loss is reduced with the limited ion currents. Table 1-2 summarizes the differences between local potentials and action potentials.

Synaptic Transmission

Synapses are specialized junctions for the transmission of impulses from one nerve cell to another. The extensive, intricate, and varied connections among nerve cells

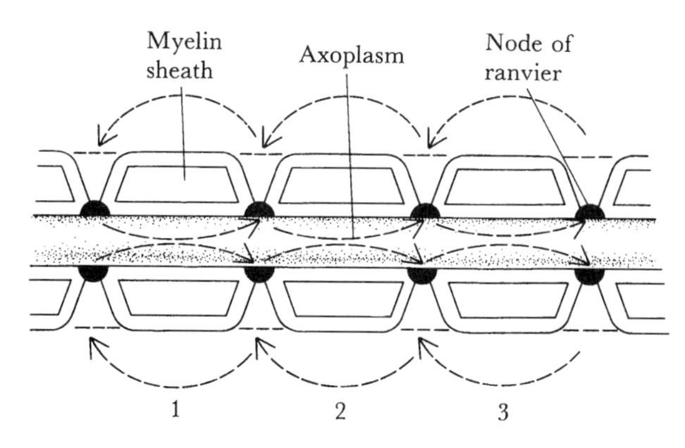

Fig. 1-11. Saltatory conduction in myelinated axons. Note that conduction is unidirectional and that transmembrane ion flow occurs only at the node of Ranvier. (From A. C. Guyton. *Textbook of Medical Physiology* [8th ed.]. Philadelphia: Saunders, 1991. P. 63.)

Table 1-2. Comparison of local and action potentials

Feature	Local potential	Action potential
Amplitude	Small (0.1–10 mV)	Large (70–110 mV)
Duration	Variable (5 msec to minutes)	Brief (1–10 msec)
Form	Graded	Spike, all or none
Polarity	Depolarizing or hyperpolarizing	Depolarizing
Propagation	Passive with local decay	Active, self-propagating
Channels	Na^+-K^+ leak channels, other special receptors (e.g., neutrotransmitters, sensory receptors)	Voltage-gated Na^+, K^+ channels

and the plasticity of the chemical synapses are essential for information transmission, processing, and storage.

Chemical Transmission

Chemical transmission involves the release by the **presynaptic cell** (PrSC) of a chemical **neurotransmitter** (NT), which diffuses across the synaptic cleft to bind to receptors on the **postsynaptic cell** (PoSC). These intermediate chemical processes result in a time delay between the arrival of the impulse in the PrSC and the generation of a postsynaptic potential. Because of the synaptic delay, conduction along a chain of neurons slows as the number of synapses in the circuit increases. The minimum synaptic delay is 0.5 msec. Known NTs include acetylcholine (Ach), several different amines (e.g., the catecholamines and serotonin), a number of amino acids, and a variety of peptides.

Neurotransmitter Release

The molecular events of neurotransmitter release are well studied and are essentially independent of the specific NT or morphologic type of synapse. When the action potential depolarizes the PrSC nerve terminal, voltage-gated Ca^{2+} channels open and allow Ca^{2+} to enter the cell down its electrochemical gradient. The Ca^{2+} influx stimulates the fusion of neurotransmitter-containing vesicles with the PrSC cell membrane. This process of **exocytosis** results in the release of NT. The number of vesicles released is directly proportional to the Ca^{2+} influx. The NT diffuses across the synaptic cleft and binds to receptors on the PoSC.

Postsynaptic Cell Response

Postsynaptic cell response varies with the NT. Moreover, the same NT can produce different responses if the PoSC receptor is different. Postsynaptic cell responses are of two general types: **Ion permeability** may be altered or **internal metabolic processes** of the PoSC may be activated. The **postsynaptic potential** (PSP) is a local graded potential (which decays with distance and time) produced by altering ion permeability in the PoSC through the inactivation of ion channels. **Excitatory PSPs** (EPSPs) result from increasing membrane permeability to Na^+, such as that which occurs with the Ach receptor. **Inhibitory PSPs** are produced by increasing membrane permeability to K^+ and Cl^-. Even if this does not hyperpolarize the cell, it can "short-circuit" excitatory stimuli, because an increase in Na^+ permeability has less of an effect on membrane potential if K^+ and Cl^- permeability are also increased. These changes in permeability are short lived and decay approximately 15 msec after the NT is removed from the receptor. Slower, longer-lasting PSPs have been described in autonomic ganglia, cardiac and smooth muscle, and cortical neurons. These potentials depend on decreasing ion permeability.

The response of the PoSC is the result of the integration of all the cells' input. A single neuron such as a spinal motor neuron receives convergent input from hundreds of sources on its soma and dendrites. No single stimulus is sufficient to induce the cell to fire or to prevent firing. Input is summed algebraically both spatially and temporally. **Temporal summation** occurs when a nerve terminal fires rapidly before the previous PSP has had a chance to decay. **Spatial summation** occurs when activity is simultaneously present at more than one of the cell's synapses (Fig 1-12). The action potential is initiated at the axon hillock (the junction between axon and cell body). This segment of the axon has a threshold for firing that is approximately 15 mV lower than that of the rest of the cell membrane because of the high concentration of voltage-gated Na^+ channels found at the axon hillock. Because PSPs are local potentials that

Fig. 1-12. Pre- and postsynaptic inhibition. Postsynaptic inhibition affects all inputs to the cell while presynaptic inhibition selectively inhibits a single input. (From E. K. Kandel and G. H. Schwartz. *Principles of Neural Science* [2nd ed.]. New York: Elsevier, 1985. P. 90.)

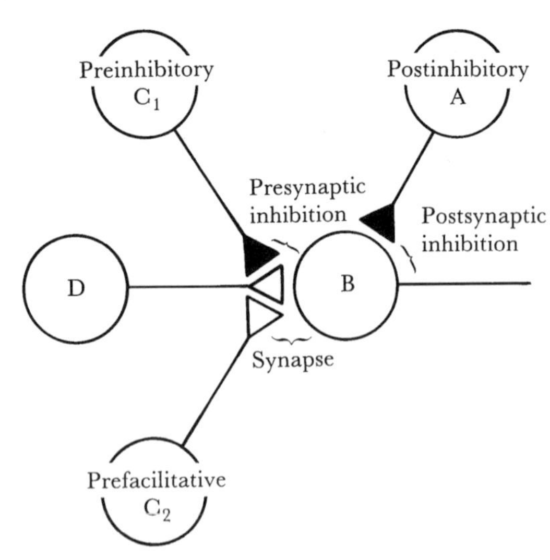

decay with distance, synapses closer to the initial segment have a greater influence on the cell's firing status than do more distant synapses. When an action potential occurs, it travels anterograde down the axon but also retrograde to the cell soma and dendrites. This retrograde firing into the soma serves to "wipe the slate clean" so that the summation process starts fresh.

Besides affecting ion channels, NTs may also activate internal metabolic processes of the PoSC such as protein kinases or receptor protein synthesis. Prolonged stimulation, or a lack thereof, may lead to down-regulation or up-regulation of receptors. Such metabolic changes produce long-lasting changes in the reactivity of the synapse. These modulations may be important in memory and learning.

Presynaptic Cell Response

In addition to excitatory and inhibitory effects on the PoSC, the PrSC can also be affected due to the presence of NT receptors in its membrane. Presynaptic inhibition or facilitation occurs by an alteration of Ca^{2+} influx and thus the amount of NT released by the PrSC. Axoaxonal synapses are involved, as shown in Fig. 1-12. The value of a mechanism such as presynaptic inhibition is that it permits selective modification of input from a specific source. In contrast, postsynaptic inhibition results in generalized depression of the PoSC. The Ca^{2+} influx can be altered by either direct modification of the voltage-gated Ca^{2+} channels or by alteration of the voltage of the action potential at the nerve terminal.

Termination of Synaptic Transmission

The action of the NT is terminated by removing the NT from the synaptic cleft by either diffusion, active re-uptake by the PrSC, or enzymatic inactivation of the NT. New transmitter is synthesized in the PrSC nerve terminal or cell body. Neurotransmitter synthesized in the cell body is transported to the nerve terminal by axonal transport.

The Neuromuscular Junction

Transmission at the neuromuscular junction (NMJ) follows the same basic mechanism as described above. Local Ca^{2+} influx at the PrSC (the spinal motor neuron) nerve terminal is induced by the action potential. Calcium in turn results in the release of the NT, acetylcholine, from the PrSC. The Ach receptor on the muscle cell binds Ach. Na^{+} permeability increases, and an end-plate potential is produced that spreads along the muscle, leading to excitation contraction coupling. A unique feature of transmission at the NMJ is that an action potential always produces a PSP sufficient to cause the muscle membrane to fire. At rest, miniature end-plate potentials of approximately 0.5 mV can be detected at the muscle end plate. These represent single vesicles (quanta) of Ach that are randomly released at rest. The action of Ach is terminated via enzymatic degradation by acetylcholinesterase, which is located in the synaptic cleft.

Skeletal Muscle

The individual muscle cell or muscle fiber is a large, multinucleate, highly organized cell, specialized for contraction. The cytoplasm contains many myofibrils, which consist of two types of myofilaments, thick and thin. These filaments are organized in a regular repeating pattern forming the basic functional unit of skeletal muscle, **the sarcomere** (Fig. 1-13).

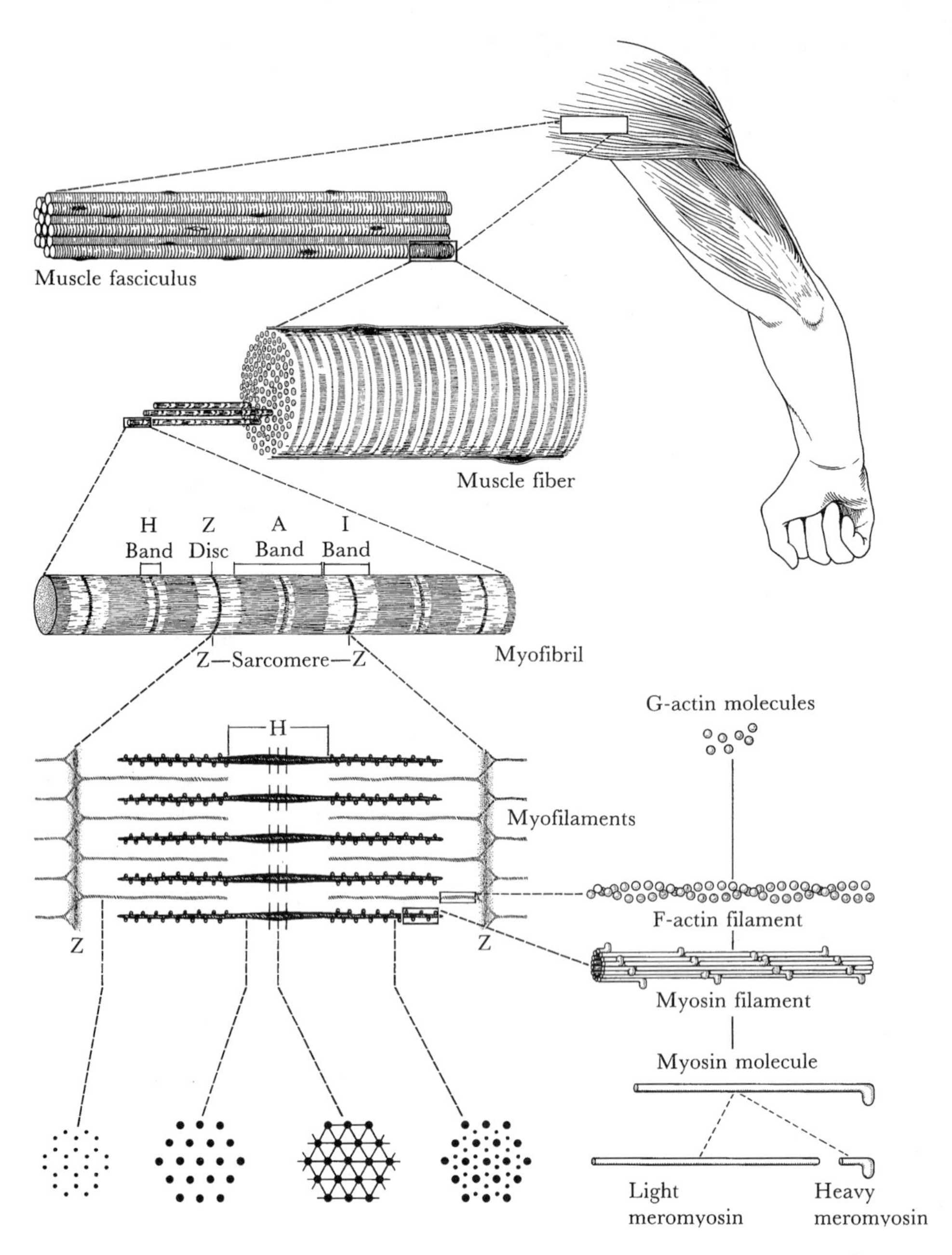

Fig. 1-13. Organization of skeletal muscle from gross to molecular level. (From D. W. Fawcett. *A Textbook of Histology* [11th ed.]. Philadelphia: Saunders, 1986. P. 282.)

Myosin and Actin

Thick filaments are composed of myosin molecules. Myosin is a protein that can be proteolytically digested into two fragments, **light meromyosin** (LMM) and **heavy meromyosin** (HMM). The LMM is rod-shaped and assembles into filaments, while HMM contains the myosin globular head (which forms the actin-myosin cross bridges) with ATPase activity and a binding site for actin. **Myosin light chain,** a smaller peptide that may modulate myosin ATPase activity, is also found in association with the globular portion of HMM. As seen in Fig. 1-13, the actin-myosin cross bridges (or myosin globular heads) are arranged in a helical array that extends outward from the H band along the myosin filament. Myosin molecules self-assemble into filaments with definite polarity. In the sarcomere the myosin thick filament reverses its polarity in the middle of the sarcomere, forming a central "bare zone" (the H band) that is devoid of myosin cross bridges (Fig. 1-14).

The principal constituent of thin filaments is actin. Actin monomer (G actin) is a 42,000-dalton molecular weight protein that has the capacity to self-assemble into a double-stranded α-helix of actin monomers (F actin). In addition to actin, thin filaments also contain **troponin** and **tropomyosin.** Tropomyosin is also a double-stranded α-helix. It is situated in the groove between the actin strands. A molecule of tropomyosin spans approximately seven actin monomers. Troponin is a complex of three separate proteins: TN-T, which binds to tropomyosin; TN-C, which binds Ca^{2+}; and TN-I, which binds to actin. There is one troponin complex for each molecule of tropomyosin (Fig. 1-15).

The Sliding Filament Theory of Muscle Contraction

During contraction, the overall length of the sarcomeres decreases, while individual thick and thin filaments remain the same length. According to the sliding filament theory of muscle contraction, during contraction thick and thin filaments slide past one another, resulting in increased overlap.

Excitation-Contraction (E-C) Coupling

The critical link between the electrical events of the muscle cell action potential and the organized interaction between actin and myosin, which causes contraction of skeletal muscle, is an increase in intracellular Ca^{2+} concentration from approximately 10^{-7} M in the relaxed state to 2×10^{-5} M or greater during contraction. Of special importance in E-C coupling are the **"triads,"** composed of two terminal cisterns located in close proximity to a T tubule. **Terminal cisterns** are specialized evaginations of the muscle cell sarcoplasmic reticulum. The **T tubules** are extensive, deep invaginations of the muscle cell membrane. In the triads, extensions of the muscle cell membranes are in close proximity to extensions of the sarcoplasmic reticulum. The muscle cell action potential is propagated down the T tubules and, by an unknown mechanism, activates the terminal cisterns and sarcoplasmic reticulum to release their stored

Fig. 1-14. A. Diagrammatic representation of the myosin molecule. B. Arrangement of myosin molecules into filaments. Note the reversal of polarity in the center, which gives rise to the central bare zone. (From L. Weiss. *Cell and Tissue Biology: A Textbook of Histology.* [6th ed.]. Baltimore: Urban and Schwarzenberg, 1988. P. 267.)

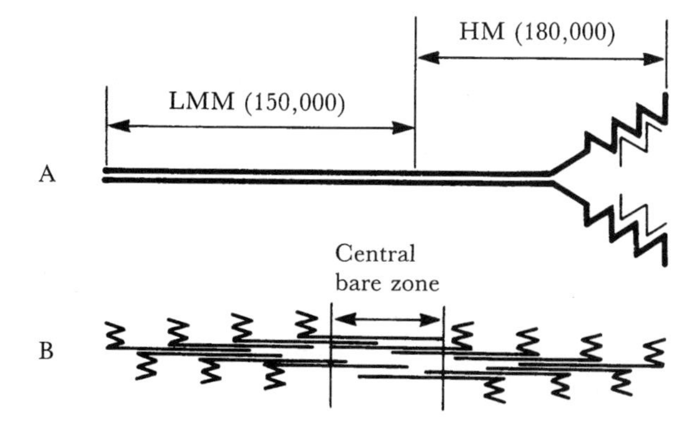

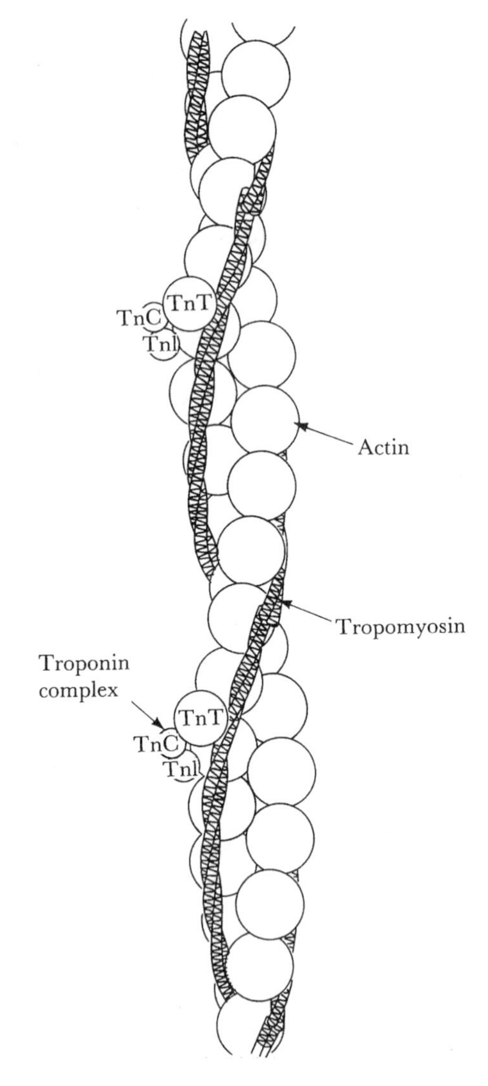

Fig. 1-15. The proposed model of a thin filament in the resting state. (After C. Cohen. Protein switch of muscle contraction. Copyright © 1975 by Scientific American, Inc. All rights reserved.)

Ca^{2+} into the cytoplasm. The sarcoplasmic reticulum has a high-capacity Ca^{2+} ATPase that can actively pump Ca^{2+} back into the sarcoplasmic reticulum and return the cytoplasmic concentration of Ca^{2+} back to resting levels in order for muscle relaxation to take place.

Molecular Events in Muscle Contraction

In the resting state, before the rise in cytoplasmic Ca^{2+} concentration, the tropomyosin molecules are located in the groove between the strands of actin filaments in such a way that they sterically block the myosin binding sites on actin. Thus, they inhibit the formation of actin-myosin cross bridges.

During activation, Ca^{2+} is released from the sarcoplasmic reticulum. The TN-C component of troponin binds Ca^{2+}. This binding causes conformational changes in the troponin complex that in turn result in conformational changes in tropomyosin. The tropomyosin then alters its position with respect to actin in such a way that binding sites on actin become available for interaction with myosin globular heads. As long as cytoplasmic Ca^{2+} concentrations remain elevated, actin and myosin undergo cycles of association and dissociation (Fig. 1-16).

In the resting state, myosin globular heads have bound ADP and inorganic phosphate (P_i). Binding of actin to myosin globular heads causes the release of ADP and P_i from myosin. This release is associated with a tilting of the myosin globular head that is thought to be the "power stroke" causing the sliding of the thick and thin filaments past one another. For the actin-myosin complex to dissociate, ATP is required. Actin has a low affinity for myosin-ATP complex, and the binding of ATP by myosin causes the detachment of actin. Myosin globular head has ATPase activity, resulting in the hydrolysis of the bound ATP to ADP and P_i. The cycle is now complete and can be repeated as long as cytoplasmic Ca^{2+} concentrations remain greater than 2×10^{-5} M.

Slow and Fast Muscle Fibers

Skeletal muscle cells are often divided into two types, slow and fast fibers. Slow fibers are adapted for sustained contraction over long periods of time, while fast fibers are adapted for intense but sporadic contractions. The characteristic differences between fiber types are summarized in Table 1-3.

Fig. 1-16. Proposed mechanism (A–D) by which the ATP-driven cycles of actin and myosin association and dissociation generate the mechanical force needed for the sliding of thick and thin filaments during contraction. P_i=inorganic phosphate. (From L. Stryer. *Biochemistry* [3rd ed.]. New York: Freeman, 1988. P. 931.)

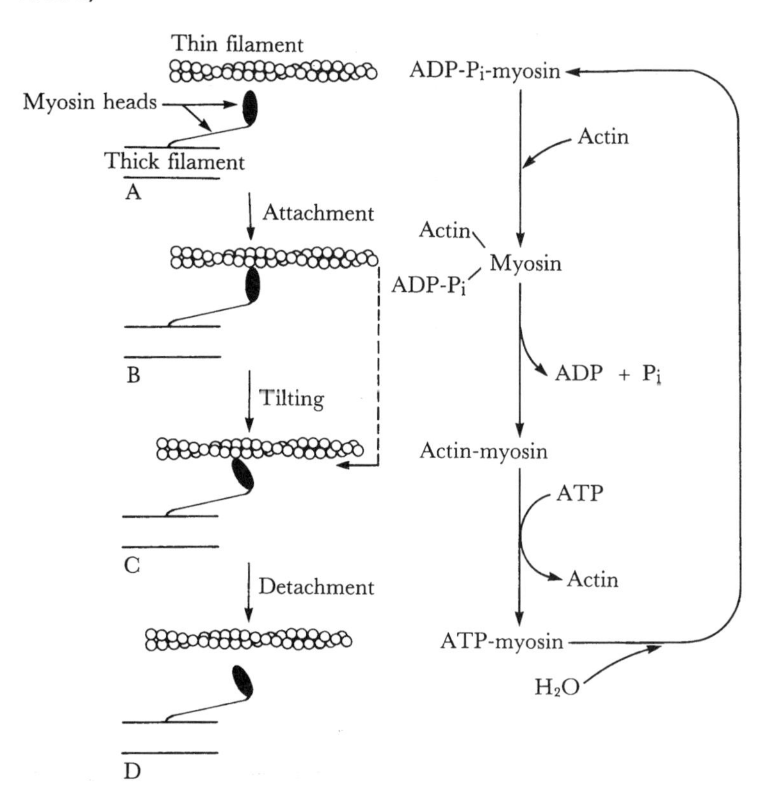

Table 1-3. Distinguishing features of slow and fast muscle fibers

Feature	Slow fiber	Fast fiber
Color	Red	White
Mitochondria content	+++*	+
Myoglobin content	+++	+
Glycolytic enzymes	+	+++
Glycogen content	+	+++
Myofibrillar ATPase	+	+++

* Number of plus signs indicate relative amount.

Slow Fibers

Slow fibers are well endowed with cellular machinery for **aerobic metabolism,** including abundant mitochondria, a high concentration of myoglobin, and a rich blood supply. The high myoglobin concentration and rich blood supply impart a red color to slow fibers.

Fast Fibers

A high concentration of glycolytic enzymes as well as large stores of glycogen are found within fast fibers that enables them to carry out **anaerobic metabolism.**

Although both slow and fast fibers are present in most muscles of the body, one or the other usually predominates. For example, in the calf muscles, which must sustain contraction for long periods of time, slow fibers predominate, while the extraocular muscles, which are adapted for rapid contraction, have a predominance of fast fibers.

Muscle Twitch

A muscle twitch is a cycle of contraction and relaxation that occurs in response to a neural stimulus in vivo or an electrical impulse passed directly into the muscle in an experimental setting. The duration of the twitch varies with different muscles. The time lag between the stimulus and the initiation of contraction is called the **latent period** (Fig. 1-17).

Refractory Period

As previously discussed (see Membrane Excitability), stimulated muscle cells exhibit an absolute and a relative refractory period during which restimulation is not possible or is relatively more difficult, respectively.

Summation and Tetanus

Let us imagine a muscle fiber that is undergoing a twitch response. What would happen if it were stimulated again, before completion of the first twitch response and before complete relaxation had taken place? Assuming that the stimulus came after the refractory period, tension would rise to a new peak that would be higher than the peak tension for the first twitch (Fig. 1-18).

Spatial Versus Temporal Summation

Each individual motor unit contracts in an "all-or-none" fashion. In other words, a suprathreshold stimulus produces a twitch, the amplitude of which is independent of

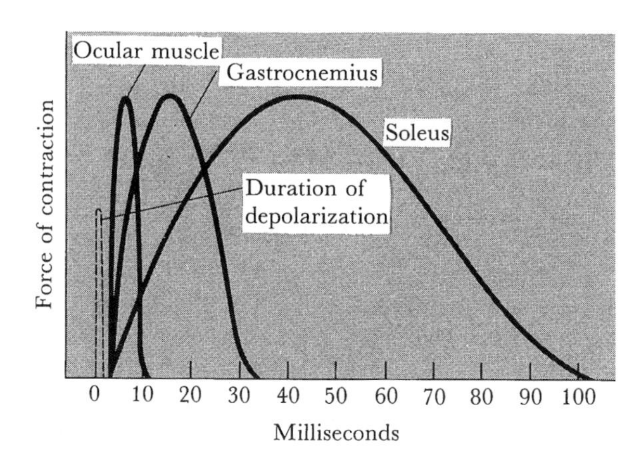

Fig. 1-17. Muscle twitch. Note the variation in the duration of muscle twitch with different muscles. Also note the latent period between action potential and muscle contraction. (From A. C. Guyton. *Textbook of Medical Physiology* [8th ed.]. Philadelphia: Saunders, 1991. P. 75.)

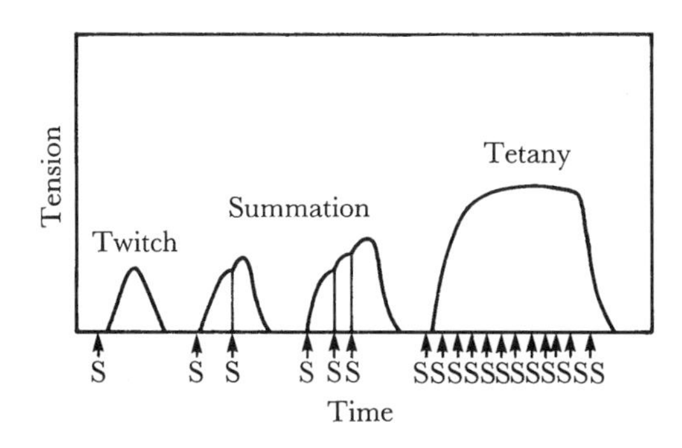

Fig. 1-18. Twitch, summation, and tetany. S = stimulus. (From J. B. West. *Best and Taylor's Physiological Basis of Medical Practice* [12th ed.]. Baltimore: Williams & Wilkins, 1991. P. 75.)

the stimulus intensity. Whole skeletal muscle, however, exhibits a graded response. Increasing intensity of stimulus produces higher amplitudes of twitch. The reason for this graded response is the fact that different motor units in a muscle have different thresholds of stimulation, and increasing stimulus intensity activates more motor units, leading to increased amplitude of response. This phenomenon is called **spatial summation. Temporal summation** refers to the fact that increasing frequency of stimulation can cause increased tension developed by each motor unit (Fig. 1-18). Thus, skeletal muscle can be made to develop increasing tension in a graded fashion in response to increases in the frequency (temporal summation) and intensity (spatial summation) of the stimulus. This phenomenon of summation (spatial or temporal) has been previously discussed (see Synaptic Transmission), and may be explained by the fact that after the second stimulus, the Ca^{2+} concentration in the sarcoplasm is higher than it was after the first stimulus, because some of the Ca^{2+} released from the sarcoplasmic reticulum by the first stimulus remained in the cytoplasm (i.e., re-uptake was not completed when the second stimulus arrived).

Tetany

As the frequency of stimulation is increased, higher and higher peak tensions are developed in the muscle, until stimulus frequency is reached where no relaxation of muscle between successive stimuli takes place. At this point a peak plateau of tension is reached, and the muscle fiber is said to be in tetany (Fig. 1-18).

Motor Unit

The functional unit in skeletal muscle is the motor unit, which consists of an **alpha motor neuron** in the anterior horn of the spinal cord and all the muscle fibers that it innervates. A motor unit may contain as few as two or three muscle fibers when precise control of muscular action is needed (e.g., laryngeal muscles), or it may consist of several hundred muscle fibers when fine control of muscle activation is not needed (e.g., gluteal muscles).

Isometric and Isotonic Contractions

The molecular events of actin and myosin interaction result in transduction of chemical energy (ATP) to mechanical energy. This mechanical energy can be in the form of tension developed in the muscle or in the form of work done by the muscle as it shortens. If two ends of a muscle are fixed such that it cannot shorten, and the muscle is stimulated to contract, tension develops in the muscle, but the length of the muscle does not change during the contraction. This is called an isometric contraction. On the other hand, if a muscle contracts such that tension remains constant while the muscle shortens, the contraction is called an isotonic contraction.

Length-Tension Relationship

The tension developed in a muscle fiber on neural stimulation depends on the extent to which the muscle was stretched before stimulation. If one were to passively stretch muscle to various lengths and measure the tension at each length, a passive length-tension curve could be plotted, as in Fig. 1-19. If the muscle were stimulated to contract isometrically at each length and the peak tension recorded for each length, one could calculate the **active tension** (tension developed as a result of contraction) at each length (Fig. 1-19).

It is apparent from Fig. 1-19 that there is an optimal length of muscle in which the greatest tension is developed on contraction. This phenomenon can be explained based on the degree of overlap between thick and thin myofilaments at each muscle length (Fig. 1-20). Based on this model, increasing overlap between thin and thick filaments produces increasing tension with stimulation. Note that between points B and C there is no change in the degree of thin and thick filament overlap because the cen-

Fig. 1-19. Resting, active, and total tension in a skeletal muscle undergoing isometric contraction at various muscle lengths. Resting tension = tension in muscle before contraction = preload. Total tension = peak total tension developed in muscle after contraction. Active tension = total tension – resting tension = tension developed in a muscle as a result of contraction.

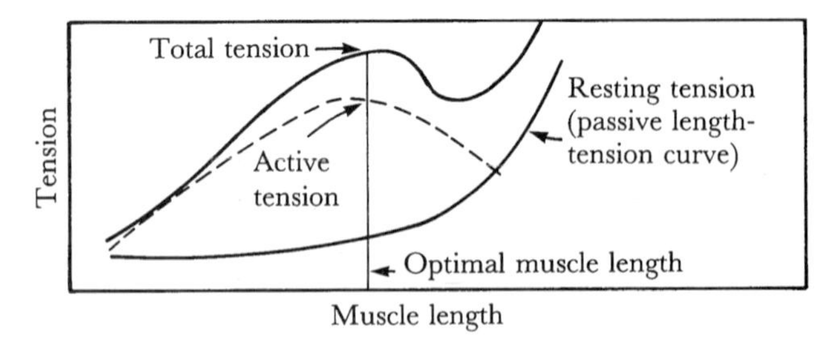

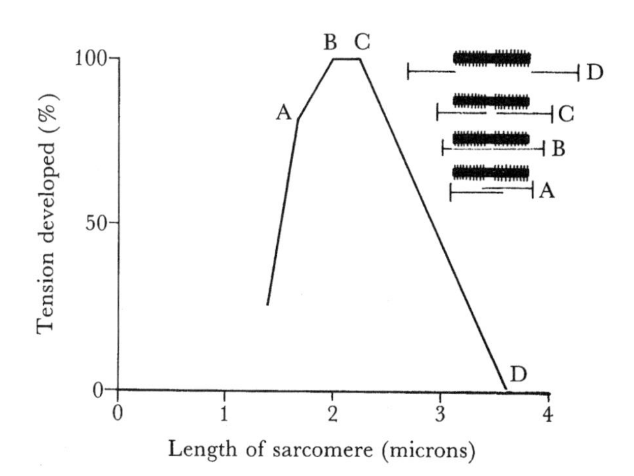

Fig. 1-20. Length-tension diagram for a single sarcomere, illustrating maximum strength of contraction when the sarcomere is 2.0 to 2.2 μ in length. At the upper right are the relative positions of thin and thick filaments at different sarcomere lengths. (From A. C. Guyton. *Textbook of Medical Physiology* [8th ed.]. Philadelphia: Saunders, 1991. P. 73.)

tral bare zone of the thick filament is devoid of myosin cross bridges and does not contribute to the overall number of actin and myosin cross links. The strength of the contraction, therefore, is constant at sarcomere lengths between B and C. If sarcomere length is decreased from B or increased from C, the degree of thick and thin filament overlap is decreased, resulting in decreased strength of contraction.

Load-Velocity Relationship

Afterload is defined as the load against which the muscle exerts its contractile force. When a muscle is undergoing isotonic contraction, the force available to cause muscle shortening is the difference between the force generated by muscle contraction and the force against which the muscle has to work, the afterload. As afterload is decreased, the force available to cause muscle shortening is increased, and therefore, velocity of contraction is increased. The maximum velocity of contraction is reached when afterload approaches zero. Conversely, as afterload is increased, the velocity of contraction is decreased. Contraction velocity approaches zero when the afterload is equal to the maximum force that the muscle can generate (Fig. 1-21).

Fig. 1-21. Force-velocity relationship for a muscle undergoing isotonic contraction with different afterloads.

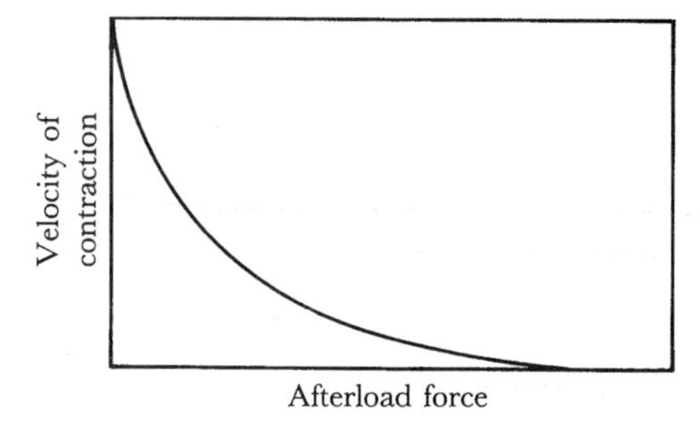

Smooth Muscle

The structural, biochemical, and mechanical properties of smooth muscle are not as well understood as those of skeletal muscle. Smooth muscles from different sites of the body vary considerably in their properties and physiologic function. There are two broad categories of smooth muscle, multiunit and single-unit smooth muscle.

Multiunit Smooth Muscle

The individual muscle cells of multiunit smooth muscle are "insulated" from one another and operate independently. Their contractile activity is controlled by neural input from the autonomic nervous system. Nonneural stimuli do not greatly influence contractile activity of multiunit smooth muscle. Examples of this type of smooth muscle include erector pili muscles in the skin and ciliary muscles in the eye.

Single-Unit Smooth Muscle

By far the most abundant type of smooth muscle, single-unit smooth muscle, is found in walls of vessels and hollow organs, such as the bladder and organs of the gastrointestinal system. In single-unit smooth muscle the individual muscle cells are connected by means of **gap junctions,** which allow the passage of ions and small molecules from one cell to the next. This essentially couples the cells electrically so that electrical stimulation of any one cell is passed on to other cells, and the muscle functions as a single unit. Contractile activity of single-unit smooth muscle is influenced not only by the autonomic nervous system, but also by nonneural stimuli, such as hormones, and by local tissue factors, such as temperature and pH.

Myosin and Actin

Smooth muscle contains thick filaments of myosin and thin filaments of actin. These interact in a similar fashion as they do in skeletal muscle. The relative ratio of these filaments is different in smooth muscle, however, as is their organization. Smooth muscle contains relatively fewer thick filaments and relatively more thin filaments. The organization of these filaments in smooth muscle is not well understood. Additionally, Z lines as observed in skeletal muscle are not defined in smooth muscle. Rather, in smooth muscle one observes **"dense bodies,"** Z-line counterparts. Dense bodies are electron-dense structures in the cytoplasm and cell membrane that anchor the thin filaments.

Excitation-Contraction Coupling

The mechanism of E-C coupling is significantly different in smooth muscle. Changes in the intracellular concentration of Ca^{2+} control the contractile state of smooth muscle just as in skeletal muscle. Smooth muscle, however, does not contain troponin (there is some tropomyosin in smooth muscle, but its function is not well understood). When Ca^{2+} concentration increases in the cytoplasm, it binds to a protein called **calmodulin.** The Ca^{2+}-calmodulin complex activates a protein kinase that phosphorylates the myosin light chain. In the resting state, the ATPase activity of the myosin head is very low. The phosphorylated myosin light chain greatly enhances the ATPase activity of the myosin head, allowing it to interact with actin in a similar fash-

Table 1-4. Distinguishing features between skeletal and smooth muscle

Feature	Skeletal muscle	Smooth muscle
Striations	Present due to the transverse register of thick and thin filaments (Z lines line up within the muscle fiber)	Not present since Z-line equivalents (or dense bodies) do not line up within the muscle fiber
Size	Large due to fusion of embryonic myoblasts	Small since myoblasts do not fuse
Shortening velocity	Fast	Slow
ATP consumption	Fast	Slow
Efficiency	High	Low
Motor end plate	Yes	No
Second messenger	Ca^{2+}	IP_3 and Ca^{2+}
Stimulus for Ca^{2+} release	Propagation of an action potential throughout sarcoplasmic reticulum	Neurotransmitter or hormone activates phospholipase C in the sarcolemma and IP_3 acts at sarcoplasmic reticulum
Ca^{2+} binding	Troponin	Ca^{2+}-dependent myosin kinase
E-C coupling	Troponin-Ca^{2+} complex causes conformational change in the thin filament, which allows cross bridge	Ca^{2+}-dependent myosin kinase phosphorylates cross bridges, allowing attachment to thin filament
Relaxation	Ca^{2+} re-uptake by active transport into sarcoplasmic reticulum and dissociation of troponin-Ca^{2+} complex	Ca^{2+} re-uptake by active transport into sarcoplasmic reticulum and dissociation of Ca^{2+}-dependent myosin kinase and dephosphorylation of cross bridges
Potential to divide	Lost	Maintained

IP_3 = inositol 1,4,5-triphosphate.

ion to the process described for skeletal muscle. The source of Ca^{2+} involved in E-C coupling is different in skeletal and smooth muscle. The Ca^{2+} released into the cytoplasm in smooth-muscle E-C coupling comes from both intracellular sources (sarcoplasmic reticulum) and extracellular sources through Ca^{2+} channels in the cell membrane. The differences between skeletal and smooth muscle are summarized in Table 1-4.

Questions

Directions: Each of the numbered items or incomplete statements in this section is followed by answers or by completions of the statement. Select the one lettered answer or completion that is best in each case.

1. All of the following factors influence membrane permeability except
   A. Concentration gradient across the membrane
   B. Interactions between membrane and particle
   C. Membrane pore size

D. Particle size
E. Temperature

2. A neutral particle, X, has a concentration (C_1) of 40 on one side of a membrane with an area of A, and a thickness of d, and a concentration on the other side (C_2) of 20. The effect of simultaneously doubling C_1, A, and d on the flux of X
   A. Cannot be determined
   B. Would result in no change
   C. Would double the flux
   D. Would triple the flux
   E. Would increase the flux 10 times
3. A cell is placed into a solution, swells, and bursts. Normal cellular osmolarity is approximately 280 mOsm. Into which solution could the cell have been placed?
   A. A 140-mM NaCl solution
   B. A 280-mM NaCl solution
   C. A 280-mM glucose solution
   D. A 580-mM glucose solution
   E. A solution of pure water
4. Which one of the following transport processes is passive (i.e., would not be affected by a disturbance in ATP synthesis)?
   A. The cotransport of an Na^+ ion and a molecule into a cell
   B. The exchange of a K^+ ion outside the cell for an Na^+ inside
   C. The exchange of an Na^+ ion outside the cell for a molecule inside
   D. The flow of H_2O through the cellular membrane
   E. The uptake of very large molecules into a cell by endocytosis
5. Protein-mediated transport allows
   A. Large and/or charged particles to be transported efficiently
   B. Transport to be regulated by the amount of protein in the membrane
   C. Transport to be regulated by voltage or ligand gating
   D. Transport to take place against a concentration gradient by coupling the process to a source of energy
   E. All of the above
6. What effect would increasing Na^+ permeability have on the membrane potential?
   A. E_m would increase (i.e., become less negative)
   B. E_m would decrease (i.e., become more negative)
   C. E_m would equal the E_{Na^+}
   D. E_m would not change
7. What would be the effect of changing the resting membrane potential, E_m, to −80 mV from −70 mV?
   A. A larger stimulus would be required to reach threshold
   B. The magnitude of the action potential would decrease
   C. The stimulus required to reach threshold would not change
   D. Threshold could not be reached
   E. Threshold would shift from −55 mV to −65 mV
8. Which of the following are known effects of neurotransmitters?
   A. Activation of internal metabolic processes
   B. Alteration of membrane permeability
   C. Alteration of receptor numbers
   D. All of the above
9. Which of the following contribute(s) to the integration of information by the postsynaptic cell?
   A. Strength of the PSP
   B. Duration of the PSP
   C. Algebraic, spatial, and temporal summation
   D. Proximity of the input to the axon hillock
   E. All of the above

10. Based on the sliding filament theory, which of the following changes would you expect in the sarcomere during contraction?
    A. The length of the A and H zones should increase
    B. The length of the A zone should increase
    C. The length of the H and I zones should decrease
    D. The length of the H zone should increase
    E. The length of the I zone should decrease
11. Ca^{2+} channel blockers would have a significant effect on muscle function in
    A. Skeletal muscle
    B. Smooth muscle
    C. Both
    D. Neither
12. Ca^{2+} binding proteins are involved in excitation contraction coupling in
    A. Skeletal muscle
    B. Smooth muscle
    C. Both
    D. Neither
13. If you were developing a drug to treat the muscle spasticity of several neurologic diseases such as cerebral palsy or multiple sclerosis, which of the following would be most useful?
    A. A drug that inhibited Ca^{2+} ATPase enzymes in the sarcoplasmic reticulum
    B. A drug that inhibited Ca^{2+} channels in the cell membrane
    C. A drug that inhibited Ca^{2+} release from the sarcoplasmic reticulum
    D. A drug that inhibited protein kinases

Directions: Each group of items in this section consists of lettered options followed by a set of numbered items. For each item, select the one lettered option that is most closely associated with it. Each lettered option may be selected once, more than once, or not at all.

Questions 14–15. Match the description concerning the development of tension within skeletal muscle to the appropriate summation process.

A. Increasing frequency of neural stimulation, thereby increasing the number of motor units that will contract
B. Increasing frequency of neural stimulation, thereby increasing the tension developed by individual contracting motor units
C. Increasing intensity of electrical stimulation, thereby increasing the tension developed by each individual contracting motor unit
D. Increasing intensity of electrical stimulation, thereby increasing the number of motor units that will contract

14. Spatial summation
15. Temporal summation

Questions 16–19. Match the letters from the figure below with the appropriate changes in K^+ concentration gradient across a cellular membrane, if the K^+ concentration gradient

16. Increases 10 times
17. Increases 100 times
18. Decreases 10 times

19. Decreases 100 times

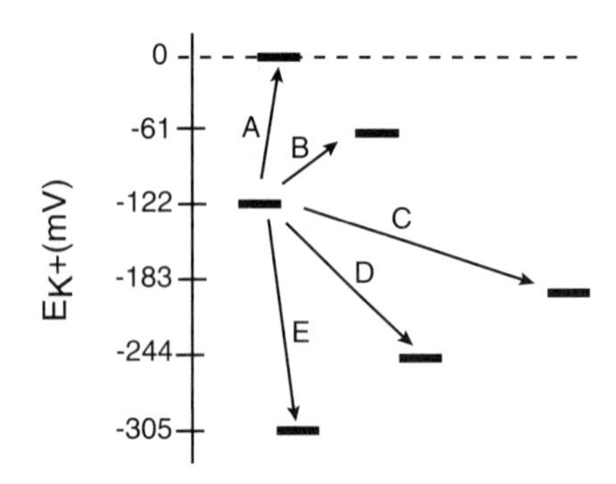

Questions 20–23. Match the letter from the figure below with the appropriate changes in ion channels and electrical potential of the membrane of a nerve cell.

20. Opening of fast-acting voltage-gated Na^+ channels, and an electrochemical potential favoring influx of Na^+
21. Open slow-acting voltage-gated K^+ channels are now slowly closing
22. Closure of voltage-gated Na^+ channels and opening of slow-acting voltage-gated K^+ channels
23. Opening of slow-acting voltage-gated K^+ channels, opening of fast-acting voltage-gated Na^+ channels, with an electric potential that no longer favors Na^+ influx

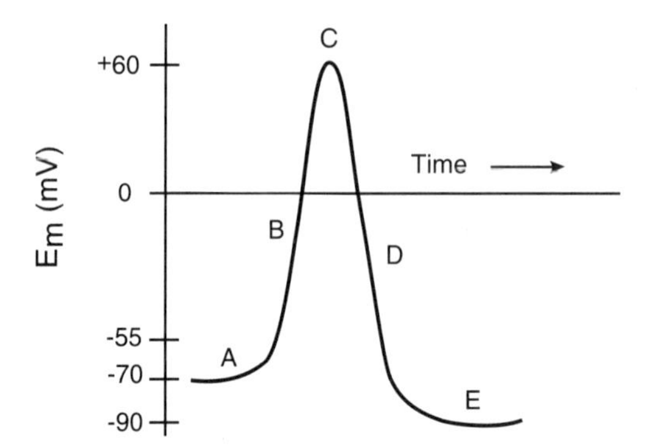

Questions 24–26. Tetrodotoxin (TTX) is a substance that specifically blocks the fast-acting voltage-gated Na^+ channels, while tetraethylammonium (TEA) is a substance that specifically blocks the slow-acting voltage-gated K^+ channels. Match the letter from the figure below with the expected response of a neuron treated with

24. TTX
25. TEA
26. TEA and TTX

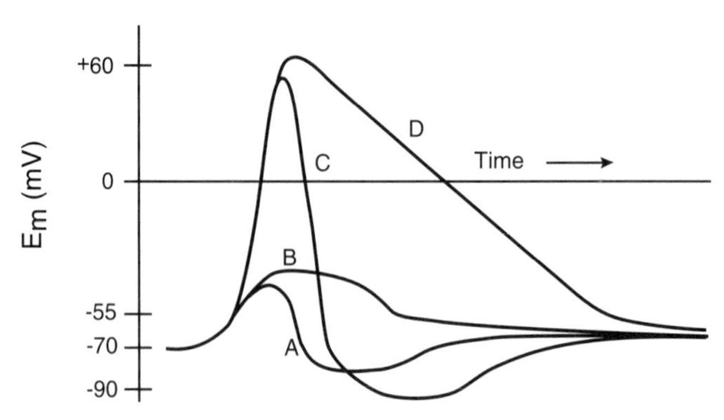

Answers

1. A A particle's size, the ambient temperature, particle-membrane interactions, and membrane pore size will all affect the permeability of a membrane to a particle. However, the concentration gradient has no influence on the properties of the membrane and only affects the flux across the membrane.
2. D The flux would increase three times. Doubling the area and the thickness together would have no effect on flux since they cancel each other in Fick's equation:

$$\text{Flux} = \frac{pA}{d}(C_1 - C_2)$$

   Increasing C_1 results in $(C_1-C_2) = 60$ (i.e., 3 times the original difference of 20). Therefore, the flux would triple.
3. E The cell would only swell in the solution of pure water. The effective osmolarity of a 140-mM NaCl solution is 280 mOsm (or isotonic with respect to the interior of the cell) since NaCl nearly completely dissolves into Na^+ and Cl^-. Therefore, there would be no change in the cell in either 140-mM NaCl or 280-mM glucose solutions. The cell would shrink when placed in hypertonic 280- or 560-mM NaCl solutions (approximately 560 and 1120 mOsm, respectively).
4. D Osmosis (simple diffusion of H_2O) would not be affected. However, primary active transport (exchange of Na^+ and K^+), the secondary active transport using the Na^+ gradient established by primary active transport, and endocytosis require energy in the form of ATP to function.
5. E Membrane protein channels may provide a conduit for large and/or charged particles. In addition, the transport of substances may be regulated by voltage or ligand gates on the protein channels. Furthermore, protein channels may be coupled to the breakdown of a chemical bond, providing energy for the transport of substances against their electrochemical gradients.
6. A E_m would increase (i.e., become less negative). E_m is equal to the average of the potentials for all ions weighted by the membrane's permeability to each ion (this is called the **Goldman equation**). Therefore, increasing the permeability of the membrane to Na^+ would shift the E_m toward E_{Na^+} (i.e., increase it). This is the basis of the action potential.
7. A The threshold of a neuron is mainly a function of the properties of the voltage gates of the Na^+ channels within the membrane. Therefore, one would not expect these to change when the E_m is decreased to –80 mV. Similarly, without changes in the voltage-gated Na^+ channels, the magnitude of the action potential would not be expected to change. Therefore, the only correct answer is that a larger depolarizing stimulus would be required to reach threshold.
8. D Neurotransmitters are known to alter membrane permeability via ligand gating, to activate metabolic processes via signal transduction within the cell, and to alter the number of receptors expressed on the cell membrane via down-regulation or up-regulation. Therefore, the answer is "all of the above."
9. E The strength and duration of the PSP; the algebraic, spatial, and temporal summation of PSPs; and the proximity of the input to the axon hillock will all contribute to the membrane potential that reaches the axon hillock. Therefore, all of the factors listed are integrated by the PoSC to determine the output.
10. C Based on the sliding filament theory and knowledge of the structure of the sarcomeres (see Fig. 1-13), one can predict that the length of the A zone depends only on the length of thick filaments, which remains constant during contraction.

The length of the H and I zones, on the other hand, depends on the degree of thick and thin filament overlap, and would therefore decrease during contraction.

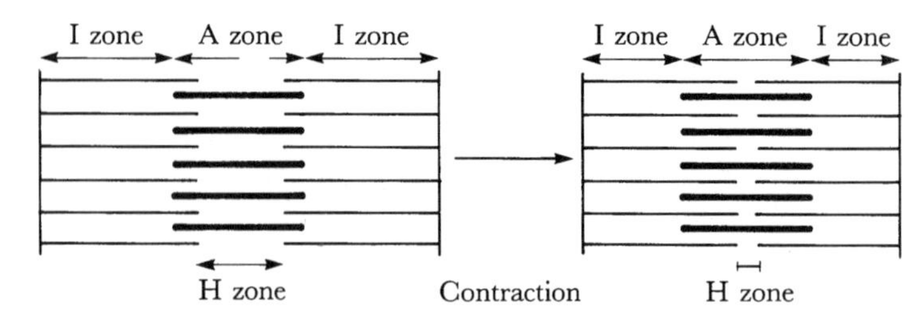

11. B Only smooth muscle depends on influx of extracellular Ca^{2+} through Ca^{2+} channels in the cell membrane for excitation contraction coupling.
12. C Both smooth and skeletal muscle contain Ca^{2+}-binding proteins that play an important role in E-C coupling. In skeletal muscle, the binding of Ca^{2+} by troponin brings about conformational changes that ultimately result in the availability of binding sites on actin for interaction with myosin. In smooth muscle, binding of Ca^{2+} by calmodulin ultimately results in enhanced ATPase activity of myosin globular heads.
13. C Inhibition of Ca^{2+} release from the sarcoplasmic reticulum would be expected to decrease spasticity in skeletal muscle. Smooth muscle function would be much less influenced by this drug because the endoplasmic reticulum Ca^{2+} stores play less of a role in smooth muscle than they do in skeletal muscle. Inhibiting the Ca^{2+} ATPase in the sarcoplasmic reticulum would result in increased intracellular Ca^{2+}. The other agents would not be as selective in inhibiting Ca^{2+} release from the sarcoplasmic reticulum.
14. D Spatial summation refers to the development of more tension within the muscle as a result of increasing stimulus intensity, which in turn activates more motor units.
15. B Temporal summation refers to the development of more tension by the individual contracting motor unit as a result of increasing stimulus frequency.
16. C Questions 17–20 require that you recognize that the inside of a cell has a high concentration of the monovalent cation, K^+, and that the simplified Nernst equation for K^+ is

$$E_x = -61.5 \log \frac{[X]_i}{[X]_o}$$

where $X = K^+$. In this instance, it becomes evident that E_{K+} will decrease by 61 mV (i.e., –61 log10).

17. D By the same reasoning as in question 16, E_{K+} will decrease by 122 mV (i.e., –61 log100).
18. B By the same reasoning as in question 17, E_{K+} will increase by 61 mV (i.e., –61 log(1/10) = 61 log10).
19. A By the same reasoning as in question 16, E_{K+} will increase by 122 mV (i.e., –61 log(1/100) = 61 log100).
20. B Once the threshold is reached, the flux of Na^+ in is greater than the flux of K^+ out. This results in a cascade of opening voltage-gated Na^+ channels, Na^+ influx, and a rapid rise in E_m.
21. E The voltage-gated K^+ channels are slow to close; therefore, the K^+ permeability remains higher than baseline for some period after an action potential. This results in a decrease in E_m and a relative refractory period to depolarization since the threshold remains at –55 mV.

22. D There is a rapid fall in E_m during the period when voltage-gated Na^+ channels are closing and voltage-gated K^+ channels are open.
23. C This is the overshoot period when the voltage-gated K^+ channels are opening and the voltage-gated Na^+ channels remain largely open.
24. A TTX would result in the blockage of the spike in the action potential since the rapid depolarization is due to the opening of voltage-gated Na^+ channels. However, the suprathreshold depolarization will still open some voltage-gated K^+ channels, resulting in a slight hyperpolarization of the membrane. This is represented by curve A.
25. D TEA would result in blockage of the rapid repolarization of the membrane seen in a normal action potential (curve C). Therefore, the E_m will return to normal under the influence of the leak channels and the Na^+-K^+ ATPase. This will take longer and will not result in a period of hyperpolarization. This is seen in curve D.
26. B TTX and TEA have specific effects on voltage-gated channels but do not affect the leak channels. Therefore, one would expect a local potential that decayed with time and distance. This simple decay with time is seen in curve B. Curve C represents a normal action potential.

Bibliography

Fawcett, D. W. *A Textbook of Histology* (11th ed.). Philadelphia: Saunders, 1986.

Ganong, W. F. *Review of Medical Physiology* (16th ed.). East Norwalk, CT: Appleton & Lange, 1993.

Guyton, A. C. *Textbook of Medical Physiology* (8th ed.). Philadelphia: Saunders, 1991.

Kandel, E. K., and Schwartz, C. H. *Principles of Neural Science* (2nd ed.). New York: Elsevier, 1985.

Stryer, L. *Biochemistry* (3rd ed.). New York: Freeman, 1988.

Vander, A. J., Sherman, J. H., and Luciano, D. S. *Human Physiology—The Mechanisms of Body Function* (6th ed.). New York: McGraw-Hill, 1994.

Weiss, L. *Cell and Tissue Biology: A Textbook of Histology* (6th ed.). Baltimore: Urban and Schwarzenberg, 1988.

West, J. B. *Best and Taylor's Physiological Basis of Medical Practice* (12th ed.). Baltimore: Williams & Wilkins, 1991.

2

Cardiovascular Physiology

Jason Goldsmith

Cardiac Electrophysiology

Cardiac Resting Membrane Potential

The Na^+-K^+ pump establishes an ion concentration gradient through expulsion of Na^+ in exchange for K^+ (see Chaps. 1 and 8). Subsequently, some of the K^+ ions diffuse out along their concentration gradient, leaving the inside of the cell negative. Cardiac resting membrane potential (V_M) can be approximated by the Nernst equation using intra- and extracellular concentrations of K^+.

$$V_M = -61.5 \log \frac{[K^+]_{in}}{[K^+]_{out}} \qquad (2\text{-}1)$$

If normal values for atrial and ventricular myocardium are used ($[K^+]_{out}$ = 4.0 mM and $[K^+]_{in}$ = 150 mM), then the approximate value obtained for V_M is –97 mV. This value is slightly more negative than the actual V_M (–85 to –95 mV) because other ions make small depolarizing contributions to the resting potential. Note that increasing the ratio of $[K^+]_{in}$-$[K^+]_{out}$ (as occurs with hypokalemia) makes the V_M more negative (hyperpolarizes the cell), and that decreasing the ratio of $[K^+]_{in}$-$[K^+]_{out}$ (as occurs when K^+ leaks out because of membrane damage) makes the V_M less negative (depolarizes the cell).

Cardiac Action Potentials

Two types of action potential (AP) are observed in the heart: (1) The **fast-response AP** is found both in the contracting cells of the atrial and ventricular myocardium and in the specialized conducting network of Purkinje fibers (Fig. 2-1). (2) The **slow-response AP** is found in the pacemaker cells of the sinoatrial (SA) and atrioventric ular (AV) nodes.

The fast-response AP is characterized by rapid Na^+ influx during phase 0 depolarization, whereas the slow-response AP shows slow Ca^{2+} influx during this period. A functional distinction between the two is that the slow AP generates the **pacemaker potential,** normally from within the SA node, whereas the fast-response AP is responsible both for transmission of the action potential along the conduction network and for depolarization of the entire myocardium, resulting in contraction.

Fast-Response Action Potentials in Cells of the Conduction Network and the Myocardium

Phase 0: **Rapid upstroke.** When the membrane is depolarized to threshold potential, **voltage-sensitive Na^+ channels** open, resulting in a rapid inward, de-

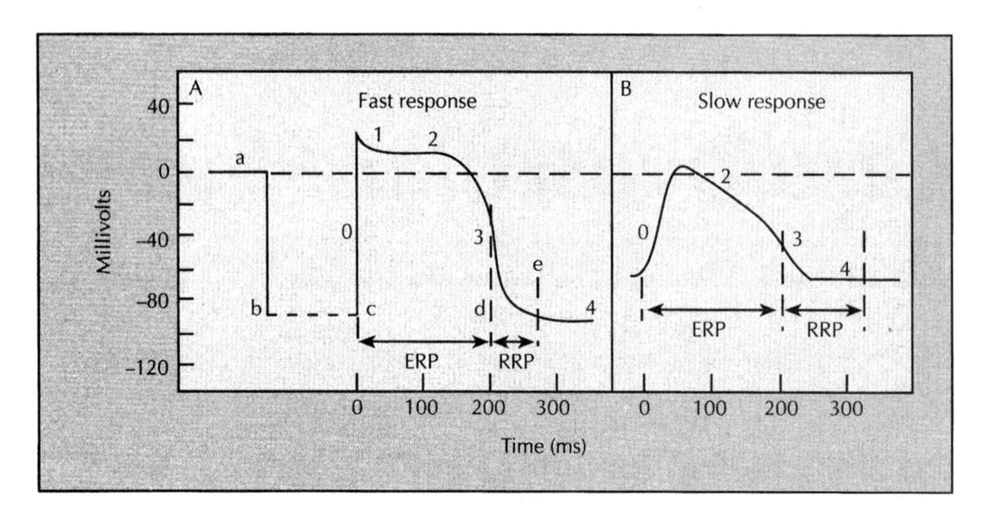

Fig. 2-1. The changes in transmembrane potential recorded from a fast-response cardiac fiber, A, and a slow-response fiber, B. Note that compared to the fast-response fiber, the resting potential of the slow fiber is less negative, the upstroke (phase 0) of the action potential is smaller, phase 1 is absent, and the relative refractory period (RRP) extends well into phase 4, after the fiber has fully repolarized. ERP = effective refractory period. (From R. M. Berne and M. N. Levy. *Cardiovascular Physiology* [6th ed.]. St. Louis: Mosby–Year Book, 1992. P. 6.)

polarizing flow of Na^+. This rapid depolarization is analogous to that seen in nerve and muscle fibers (see Chap. 1). The rate of depolarization is the upstroke velocity, or V_{max}.

Phase 1: **Partial repolarization.** During this phase, the fast Na^+ channels are inactivated and depolarization ceases; partial repolarization follows because of a small efflux of K^+. The cell is then in the **absolute refractory period,** during which its channels are refractory to further stimulation until "reset" by additional repolarization during phase 3.

Phase 2: **Plateau.** The slow efflux of K^+ in phase 1 is temporarily counteracted by the **voltage-gated inward Ca^{2+} current** (a transient slow influx of Ca^{2+}), which is part of the **excitation-contraction coupling** process (described below). These Ca^{2+} channels are blocked by the Ca^{2+} channel blocking drugs, such as nifedipine, verapamil, and diltiazem.

Phase 3: **Rapid repolarization.** The ionic bases for this phase are the gradual increase in K^+ efflux due to the activation of **voltage-gated K^+ channels** and the inactivation of the voltage-gated Ca^{2+} channels of phase 2. Repolarization resets the voltage-gated Na^+ channels of phase 0. During this phase, the cell enters the **relative refractory period** (RRP), when it becomes subject to premature activation. At the end of phase 3, the cell reaches its resting potential and exits the RRP.

Phase 4: **Diastole.** The fully recovered cell remains at the resting potential until a depolarization event that occurs in an adjacent cell is transmitted through a **gap junction.**

In the fast-response AP, Na^+ influx depolarizes the cell, and K^+ efflux repolarizes the cell, while Ca^{2+} contributes to the plateau and to excitation-contraction coupling. Note that the Na^+-K^+ pump is operating continuously to reestablish the ion concentration gradient.

Slow-Response Action Potentials in Pacemaker Cells

Phase 0: **Slow upstroke.** In SA and AV nodal cells, in which there are no fast Na^+ channels, the slower AP upstroke is mediated mainly by **Ca^{2+} channels.**

When spontaneous diastolic depolarization reaches the threshold potential (see phase 4, below), the influx of Ca^{2+} accelerates, resulting in phase 0 of the "slow" AP. The amplitude and rate of depolarization are less than in the fast-response AP.

Phase 1: This phase is absent in slow-response cells.

Phase 2: **Plateau.** The plateau is less prominent than in the conducting and contracting cells of the myocardium.

Phase 3: **Rapid repolarization.** The ionic basis for this phase is the same as for the fast-response AP.

Phase 4: **Diastolic depolarization.** The sinoatrial and atrioventricular nodes, and to some extent the Purkinje fibers, are capable of **spontaneous depolarization** due to the interplay of at least three ion flows (increased Ca^{2+} and Na^+ influx and decreased K^+ efflux), of which Ca^{2+} influx concerns us here. Spontaneous depolarization accounts for **automaticity,** the intrinsic ability of these specialized cardiac pacemaker cells to initiate rhythmic beats. This is also referred to as the **pacemaker potential.**

The slow-response AP is initiated from a reduced membrane potential (due to spontaneous depolarization), and has a low V_{max} (upstroke velocity), a very slow conduction velocity, and a prolonged relative refractory period (see the section on conduction).

Mediators of Heart Rate

The rate at which spontaneous depolarization proceeds from the resting to the threshold potential determines the heart rate (Fig. 2-2). It follows, then, that the slope of phase 4 depolarization and the levels of the resting and threshold potentials influence the rate at which the pacemaker cells fire an AP. These parameters are controlled by the autonomic nervous system through release of acetylcholine by the vagus nerve, norepinephrine by sympathetic nerves, and epinephrine by the adrenal glands.

Acetylcholine lowers the heart rate by opening specific K^+ channels that result in K^+ efflux. This, in turn, hyperpolarizes the pacemaker cell (lowering the resting potential) and reduces the slope of the pacemaker potential. Both effects lengthen the time it takes for spontaneous depolarization to reach threshold, thus slowing the heart rate.

Catecholamines (norepinephrine and epinephrine) increase the heart rate by increasing the slope of the pacemaker potential. Catecholamine receptors are coupled to adenylate cyclase, which, through a signal amplification cascade that includes cyclic adenosine monophosphate (cAMP) and a protein kinase, leads to the opening of one type of Ca^{2+} channel. Increased Ca^{2+} flow during phase 4 increases the rate of spontaneous depolarization. Increased Ca^{2+} influx in response to catecholamines also occurs in the fast-response fibers of the myocardium, resulting in more forceful contraction (see the section on excitation-contraction coupling, below). Thus, heart rate and contractility increase in response to catecholamines.

Refractory Periods

Effective Refractory Period

The effective refractory period (ERP) is the interval in which a second AP cannot be elicited. It extends from phase 0 to the middle of phase 3 (see Fig. 2-1). It is also referred to as the absolute refractory period.

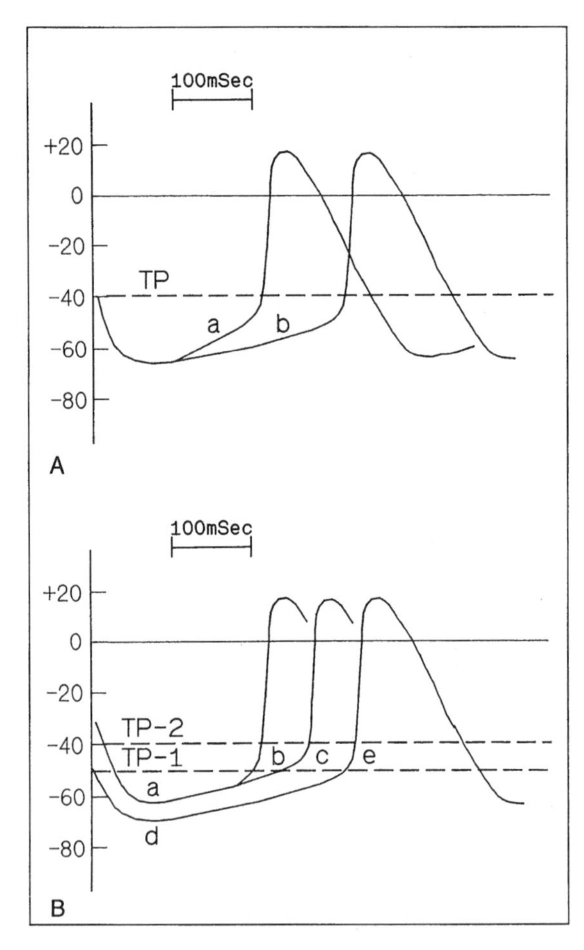

Fig. 2-2. The rate at which spontaneous depolarization proceeds from the resting to the threshold potential determines the rate of the heartbeat. Acetylcholine opens specific K^+ channels, which (1) reduce the slope of the pacemaker potential (from a to b in A) and (2) hyperpolarize the pacemaker cell (from a to d in B). Raising the threshold potential (from TP-1 to TP-2) may also decrease heart rate. (From I. R. Josephson and N. Sperelakis. Initiation and Propagation of the Cardiac Action Potential. In N. Sperelakis and R. O. Banks, *Physiology: Essentials of Basic Science*. Boston: Little, Brown, 1993. P. 266.)

Relative Refractory Period

The relative refractory period (RRP), which follows the ERP, is the interval during which a second AP may be elicited, although not at full magnitude. This has different implications for the slow- and fast-response cardiac fibers:

Fast-response AP. An AP of full magnitude cannot be elicited until the membrane is completely repolarized at the beginning of phase 4. The contractile response of the myocardial fibers is more than half over by the time a second response can be elicited. Thus, the refractory period allows time before the next contraction for the myocardial fibers to relax and for the ventricles to refill with a sufficient volume of blood.

Slow-response AP. The RRP of the slow-response cell is longer than that of the fast-response cell. This fact prevents slow-response fibers (pacemaker cells) from initiating premature full-magnitude beats before the fast-response cells are completely recovered. Although fast-response fibers can potentially initiate full-magnitude action potentials before slow-response fibers (because their RRPs are shorter), they are unable to do so because they do not undergo spontaneous depolarization. Therefore, the slow-response fibers are the normal pacemakers of the heart.

Normal and Ectopic Pacemakers

The **SA node** is the normal (dominant) pacemaker of the heart, as it undergoes spontaneous depolarization at the highest rate (as compared to the other natural pacemakers). Depolarization spreads from the SA node to the rest of the myocardium before any other potential pacemakers are able to discharge spontaneously, a phenomenon known as **overdrive suppression** of latent (or subsidiary) pacemakers. As **AV nodal cells** discharge at the second highest rate, followed by the **His-Purkinje fibers,** either can serve as pacemakers if damage occurs to the SA node or its conduction pathway.

In the event of myocardial injury or disease, damaged cells may take over as pacemakers. For instance, if a damaged cell becomes leaky to ions, it may spontaneously depolarize at a faster rate than the SA node, thus becoming an **ectopic pacemaker.** This is one example of a dysrhythmia (see below).

Excitation-Contraction Coupling

Excitation-contraction coupling in cardiac muscle is similar to that in skeletal muscle in that excitation (depolarization) is coupled to contraction via calcium. The mechanisms, however, are slightly different. In skeletal muscle, depolarization (caused by Na^+ influx) results in Ca^{2+} release from the sarcoplasmic reticulum. In cardiac muscle, Ca^{2+} that enters during the plateau phase mediates additional release of Ca^{2+} from the sarcoplasmic reticulum and also directly interacts with troponin (Fig. 2-3). The Ca^{2+}-troponin complex binds tropomyosin and facilitates the interaction between actin and myosin that produces contraction. The Ca^{2+} is removed by re-uptake into the sarcoplasmic reticulum and by extrusion via a Ca^{2+}-Na^+ pump located in the plasma membrane, resulting in relaxation. Finally, the Na^+-K^+ ATPase restores the Na^+ and K^+ balance.

Conduction

In the normal heart, excitation originates in the SA node and then continues sequentially through the atrial myocardium, AV node, bundle of His, Purkinje system, and ventricular muscle.

An AP propagates down individual myocardial fibers by generating **local currents** (the same mechanism used in nerve and skeletal muscle). Excitation spreads from fiber to fiber through gap junctions, a process not found in skeletal and smooth muscle, where impulses in each fiber are individually initiated by direct innervation from motor neurons.

Conduction velocity in fast-response contracting and conducting cells increases with (1) increased rate of change of potential during phase 0 (upstroke velocity, or V_{max}), (2) increased amplitude of the AP, and (3) greater negativity of the resting membrane potential (V_M). A reduced V_M slows conduction velocity and may predispose to arrhythmia. Impulses propagate at the highest speed in the Purkinje system; at intermediate speed in the internodal pathways, His bundle, and ventricular muscle; and at the slowest speed in the SA and AV nodes.

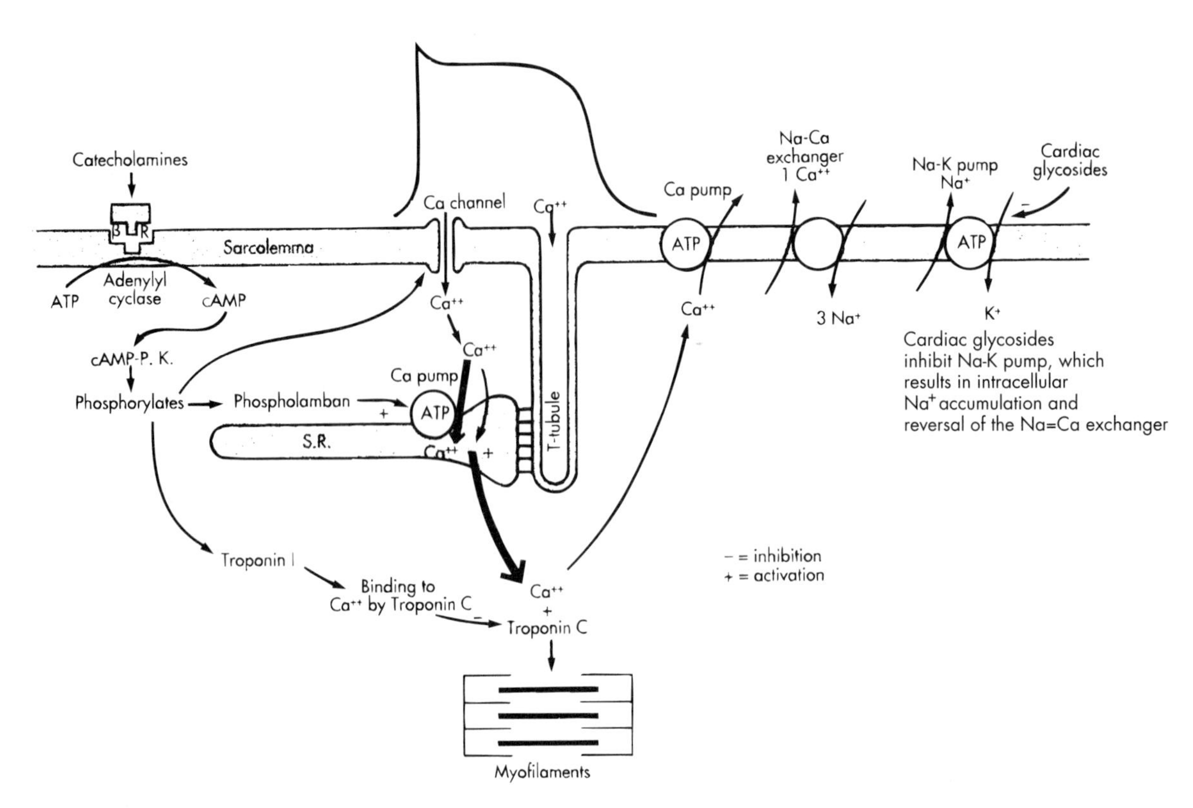

Fig. 2-3. Schematic diagram of the movements of calcium in excitation-contraction coupling in cardiac muscle. The influx of Ca^{2+} from the interstitial fluid during excitation triggers the release of Ca^{2+} from the sarcoplasmic reticulum (S. R.) The free cytosolic Ca^{2+} activates contraction of the myofilaments (systole). Relaxation (diastole) occurs as a result of uptake of Ca^{2+} by the sarcoplasmic reticulum extrusion of intracellular Ca^{2+} by Na^{+}-Ca^{2+} exchange and to a limited degree by the Ca pump. ßR = beta-adrenergic receptor; cAMP = cyclic adenosine monophosphate; cAMP-P.K. = cyclic AMP–dependent protein kinase; ATP = adenosine triphosphate. (From R. M. Berne and M. N. Levy. *Cardiovascular Physiology* [6th ed.]. St. Louis: Mosby–Year Book, 1992. P. 60.)

Conduction through the AV node does not adhere to these general principles. Conduction velocity *slows* with greater negativity of the resting membrane potential (resulting from vagal stimulation, as discussed previously). In addition, the long relative refractory period of the AV node may block one half to one third of high-frequency atrial impulses, allowing adequate time for ventricular filling.

Dysrhythmias

Most dysrhythmias result from either (1) an **abnormality in impulse generation** that arises either in the SA node or some ectopic pacemaker and which affects the rate or regularity of the heartbeat, or (2) an **abnormality in impulse conduction** that alters the normal sequence of activation of the atria and ventricles, resulting in reentrant dysrhythmia.

Events that damage myocardial cells and make the membrane potential (V_M) more positive (depolarizing the cells) may be proarrhythmic. Precipitating events include myocardial hypoxia, potassium imbalance (either increased extracellular K^+ or decreased intracellular K^+), and toxic drug levels (e.g., digoxin poisons the Na^+-K^+ pump, thus depolarizing V_M). The more positive V_M reduces membrane excitability and conduction velocity (predisposing to blockage of impulse transmission and reentrant arrhythmia) and may induce abnormal automaticity.

The Electrocardiogram

By recording the conduction of action potentials through the heart, the ECG measures changes in rate, rhythm, conduction, and, frequently, size and position of the heart.

The initial discharge from the SA node (located in the right atrial wall) is too small to be seen by ECG. The excitation spreads via the atrial tracts through the atrial myocardium and produces the P wave (Fig. 2-4). Excitation then enters the AV node (located in the interatrial septum) via internodal pathways. Meanwhile, the atrial myocardium completes its depolarization, leading to atrial contraction and ejection of blood into the ventricles. From the AV node, the excitation moves to the bundle of His, located in the interventricular septum. The conduction delay in the AV node and His bundle allows time for atrial contraction and ventricular filling. The His bundle divides into right and left bundle branches, which eventually run into the Purkinje system, fibers that ramify through the myocardium of both ventricles. These events are recorded in the ECG as follows:

P wave	Depolarization of the atria (note that atrial repolarization is masked by the QRS complex)
PR interval	Time from atrial to ventricular activation Includes conduction delay through AV node and His bundle Normally 0.12 to 0.20 seconds Prolonged in disturbances of AV conduction
QRS complex	Depolarization of the ventricular myocardium Normally 0.06 to 0.10 seconds Prolonged in disturbances of ventricular conduction pathway
QT interval	Ventricular contraction Duration varies inversely with heart rate

Fig. 2-4. Electrocardiogram showing the important deflections and intervals. A ventricular myocardial action potential is shown above it to demonstrate the relative timing between the two.

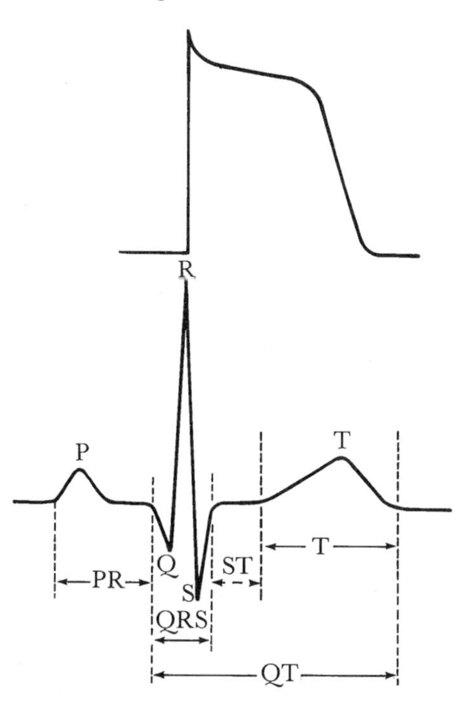

ST segment	Isoelectric line, entire ventricle depolarized Elevation may indicate myocardial injury or infarction Depression may indicate myocardial ischemia
T wave	Ventricular repolarization Normally deflected in the same direction as the QRS complex Abnormal deviation or amplitude may result from ischemia, myocardial damage, electrolyte disturbances, or cardiac hypertrophy; potentially reversible

Surface ECG waves are the result of extracellular currents produced when electrical excitation propagates down an excitable fiber. The magnitude of the current is proportional to the rate of membrane potential change. In ventricular myocardial cells, this rate of change is greatest during phase 0 (rapid depolarization) and phase 3 (rapid repolarization) of the AP. Hence, these phases correlate, respectively, with the QRS complex and the T wave (Fig. 2-4). In the normal heart, repolarization proceeds in the opposite direction of depolarization, thus accounting for the findings that R waves (upright portion of the QRS complex) and T waves are either both upright or both inverted in the same leads despite opposite charges.

In **first-degree heart block,** all atrial impulses reach the ventricles, but are slowed, producing a long PR interval. In **second-degree block,** not all atrial impulses reach the ventricles (there are more P waves than QRS complexes). **Third-degree block** is a complete block, in which the ventricles beat independent of and usually at a slower rate than the atria.

Ventricular tachycardia and fibrillation are more serious than their atrial counterparts because cardiac output is compromised. Atrial fibrillation is more benign because atrial pumping accounts for only a fraction of ventricular filling in the normal, compliant heart. However, a hypertrophied, noncompliant ventricle may depend on atrial contraction for adequate presystolic filling; in this case atrial fibrillation may have more serious consequences.

Myocardial infarction occurs when one of the coronary arteries becomes totally occluded in the face of insufficient anastomotic flow. There are three characteristic findings in myocardial infarction: (1) history of characteristic chest pain, (2) altered ECG pattern, and (3) elevation of cardiac serum enzymes.

Cardiac Cycle, Output, and Mechanics

Cardiac Cycle

The valvular events, heart sounds, pressures, volumes, and ECG of the cardiac cycle are shown in Fig. 2-5. Ventricular contraction (QT interval on ECG) coincides with a rise in left ventricular pressure (see top set of curves). When the aortic valve opens, aortic pressure rises as aortic blood flow increases from baseline (second curve from top). Note that increased aortic flow coincides with decreased ventricular volume (third curve).

Atrial Events

As no valves separate the superior vena cava from the right atrium, right atrial pressure transmits directly into the venous jugular vein. Thus, the venous pulse curve in Fig. 2-5 reflects right atrial pressure. Atrial contraction produces the venous **a wave.** Ventricular contraction produces the venous **c wave,** as the tricuspid valve closes and bulges into the right atrium. Venous delivery into the atrium continues during ven-

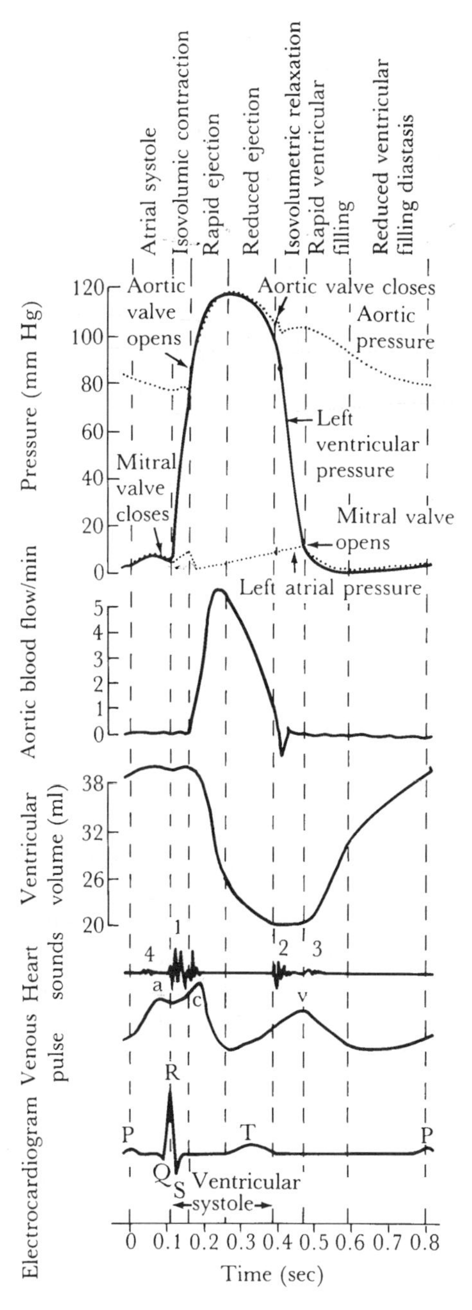

Fig. 2-5. The cardiac cycle, showing left atrial, aortic, and left ventricular pressure pulses correlated in time with aortic flow, ventricular volume, heart sounds, venous pulse, and electrocardiogram. (From R. M. Berne and M. N. Levy. *Cardiovascular Physiology* [6th ed.]. St. Louis: Mosby–Year Book, 1992.)

tricular systole and the progressive increase in atrial pressure gives rise to the venous **v wave.**

Ventricular Events

In the following discussion, note the labels at the top of Fig. 2-5. Systole begins with ventricular contraction and closure of the AV valves (mitral and tricuspid). In the first part of ventricular systole, **isovolumic contraction,** ventricular pressures rise rapidly but the volumes remain constant. When the ventricular pressures exceed those in the great arteries, the semilunar valves (aortic and pulmonic) open, and the systolic **ejection phase** begins. Ejection continues until the ventricular pressures fall below those of the great arteries. At that point, the semilunar valves close and diastole begins. The "rebound" of these valves produces the **dicrotic notch,** or **incisura,** in the aortic and pulmonic pressure curves.

In the first part of diastole, **isovolumic relaxation,** the ventricles relax and the pressures drop rapidly, yet the ventricular volumes remain constant because the AV and tricuspid valves are still closed. When the ventricular pressures fall below those of the atria, the AV valves open, marking the beginning of the **rapid filling phase.** Blood flows passively from the venous system, filling the ventricles rapidly at first and then slowly by mid-diastole. Atrial contraction provides the final measure of ventricular filling prior to systole.

Systolic pressure is the peak pressure (approximately 120 mm Hg) in the aorta during systolic ejection. **Diastolic pressure** is the lowest aortic pressure (approximately 80 mm Hg) during diastole.

Mean arterial pressure is approximately equal to the diastolic blood pressure plus one-third of the pulse pressure. **Pulse pressure** is the difference between the systolic and diastolic aortic pressures, and is a function of the volume of blood per beat, the aortic compliance, and the total peripheral resistance.

Heart Sounds

The **first heart sound (S1)** is due to closure of the AV valves (see Fig. 2-5). The mitral valve closes slightly before the tricuspid valve; an audible split is normal but not always heard. The **second heart sound (S2)** results from closure of the semilunar valves. The aortic valve closes slightly before the pulmonic valve; the split is more apparent with inspiration. The opening of normal valves is silent. A **third sound (S3)** may be heard during early diastole in children, in well-conditioned athletes (because of rapid ventricular filling), and in heart failure patients with reduced left ventricular compliance. A **fourth heart sound (S4),** in late diastole, may occur when the atrium contracts and propels blood into hypertrophied ventricles that have low compliance.

Cardiac Output and Mechanics

Cardiac output (CO) is the product of heart rate (HR) and stroke volume (SV), the latter being the volume of blood ejected in one beat.

$$\text{Cardiac output} = \text{heart rate} \times \text{stroke volume} \qquad (2\text{-}2)$$

At rest, the heart rate is about 70 beats per minute, SV is about 80 ml, and CO is about 5.6 liters per minute.

Fick's principle for the measurement of cardiac output is

$$\text{Cardiac output} = \frac{\text{oxygen consumption rate by the body}}{\text{central venous} - \text{aortic oxygen concentration}} \qquad (2\text{-}3)$$

Under normal conditions this corresponds to

$$\frac{250 \text{ ml } O_2/\text{min}}{190 - 140 \text{ ml } O_2/\text{liter blood}} = \frac{250}{50}$$

$$= 5 \text{ liters blood/min}$$

Fick's equation can be used to measure blood flow through any organ provided its oxygen consumption rate and ateriovenous oxygen difference are known.

Cardiac output can also be measured by the **indicator dilution method.** Dye, when injected into the vena cava or right heart, is diluted as it passes through the lungs and left heart. The degree of dilution is a function of the CO. Thus, if dye concentration is measured in the aorta and plotted against time, the area under the curve is a function of CO. Alternatively, cold saline can be substituted for the dye; a thermosensor, normally placed in the pulmonary artery, can measure the dilution of cold saline injected into the right atrium, and hence CO as a function of temperature change.

Factors that impact on cardiac output include heart rate, stroke volume, preload, afterload, and contractility.

Heart rate is regulated mainly by the autonomic nervous system, as described previously.

Stroke volume is equal to ventricular end-diastolic volume minus the ventricular end-systolic volume, and is a function of contractility, preload, and afterload.

Preload, the ventricular end-diastolic volume (or pressure), reflects how much the heart is stretched before contraction. Preload increases with increased venous return.

Afterload, the pressure against which the heart must pump to eject blood, is a function of the **total peripheral resistance.** At constant preload and contractility, stroke volume is inversely related to afterload. Stroke volume increases with decreased peripheral resistance (e.g., with a vasodilator). Afterload increases with increased blood pressure (e.g., when peripheral arterioles constrict in response to epinephrine).

Contractility, the force the heart is capable of developing as it contracts, can be measured by the maximum rate of pressure (or force) that develops within the ventricle.

Ejection fraction is equal to stroke volume divided by ventricular end-diastolic volume, and is an index of ventricular function. A more forceful contraction propels more blood into the arteries, leaving the ventricles at a smaller end-systolic volume. Hence, stroke volume and ejection fraction increase with contractility.

Factors that increase the contractile state of the heart are called **positive inotropic factors.** These include β_1-sympathetic stimulation (Table 2-1), increased extracellular Ca^{2+} concentration, decreased extracellular Na^+ concentration, digitalis, and increased heart rate. Many of these effects are ultimately mediated by increased Ca^{2+} influx into myocardial fibers. As discussed before, sympathetic stimulation exerts its inotropic effects by activating the adenylate cyclase–cAMP system; this opens Ca^{2+} channels, leading both to increased contractility and to increased rate of spontaneous depolarization (increasing heart rate).

Digitalis inhibits the Na^+-K^+ pump. This causes Na^+ to accumulate in the cell and forces the Na^+-Ca^{2+} pump to reverse direction (see Fig. 2-3). A rise in the extracellular Ca^{2+} also stimulates the Na^+-Ca^{2+} pump to reverse. Both mechanisms result in increased intracellular Ca^{2+} and, thus, increased contractility.

Negative inotropy results from pharmacologic blockage of the catecholamine receptors (beta-blockers), heart failure, acidosis, hypoxia, and hypercapnia.

Starling's Law

The relationship between preload and stroke volume is described by Starling's law, which states that stroke volume is a function of the degree to which heart fibers are prestretched. The ventricular function curve (Fig. 2-6), in which stroke volume is plot-

Table 2-1. Autonomic innervation of the heart and blood vessels

	Adrenergic response	Receptor involved	Intracellular effects	Cholinergic response	Intracellular effects
Heart					
Rate	Increases heart rate	β_1	⇑cAMP, which increases Ca^{2+} influx	Decrease	Opens K+ channels, which hyperpolarizes cell
Contractility	Increases contractility	β_1	⇑cAMP, which increases Ca^{2+} influx	None	None
Blood Vessels*					
Arteries (most)	Constriction	α_1	⇑Ca^{2+} release from ER via IP_3	Limited to a few sites (see text)	?
Skeletal muscle arteries	Dilation	β_2	⇑cAMP	No innervation	None
Veins	Constriction	α_2	⇑Ca^{2+} via opening of Ca^{2+} channel	No innervation	None

ER = endoplasmic reticulum; IP_3 = inositol triphosphate.

*In general, events that lead to an increase in Ca^{2+} in smooth muscle result in vasoconstriction; events that lead to an increase in cAMP or cyclic guanosine monophosphate result in vasodilation.

ted against **left ventricular end-diastolic volume** (LVEDV), illustrates Starling's law. Stroke volume increases with preload up to the point of optimal initial sarcomere length; this is when actin and myosin filaments are most ideally situated to form the maximum number of force-generating sites. When muscle fibers are stretched beyond their physiologic limits, however, cross bridges between the actin and myosin filaments cannot form and, thus, contractile force decreases; this is shown by the descending limb of Starling's curve (dashed line). The descending limb is not seen under normal physiologic conditions.

Left ventricular end-diastolic volume increases with increased **venous return** (blood volume returning to the heart per unit time). Venous return is increased by exercise, sympathetic venoconstriction, and overtransfusion and is decreased by ve-

Fig. 2-6. Ventricular function curve, where LVEDV is left ventricular end-diastolic volume and the dashed line represents the descending limb of Starling's curve.

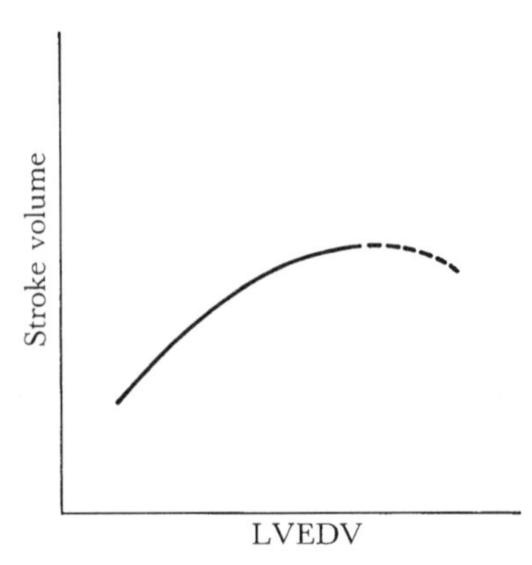

nodilation, hemorrhage, and diuretics. LVEDV fails to increase normally with decreasing **compliance** (capacity of the heart to increase in volume with increased pressure). Thus, a hypertrophied ventricle, which may have little compliance, tends to limit ventricular filling.

The ventricular function curve illustrates the relationship between contractility and stroke volume (Fig. 2-7). The middle curve represents the normal contractile state of the heart. When contractility is increased (e.g., by sympathetic stimulation), the curve shifts upward; when contractility is decreased (e.g., during heart failure), the curve shifts downward. Thus, for a constant preload, stroke volume can be adjusted by changing contractility; for a given contractile state, stroke volume can be adjusted by altering preload (see Starling's law above).

Cardiac failure is defined as inadequate CO to meet the metabolic needs of the body. Cardiac failure may result from systolic or diastolic dysfunction. **Systolic dysfunction** occurs when the heart is unable to sustain an adequate ejection fraction and is frequently the result of reduced contractility. Many pathologic entities, including ischemia and conduction disorders, predispose to systolic dysfunction. Cardiac dilatation frequently follows (see Compensatory Mechanisms, below). In **diastolic dysfunction,** the noncompliant heart—commonly the result of hypertrophy due to long-standing hypertension—does not adequately relax during diastolic filling, limiting end-diastolic volume. Although the ejection fraction may be normal, the CO is reduced.

Compensatory Mechanisms

Cardiac failure induces several interrelated compensatory mechanisms, all of which may increase cardiac output. Eventually, however, the compensatory mechanisms may exacerbate failure by setting up a vicious cycle of increased preload and afterload. Compensatory mechanisms increase CO but may overwork the heart, thereby contributing to possible hypertrophy, angina, and myocardial infarction.

Increased sympathetic activity results in increased heart rate, contractility, and stroke volume (due to α_2-mediated venoconstriction, which results in increased venous return). In addition, however, concurrent α_1-adrenergic stimulation results in arteriolar constriction and increased peripheral resistance. All these responses increase the work of the heart and may worsen the degree of failure.

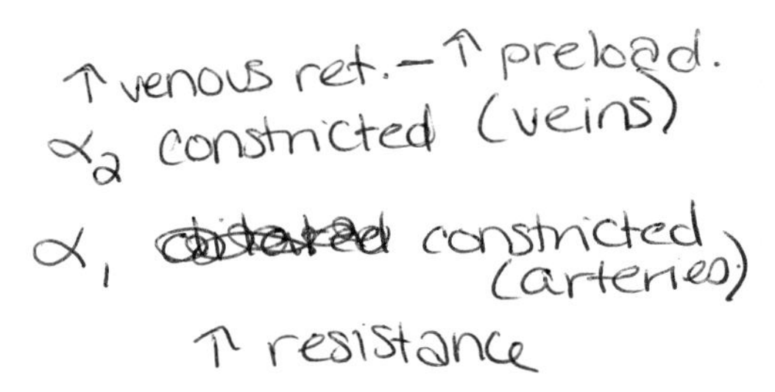

Fig. 2-7. Ventricular function curves and different contractility states.

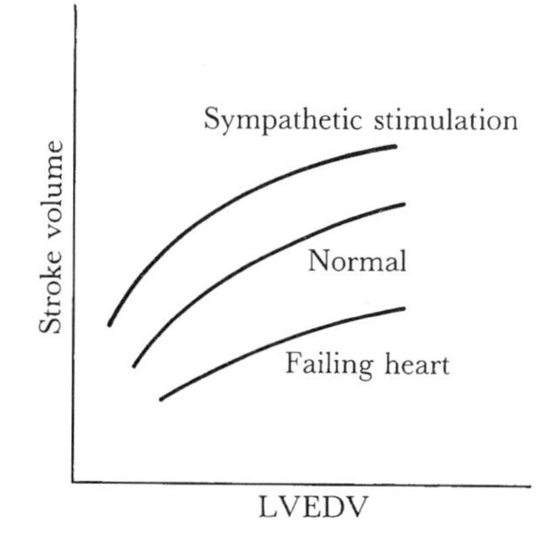

Fluid retention occurs as a result of reduced renal perfusion and activation of the renin/angiotensin/aldosterone axis. Increased venous pressure increases preload (and, thus, CO via Starling's mechanism), but also contributes to peripheral and pulmonary edema. In addition, the work of the heart increases. Dilation of the ventricles (a frequent manifestation of heart failure) may be a consequence of fluid retention. Increased ventricular volume results in increased wall tension, which, in turn, increases the myocardial oxygen requirement. Thus, an increase in ventricular volume may precipitate a myocardial infarction. Administration of a positive inotropic agent increases contractility and heart rate to produce a comparable CO at a lower end-diastolic volume, thus avoiding the deleterious consequences of fluid retention.

Myocardial hypertrophy occurs as a result of increased work. Eventually, decreased compliance may follow, resulting in decreased stroke volume.

The sequence of events in heart failure resulting from systolic dysfunction and the body's **compensatory responses** are illustrated in Fig. 2-8. Systolic heart failure results in a decrease in the contractile state of the heart, which shifts the Starling curve downward. Symptoms of congestive heart failure may develop, including pulmonary edema and dyspnea with left ventricular failure or peripheral edema with right ventricular failure. One compensatory response is fluid retention (further exacerbating edema), which increases stroke volume via the Starling mechanism. The increase in stroke volume is slight, however, because the Starling curve for failing hearts is relatively flat due to decreased ejection fraction. Additionally, if the ventricles are hypertrophied and noncompliant, they will not dilate significantly with increased venous return. Compensation also occurs through **increased sympathetic discharge,** which increases contractility and venous return, thus shifting the heart to a higher Starling curve. Note that this higher curve is steeper and thus more responsive to increased LVEDV. Treatment with digitalis increases contractility even further, thus moving the heart to a still higher Starling curve; contractility rises to a level at which sympathetic activity decreases, resulting in reduced venoconstriction and peripheral resistance. Thus, the heart moves leftward along the Starling curve to a point where cardiac output is similar to that of the normal heart (although the treated heart is functioning at a higher LVEDV).

Other therapeutic modalities for the treatment of heart failure include (1) β_1-adrenergic agonists, such as dopamine or dobutamine, which increase contractility by in-

Fig. 2-8. Congestive heart failure (CHF), compensatory steps, and digitalis therapy. A. Pump failure. B. Fluid retention. C. Sympathetic stimulation. D. Digitalis treatment. E. Decreased sympathetic stimulation.

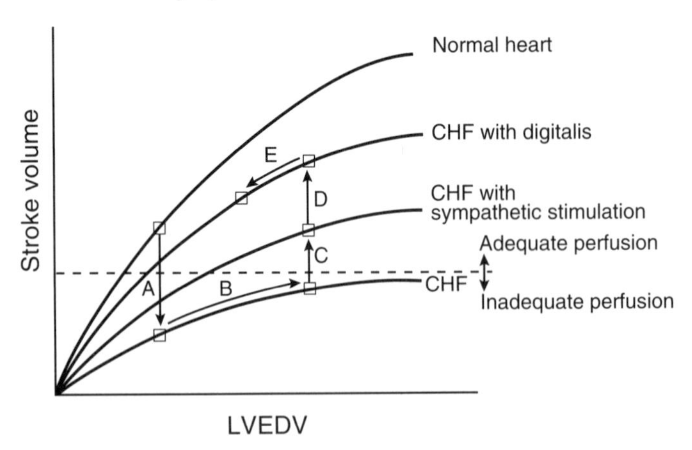

creasing Ca^{2+} flow into cardiac cells via the cAMP mechanism discussed above; (2) phosphodiesterase inhibitors, such as amrinone or milrinone, which increase contractility by inhibiting the breakdown of cAMP; (3) vasodilators, such as nitroprusside, which produce nitric oxide (discussed below) and decrease peripheral resistance and venous return; (4) selective venodilators, such as low-dose nitroglycerin (it also produces nitric oxide), which reduce venous return; and (5) angiotensin I–converting enzyme (ACE) inhibitors, such as captopril, which inhibit conversion of inactive angiotensin I to the potent vasoconstrictor, angiotensin II, and, thus, are effective vasodilators.

Pressure-Volume Curve and Cardiac Work

The changes in ventricular pressure and volume during the cardiac cycle can also be seen in the pressure-volume diagram (Fig. 2-9). Increase in the contractile state is indicated by dashed lines. **Work** done by the heart during the ejection phase is represented by the area within the pressure-volume curve (this is approximated by stroke volume multiplied by systolic pressure). Work increases with increased preload, afterload, contractility, heart rate, and wall stress. The greater the work performed, the greater is the oxygen demand and the more likely the heart is to develop hypertrophy when the workload is sustained.

In **aortic stenosis,** ejection must be carried out at higher pressures (for pressure overload, see Fig. 2-10), which requires more work (i.e., greater area within the pressure-volume loop). The heart may undergo hypertrophy and eventually fail. In **aortic insufficiency,** during diastole, blood leaks through the aortic valve into the left ventricle and causes an increase in LVEDV (volume overload). Stroke volume and ventricular work increase because of increased preload, which may lead to cardiac hypertrophy and eventual failure. Pulse pressure is narrower in aortic stenosis due to the slower rate of rise of aortic pressure during ejection; it is wider in aortic insufficiency, resulting both from a rise in systolic pressure (due to increased preload and SV) and from reduced diastolic pressure (due to backflow through the incompetent aortic valve).

Fig. 2-9. Pressure-volume loop of left ventricle for a single cardiac cycle showing MC (mitral closure), AO (aortic opening), AC (aortic closure), and MO (mitral opening). The dashed line represents the loop in a state of increased contractility. ESV = end-systolic volume; EDV = end-diastolic volume.

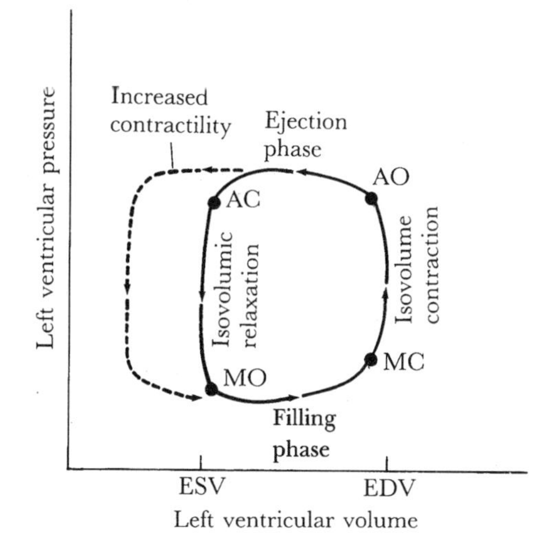

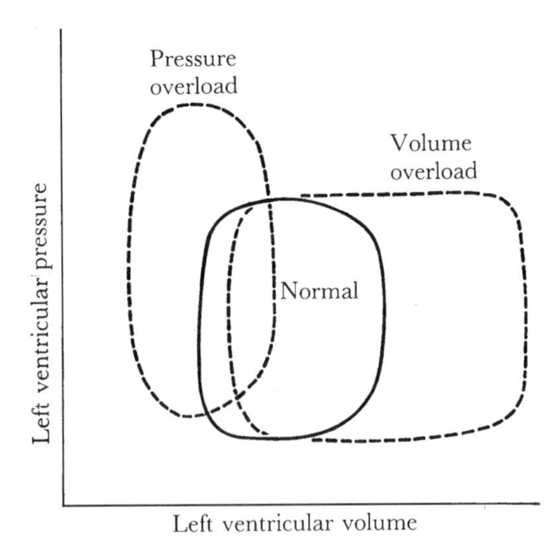

Fig. 2-10. Pressure-volume loops for normal and pathologic states. The pressure overload loop represents aortic stenosis, in which ejection must be carried out at higher pressures and, therefore, requires more work (i.e., greater area within the pressure-volume loop.) The volume overload loop represents aortic insufficiency: Aortic regurgitation causes an increase in LVEDV, leading to increased preload, stroke volume, and ventricular work.

Hemodynamics and the Peripheral Circulation

Flow, Pressure, and Resistance

The rate of flow through a single vessel is equal to the pressure difference between the ends of the vessel divided by its resistance.

$$\text{Blood flow (Q)} = \frac{\text{driving pressure}}{\text{resistance}} = \frac{\Delta P}{R} \quad (2\text{-}4)$$

Flow, Velocity, and Vessel Diameter

The rate of flow through a single vessel is also equal to the product of velocity and cross-sectional area. Consider flow through a vessel that decreases in diameter: Since flow is constant, as the diameter (or cross-sectional area) decreases, the velocity must increase. This is described by the **continuity equation** for flow, velocity (V), and cross-sectional area (A).

$$\text{Flow} = V_1A_1 = V_2A_2 \quad (2\text{-}5)$$

Thus, flow is constant and velocity varies inversely with the cross-sectional area.

The **resistance** of a vessel is directly proportional to length (L) and viscosity (η), and inversely proportional to the fourth power of the radius (r):

$$R = \frac{8\eta L}{\pi r^4} \quad (2\text{-}6)$$

Therefore, if the radius of a vessel is halved, resistance increases 16 times. This fact makes the circulation very sensitive to small changes in caliber. Resistance is also affected by blood viscosity. Viscosity is decreased in severe anemia and may be elevated in severe polycythemia, macroglobulinemia, and hereditary spherocytosis.

When Equations 2-4 and 2-6 are put together, they yield the **Poiseuille-Hagen formula:**

$$Q = \frac{\pi \Delta P r^4}{8 \eta L} \quad (2\text{-}7)$$

Conservation of Energy

Bernoulli's principle describes the interrelationship between kinetic energy ($\frac{1}{2}\rho v^2$), gravitational potential energy (ρgh), and transmural pressure (P), which is the lateral pressure against the vessel wall):

$$E = \frac{1}{2}\rho v^2 + \rho gh + P \quad (2\text{-}8)$$

in which E is energy (energy per unit volume), ρ is density of the fluid, v is flow velocity, g is acceleration due to gravity, h is the relative height of the fluid, and P is the transmural pressure. Bernoulli's principle holds true for nonviscous fluids moving in a solid tube. Although blood is viscous and blood vessels are distensible, this formula may be applied as a rough approximation.

At any given point in the circulatory system, energy is constant (minus losses due to friction) yet can be interconverted between the forms of transmural pressure (which is a form of potential energy), kinetic energy, or gravitational potential energy. For example, as blood passes through a constriction, it speeds up (obeying the continuity equation, above) as the transmural pressure is converted to kinetic energy. As another example, consider the smoothing out of pulsatile flow. Some of the kinetic energy of flow is stored as transmural pressure when elastic arteries distend with each heartbeat, and is reconverted to kinetic energy during diastole, thus filling in the diastolic gaps and smoothing out flow (note that high-resistance arterioles also dampen flow).

Another application of Bernoulli's principle can be seen in the effect of gravity on blood. If all other variables in Equation 2-8 are held constant, but height of the fluid varied, transmural pressure changes accordingly because transmural pressure and gravitational potential energy are interconvertible. This explains why blood pressures above the level of the heart are lower than those below. In fact, in the upright person, pressure in the dural sinuses is subatmospheric (the sinuses do not collapse because they are attached to the calvarium).

Bernoulli's principle can also be applied to aortic aneurysms (dilatations due to weakness in the wall). At the aneurysm, the cross-sectional area is enlarged. Consequently, velocity slows and pressure increases. The increased pressure pushes the wall out further, and the velocity decreases still more; a vicious cycle ensues that may ultimately rupture the weakened aortic wall.

An interesting point can be made regarding the orientation of pressure transducers. As shown in tube P_1 in Fig. 2-11, if the opening of a pressure transducer faces upstream, it will record a higher pressure than a tube with a tangential orientation (P_2) in the same location. This is because tube P_1 measures both transmural pressure and kinetic energy, whereas tube P_2 only measures the former. When velocity increases at the point of constriction, the transmural pressure measured in tangential tube P_4 decreases (as described above). Tube P_3, however, measures both the decreased transmural pressure and the increased kinetic energy and therefore remains unchanged. Aortic valve stenosis illustrates this concept. Velocity increases through the stenotic valve; thus, kinetic energy increases and transmural pressure decreases. The orifices

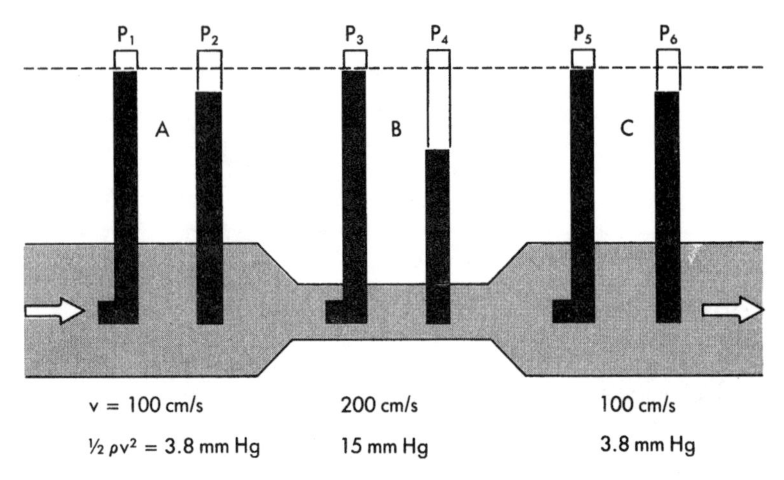

Fig. 2-11. In a narrow section (B) of a tube, the linear velocity (v) and hence the dynamic component of pressure ($1/2pv^2$) are greater than in the wide sections (A and C) of the same tube. If the total energy is virtually constant throughout the tube—that is, if the energy loss because of viscosity is negligible—the total pressures (P_1, P_3, and P_5) will not be detectably different, but the lateral pressure (P_4) in the narrow section will be less than the lateral pressures (P_2 and P_6) in the wide sections of the tube. (From R. M. Berne and M. N. Levy. *Cardiovascular Physiology* [6th ed.]. St. Louis: Mosby–Year Book, 1992.)

of the right and left coronary arteries are located just distal to the aortic valve and are oriented at right angles to the direction of aortic flow. Thus, the orifices experience the same pressure drop as the tangential pressure transducers in tube P_4. In fact, in aortic stenosis, the transmural pressure at the coronary orifices drops so low (due to increased velocity through the stenotic valve) that the direction of coronary flow often reverses in systole; that is, blood flows toward the aorta!

Wall Tension

Blood pressure across the heart and vessels is the transmural pressure, as described previously. The **wall tension** is the force that opposes this pressure and keeps the vessel intact. Wall tension is described by different formulas for the heart and vessels:

In **vessels,** wall tension (T) equals the product of pressure (P) and radius (r):

$$T = Pr \qquad (2\text{-}9)$$

In blood vessels with a smaller radius, less wall tension is needed to balance the distending transmural pressure. For this reason, capillaries with only endothelial cell walls normally do not rupture.

In a **heart** chamber, the wall tension is

$$T = Pr/2h \qquad (2\text{-}10)$$

where T is the intramyocardial tension, P is intraventricular pressure, r the radius of the ventricle, and h the thickness of the ventricular wall. When the heart dilates, wall tension increases. Ventricular wall hypertrophy may follow (increasing h), thus alleviating the wall tension of the dilated chamber(s). Hypertrophy increases the myocardial oxygen requirements, however, and thus may precipitate a myocardial infarction.

Hemodynamics of the Systemic Circulation

The relationship between pressure, velocity, cross-sectional area, and capacity of blood vessels of the systemic circulation is shown in Fig. 2-12. Note that velocity falls as cross-sectional area increases (as predicted by the continuity equation, 2-5). However, the finding for the relationship between velocity and pressure is not an inverse one, as predicted by the Bernoulli equation. Instead, pressure falls with velocity because significant amounts of energy are lost to friction as blood passes through the high-resistance arterioles. Note that the majority of the blood volume is located on the venous side of the circulation.

Resistance to flow is influenced most significantly by vessel radius. Arterioles are the major source of resistance in the systemic circulation; because their smooth muscle tone is regulated, they are a major site for regulation of mean arterial pressure as well as for the differential regulation of blood flow to meet local metabolic needs. As seen in Fig. 2-12, arterioles are the site of the largest change in resistance, pressure, and velocity.

Resistance increases in series and decreases in parallel:

$$\text{Blood vessels in series: } R_{total} = R_1 + R_2 + \ldots R_n \quad (2\text{-}11)$$

$$\text{Blood vessels in parallel: } 1/R_{total} = 1/R_1 + 1/R_2 + \ldots 1/R_n \quad (2\text{-}12)$$

For example, five identical resistors in series yield a total resistance of 5R, whereas five resistors in parallel yield a total resistance of 1/5R, a value less than any of the

Fig. 2-12. Pressure, velocity of flow, cross-sectional area, and capacity of the blood vessels of the systemic circulation. (Modified from R. M. Berne and M. N. Levy. *Cardiovascular Physiology* [6th ed.]. St. Louis: Mosby–Year Book, 1992.)

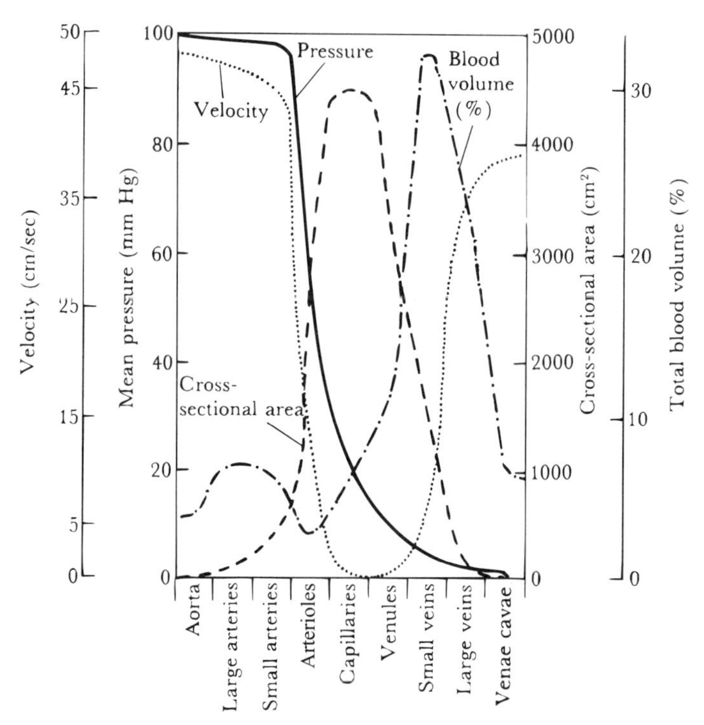

individual resistances. Don't confuse this concept with what happens when an artery ramifies into arterioles of smaller radii (and hence nonidentical resistors). Here, total resistance increases even though cross-sectional area increases, until the point that cross-sectional area is four times its pre-arteriolar value.

Arterial elasticity describes the tendency of arteries to return to a resting state after each pulse of blood flow. Elasticity, which is derived mostly from collagen, provides the **recoil** that propels blood flow during diastole. Arteries, because of high elastic content, become less compliant as they fill; thus, pressure within them rises rapidly during systole. Arterial compliance decreases with age.

Veins, in contrast, have low elastic content and are extremely compliant. They are **capacitance vessels,** as they can accommodate large increases in volume with slight increases in pressure.

Transcapillary Flow and Starling's Hypothesis

The forces that drive fluid across the capillary wall are oncotic and hydrostatic pressures. The net flow of fluid across the capillary wall (transcapillary flow) results from the balance of these forces and is described by **Starling's hypothesis:**

$$\begin{aligned}\text{Transcapillary flow} &= \text{outward forces} - \text{inward forces} \\ &= k\,[(P_{cap} + \pi_i) - (P_i + \pi_{cap})]\end{aligned} \qquad (2\text{-}13)$$

where P_{cap} is capillary hydrostatic pressure, π_i is interstitial fluid oncotic pressure, P_i is interstitial fluid hydrostatic pressure, π_{cap} is capillary (plasma) oncotic pressure, and k is the filtration constant for the capillary wall.

The capillary hydrostatic pressure (P_{cap}), the principal force in capillary filtration, forces fluid from the capillary into the interstitial (tissue) space. P_{cap} decreases from the proximal to the distal end of the capillary as a result of resistance; this gradient in hydrostatic pressure along the capillary results in filtration at the arterial end and absorption at its venous end. P_{cap} rises with increasing arterial or venous blood pressure, and may result in edema formation. For example, in the event of left-sided heart failure, increasing pulmonary venous pressure may result in pulmonary edema.

P_{cap} is opposed by P_i, the interstitial fluid hydrostatic pressure (approximately 5 mm Hg) and by π_{cap}, the plasma protein oncotic pressure (about 25 mm Hg), which provides an inward oncotic force. The interstitial fluid oncotic pressure (about 1 mm Hg), which results from the small amount of albumin that escapes from the capillary, counters the inward oncotic force.

The balance of fluid exchange is maintained by two mechanisms: (1) regulation of the capillary pressure, P_{cap}, by altering the ratio of precapillary to postcapillary resistance via adjustments in arteriolar and venous resistance, and (2) excess interstitial fluid being taken up by the lymphatic circulation.

Lymphatic Circulation

The lymphatic circulation functions to remove tissue fluid that has not been reabsorbed in the capillaries and to bring this fluid into the circulation. Lymph flow is due to four factors: (1) pumping action of skeletal muscle movement, (2) negative intrathoracic pressure during inspiration, (3) suction effect of the high-velocity blood flow in the great veins into which the lymphatics drain (Bernoulli's principle), and (4) rhythmic contractions of the walls of the large lymph ducts. Both veins and lymph vessels have valves that prevent backflow.

Cardiovascular Regulation

Goals

The primary goal of cardiovascular regulation is to maintain an adequate pressure head to perfuse and meet the metabolic needs of the tissues. This is accomplished by adjustments in CO and total peripheral resistance (TPR). Regional blood flows are differentially regulated to meet local needs, while protecting the critical cerebral and coronary blood flows. The rate of transcapillary flow (described above) is regulated by altering the ratio of precapillary to postcapillary resistance. Venous return, which impacts CO via Starling's law, is regulated through adjustments in venous capacitance.

Regulation of Mean Arterial Pressure

Two factors regulate mean arterial pressure (MAP) and, thus, flow to all vascular beds in the body: cardiac output (CO) and total peripheral resistance (TPR). This relationship is described by the equation

$$\text{MAP} = \text{CO} \times \text{TPR} \tag{2-14}$$

As described previously, CO is equal to the product of stroke volume (SV) and heart rate (HR). Thus,

$$\text{MAP} = \text{SV} \times \text{HR} \times \text{TPR} \tag{2-15}$$

Integrated Control of Arterial Pressure by the Central Nervous System

The CNS regulates cardiovascular function by influencing sympathetic and parasympathetic output and by regulating the release of humoral factors. In addition, the CNS receives and integrates a great deal of information regarding cardiovascular function—information regarding pressure from the baroreceptors, arterial oxygenation from the carotid and aortic bodies, and a variety of other physiologic signals, many of which arrive in the form of humoral factors.

Cardiovascular Regulation by the Autonomic Nervous System

The parasympathetic system primarily controls heart rate. The sympathetic system controls heart rate, heart contractility, arteriolar resistance, and venous capacitance. Neural controls for the two systems lie within specific nuclei in the medulla. The medullary activities, in turn, are modulated by peripheral baroreceptors and chemoreceptors and by inputs from higher brain centers.

para – controls HR

sym - controls HR
HC
AR
VC

Sympathetic regulation of the heart is more extensive than parasympathetic regulation. The major sympathetic neurotransmitter is norepinephrine, although bloodborne epinephrine from the adrenal medulla also plays a role. The cardiac adrenergic receptors are of the β_1-adrenergic type. Responses to sympathetic stimulation include: (1) SA node stimulation with increased heart rate, (2) atrial stimulation with increased contractility and conduction velocity, (3) AV node and His-Purkinje system stimulation with increased automaticity and conduction velocity, and (4) ventricular stimulation with increased contractility and conduction velocity. The overall effect of sympathetic stimulation on the heart is a rise in cardiac output due to a faster rate and a larger stroke volume.

Parasympathetic regulation of the heart is via the vagus nerve, which acts predominantly on the SA and AV nodes and the atria. Increased vagal activity slows the heart rate by hyperpolarizing the pacemaker cell and by reducing the slope of the pacemaker potential (as discussed previously). At rest, the **cardioinhibitory center** (or **vagal center**) in the medulla initiates a **tonic vagal discharge,** thereby maintaining a slower heart rate than would occur in a denervated heart. Less important parasympathetic effects on the heart include decreases in both atrial contractility and conduction velocity in the AV node.

Regulation of the peripheral circulation is under dual control by the autonomic nervous system and by bloodborne agents, many of which are locally produced and influence nearby blood vessels. In addition, vascular beds exhibit **autoregulation,** which is a myogenic response to stretch; changes in perfusion pressure are met by changes in vascular resistance such that a constant blood flow is maintained.

Autonomic Control of Vascular Smooth Muscle Tone

Unlike most organ systems, which receive both sympathetic and parasympathetic innervation, nearly all blood vessels of the body are supplied solely with sympathetic nerve fibers. As summarized in Table 2-1, whether a given vascular bed dilates or constricts in response to sympathetic stimulation depends on its endowment of postsynaptic receptors.* For example, skeletal muscle, coronary arteries, and liver arterioles, which are richly endowed with β_2 receptors, vasodilate in response to sympathetic stimulation. On the other hand, skin vascular beds, as well as those in many other locations, are endowed with α_1 receptors that induce vasoconstriction in response to sympathetic stimulation. Thus, in general, β_2 stimulation causes vasodilation and α_1 stimulation causes vasoconstriction. For example, systemic vascular release of epinephrine by the adrenal medulla during "fight or flight" situations results in dilation of skeletal muscle vascular beds (β_2 stimulation) and constriction of skin vascular beds (α_1 stimulation).

There is also a **tonic sympathetic discharge** from the **vasomotor center** in the medulla, which, on balance, stimulates a larger number of α_1 and α_2 receptor–regulated vascular beds than those regulated by β_2 receptors, thus maintaining a state of partial vasoconstriction. The result of this tonic sympathetic activity is increased TPR and elevated blood pressure due to α_1-mediated vasoconstriction, as well as greater venous return due to α_2-mediated venoconstriction.

All of these responses are brought about by various intracellular processes, as summarized in Table 2-1 and Fig. 2-13. In general, it is believed that vasoconstriction is initiated by intracellular events that lead to increased intracellular Ca^{2+}, and that vasodilation is initiated by intracellular events that result in increased levels of cAMP or cyclic guanosine monophosphate (cGMP), which, in turn, lead to a decrease in intracellular Ca^{2+}.

Systemic and Local Regulatory Factors

The vasculature responds to various bloodborne **systemic regulatory factors,** including the vasoconstrictive agents angiotensin and vasopressin and the vasodilatory agents bradykinin, atrial natriuretic peptide, and histamine. In addition, **local tissue metabolites** regulate flow in many tissues. For example, in skeletal muscle, the vasodilatory effects of local metabolites produced or depleted during exercise are more

* Parasympathetic innervation does exist to the vasculature of the head, viscera, genitalia, bladder, and large bowel. However, because these represent a small proportion of the resistance vessels of the body, their effect on total peripheral resistance is small.

important than adrenergic stimulation. Metabolites that mediate vasodilation include lowered oxygen tension and increased carbon dioxide and hydrogen ion, as well as increases in adenosine triphosphate (ATP) and nitric oxide.

Many of these vasodilatory agents (e.g., histamine and bradykinin) are thought to act via **nitric oxide,** which may be a common pathway leading to vasodilation (see Fig. 2-13). It is believed that nitric oxide is produced by the vascular endothelium; hence, nitric oxide is also referred to as endothelium-derived relaxation factor, or EDRF. The endothelium converts arginine to nitric oxide, which then diffuses into vascular smooth muscle and combines with the heme group in guanylate cyclase, resulting in cGMP formation. A cascade of intracellular events then leads to relaxation of vascular smooth muscle.

Various Bloodborne, Systemically Active Factors That Regulate Cardiovascular Function

Vasopressin (antidiuretic hormone) increases blood pressure by stimulating renal tubules to retain water, and by direct action on blood vessels, leading to vasoconstriction. Vasopressin is released by the posterior pituitary in response to (1) decreased blood pressure (stimulates baroreceptors), (2) dehydration and high sodium intake (stimulate hypothalamic osmoreceptors), (3) surgery and anesthesia, and (4) trauma and stress.

Angiotensin II increases blood pressure by (1) direct action on blood vessel smooth muscle (increases intracellular Ca^{2+}), (2) stimulating the central and peripheral nervous systems to increase sympathetic nervous activity, (3) stimulating the CNS to release vasopressin, and (4) inducing the release of aldosterone by the adrenal gland, which stimulates the kidney to retain Na^+ and H_2O. Angiotensin II is produced via a biosynthetic cascade that begins with the production of renin by the kidney in response to low renal arterial pressure or low concentration of filtered sodium in the kidney.

Atrial natriuretic peptide reduces blood pressure by stimulating Na^+ and H_2O loss (natriuresis) and by inducing relaxation of vascular smooth muscle. Atrial natriuretic peptide is formed in the myocardium and is released in response to increased atrial pressure, as occurs with volume overload and a sustained increase in venous return. Its vascular effect is via receptor-regulated production of nitric oxide.

Kinins, such as bradykinin, are extremely potent vasodilators that act via nitric oxide (Fig. 2-13). Bradykinin relaxes arterioles, increases venular permeability, and induces extravascular smooth muscle contraction. Tissue injury activates the kinin system, through which high-molecular-weight kininogen activates a cascade of bloodborne factors that lead to bradykinin formation.

Histamine has a wide tissue distribution in mast cells, basophils, and platelets. It is released by mast cells in response to a variety of stimuli, including physical injury, such as trauma or heat, and immune- and complement-mediated reactions. Histamine causes arteriolar dilation via nitric oxide production, increases venular permeability, and induces extravascular smooth muscle contraction (anaphylactic constriction of bronchiolar smooth muscle).

Regulation of Total Peripheral Resistance

In summary, four factors have been discussed that impact on TPR: (1) nerves (mainly sympathetic, some of which mediate constriction [α_1 receptors] and some dilation [β_2 receptors]); (2) bloodborne factors; (3) autoregulation, a myogenic response to stretch; and (4) vasodilatory actions of tissue metabolites, which counteract the basal tone imposed on most vessels by the centrally directed sympathetic vasoconstrictor influence.

TPR maintained by:
1) sym. nerves
2) blood borne (ADH etc)
3) auto reg.
4) vasodilators.

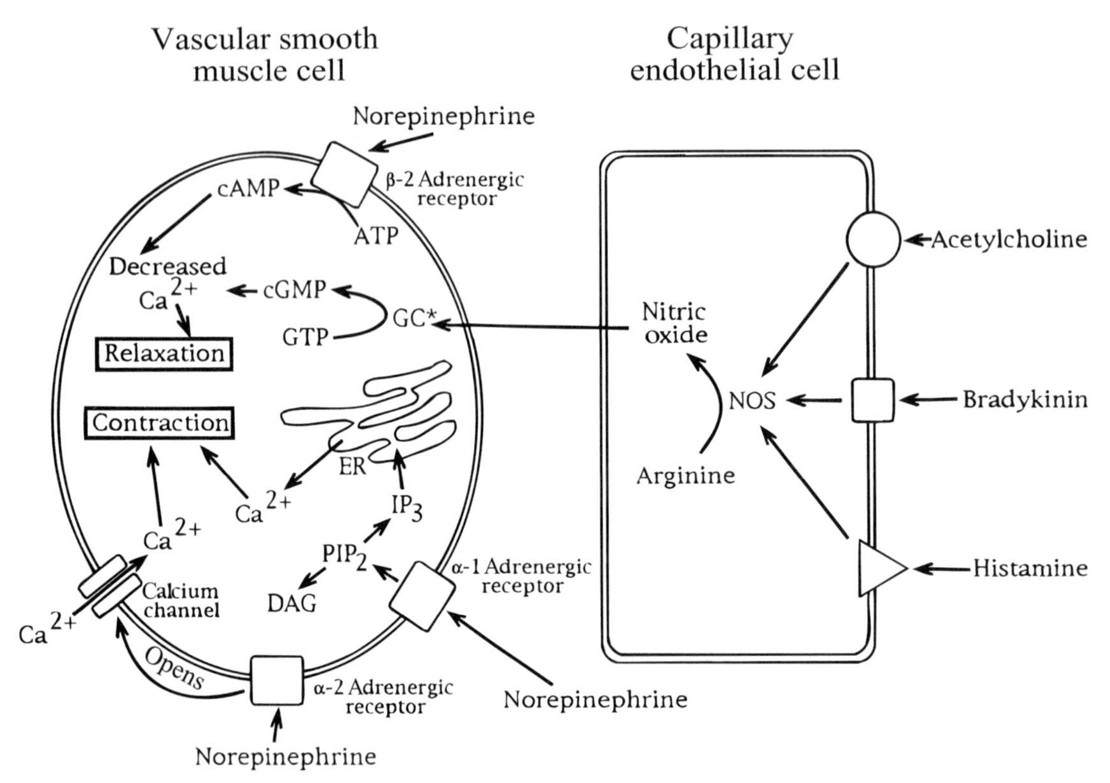

Fig. 2-13. Regulation of vascular smooth muscle tone by the autonomic nervous system and the vascular endothelium, both of which regulate the level of intracellular Ca^{2+}. Sympathetic stimulation results in vasodilation or constriction, depending on the endowment of receptors on a particular fiber. In response to stimulation by various agents, the endothelium converts arginine to nitric oxide, which then diffuses into vascular smooth muscle and combines with the heme group in guanylate cyclase (GC), resulting in cyclic guanosine monophosphate (cGMP) formation. A cascade of intracellular events then leads to relaxation of vascular smooth muscle. NOS = nitric oxide synthase; PIP_2 = phosphatidylinositol biphosphate; IP_3 = inositol triphosphate; DAG = diacylglycerol; GTP = guanosine triphosphate; * = GC activated by binding with nitric acid; ER = endoplasmic reticulum.

Cardiovascular Reflexes

The medullary cardiovascular centers receive inputs from special baro- and chemoreceptors in the periphery. **Peripheral arterial baroreceptors** have an important role in short-term regulation of blood pressure, whereas long-term regulation is the responsibility of the kidney via its regulation of fluid balance. There are no central baroreceptors. Arterial baroreceptors, located at the bifurcation of each common carotid artery (carotid sinus) and in the aortic arch, are sensitive to stretch and send afferent fibers via cranial nerves IX and X to the medulla. The vasomotor center in the medulla normally transmits tonic vasoconstriction. Increased arterial pressure stretches the peripheral baroreceptors, which reflexively excite the cardioinhibitory (vagal) center and simultaneously inhibit the vasomotor (sympathetic) center. These responses slow the heart and decrease sympathetic discharge to the vessels, resulting in arteriolar and venous dilation. The net effect is a reduction in cardiac output, peripheral resistance, and blood pressure (see Equation 2-14). A decline in arterial blood pressure produces opposite effects: A decrease in parasympathetic and an increase in sympathetic discharge induce a faster heart rate and vasoconstriction. In this manner, peripheral baroreceptors regulate blood pressure on a moment-to-moment basis.

Central chemoreceptors are located in the ventral medulla. They are important in minute-to-minute control of respiration as they respond to decreased arterial pH (reflecting increased arterial CO_2 levels). **Peripheral chemoreceptors** (the **aortic and carotid bodies**) respond primarily to arterial hypoxemia, but also to carbon dioxide tension and pH. Stimulation of both the central and peripheral chemoreceptors primarily increases respiration, but also induces vasoconstriction and an increase in heart rate.

Higher brain centers can directly activate the medullary cardiovascular centers. Excitement, fear, anger, and painful stimuli accelerate heart rate and raise blood pressure; other stimuli, such as grief, may slow the rate and lower pressure.

Several other important factors also affect the heart and blood vessels. Inspiration increases heart rate and blood pressure; expiration does the opposite. A rise in intracranial pressure reduces blood flow to the medullary vasomotor center, which then increases its rate of discharge. The resultant vasoconstriction raises the blood pressure, which in turn slows the heart rate. An increase in body temperature is sensed by the hypothalamus, which reduces α_1-mediated sympathetic stimulation of skin blood vessels, leading to vasodilation, increased peripheral flow, and radiation of excess heat.

The integration of these regulatory mechanisms can be seen with movement from the supine to the upright position. Gravity causes blood to pool in the venous capacitance vessels of the lower extremities, which reduces venous return. This reduces stroke volume and thus cardiac output, which in turn lowers the blood pressure. The initial drop in blood pressure is sensed by baroreceptors in the carotid sinus and aortic arch; the compensatory responses are an increased heart rate and vasoconstriction. Increased total peripheral resistance returns arterial blood pressure to normal. Without such compensatory responses, the reduction in cardiac output would compromise cerebral blood flow, which might result in fainting. If the individual moves around, the pumping action of muscle also assists the venous return.

Exercise also illustrates cardiovascular regulation. During exercise, the heart increases its output by increasing both rate and stroke volume. The increase in rate, due to decreased parasympathetic and increased sympathetic discharge, accounts for most of the increase in cardiac output. The larger stroke volume is attributable to the higher contractile state as a result of sympathetic stimulation; the increased venous return is due to sympathetic venoconstriction and the pumping action of muscle during exercise. In the periphery, cardiac output is redistributed to favor working muscles. Local metabolites released by working muscle induce arteriolar vasodilation and increase regional blood flow. In nonworking organs, sympathetic vasoconstriction reduces blood flow. The net result is that systolic blood pressure is raised modestly, while diastolic blood pressure remains relatively constant. The mean blood pressure is constant or slightly elevated and the systemic vascular resistance is significantly lowered. The muscle tissue also extracts a greater amount of oxygen per unit blood volume during exercise, producing a wider arteriovenous oxygen difference. The body is capable of increasing oxygen consumption up to 20 times during exercise. To meet this demand, the heart rate may increase by a factor of 4, the stroke volume by 1.5, and the arteriovenous oxygen difference by 3.3. Note that the increase in CO observed during exercise is primarily due to an increase in heart rate, for diastole shortens in duration, thus limiting presystolic filling and, hence, stroke volume.

Circulation Through Specific Organs

Blood flow and oxygen consumption rate per gram of tissue are exceptionally high in the heart, kidney, liver, and brain, organs that are metabolically highly active. On a per gram basis, the kidneys receive by far the largest amount of blood of any organ (except the carotid body) and the heart consumes oxygen at the highest rate.

Cerebral Circulation

The blood flow through the brain is about 750 ml per minute. Because the brain is extremely sensitive to ischemia, its circulation is regulated in such a way that total cerebral blood flow (CBF) remains constant despite changes in blood pressure. Applying Equation 2-4 to the cerebral circulation, overall CBF can be defined in terms of mean arterial pressure (MAP), intracranial pressure (ICP), and cerebrovascular resistance (CVR).

$$CBF = \frac{MAP - ICP}{CVR} \qquad (2\text{-}16)$$

MAP depends on cardiac output and total peripheral resistance. CVR is a function of intracranial pressure, cerebral arteriolar caliber, and blood viscosity. Because the cranium has a fixed volume, an increase in intracranial pressure compresses the cerebral vessels, which increases the ICP, thus decreasing the numerator in Equation 2-16, and the rate of cerebral blood flow.

The roles of sympathetic innervation and circulating vasoactive substances are minimal in cerebral vessels. In fact, it can be argued that cerebral circulation is controlled by the regulation of the rest of the circulation so that cerebral pressure remains constant. Cerebral flow does respond, however, to metabolic states of the body. For instance, cerebral vessels dilate in response to increased carbon dioxide or decreased oxygen concentration in the blood, and flow is restricted with decreased carbon dioxide concentration, as occurs in hyperventilation. Cerebral flow does not change with the pH of the blood, as the endothelial component of the blood-brain barrier prevents H^+ ions from diffusing into arteriolar smooth muscle; small nonpolar molecules, such as oxygen and carbon dioxide, can, however, cross the barrier. Increased vascular carbon dioxide (hypercapnea) reduces the pH of the cerebrospinal fluid (due to carbonic acid formation), which in turn stimulates increased respiration but probably has no effect on cerebral vascular resistance.

The vascular diameter is regulated at regional levels within the brain by locally produced metabolic factors. Regional blood flow correlates well with regional metabolic activity. For instance, stimulation of the retina with light increases flow to the visual cortex. Several substances are under consideration for this metabolic regulation, including K^+, adenosine, and pH (H^+ ion produced on the parenchymal side of the blood-brain barrier can interact with arteriolar smooth muscle).

Cerebral circulation also utilizes autoregulation (probably via a myogenic mechanism) to maintain a constant CBF. Although MAP is normally 90 mm Hg, because of autoregulation, it can rise to 150 mm Hg or fall to 60 mm Hg without compromise of CBF.

Under unusual circumstances, the brain regulates its blood flow by changing systemic blood pressure. Elevated intracranial pressure, as may result from a cerebral tumor or trauma, reduces flow. The vasomotor center (in the medulla) responds to cerebral ischemia by increasing systemic blood pressure, in a response known as **Cushing's phenomenon.**

Pulmonary Circulation

While a constant flow is maintained in the cerebral circulation, the pulmonary circulation maintains a constant pressure. Total pulmonary flow is not regulated; all of the output from the right ventricle flows through the lungs and returns to the left heart so that flows in the right and left ventricles are equal. There is, however, regional regulation of pulmonary flow via **hypoxic vasoconstriction,** whereby blood flow is routed from poorly to well-ventilated areas of the lung, thus maintaining optimum ventilation-perfusion ratios (see Chap. 4). In fact, pulmonary arterial smooth muscle is distinguished from all other vascular smooth muscle in that it responds to hypoxia with constriction and to high oxygen content with dilation.

Unlike systemic blood vessels, pulmonary vessels have thin walls, scant smooth muscle, and little neural regulation. The pulmonary circulation is a low-pressure system (systolic, 20 mm Hg; diastolic, 10 mm Hg) and a low-resistance system (2 resistance units compared to 20 resistance units in the systemic vasculature). The vessels are capable of autoregulation in that resistance can be decreased in the face of an increased cardiac output. This mechanism is so effective that there is no change in pulmonary artery pressure until cardiac output increases 2.5 times. This is accomplished by recruitment of additional pulmonary capillaries and distention of capillary beds.

Flow increases from the top to the bottom of the lung. This occurs because the higher blood pressures in the lower lung fields can overcome the alveolar pressure, which limits flow in the upper lung fields by forcing the capillaries and veins to collapse (see Chap. 4). Applying Equation 2-4 to the pulmonary circulation, pulmonary flow—or right ventricular CO—can be related to pulmonary artery pressure (PAP), left atrial pressure (LAP), and pulmonary vascular resistance (PVR).

$$\text{CO} = \frac{\text{PAP} - \text{LAP}}{\text{PVR}}$$

Rearrangement gives

$$\text{PAP} = (\text{CO} \times \text{PVR}) + \text{LAP} \qquad (2\text{-}17)$$

Equation 2-17 indicates that pulmonary hypertension (pathologically increased PAP) may be caused by an increase in either CO, PVR, or LAP. An increase in PVR can result from pulmonary arteriolar vasoconstriction (e.g., secondary to alveolar hypoxia), vessel wall fibrosis due to diffuse lung disease, or pulmonary emboli. A common cause of pulmonary hypertension is a rise in the LAP as a result of left heart failure or mitral stenosis. In left ventricular failure, the left ventricular end-diastolic volume and pressure increase. The latter is transmitted to the left atrium and then to the pulmonary capillaries. The increase in pulmonary capillary pressure causes fluid to move into the interstitial and alveolar spaces, causing pulmonary edema.

Coronary Circulation

↑ HR, diastole ↓

When the heart contracts, the blood vessels that supply the myocardium are compressed. Because left ventricular pressure is slightly greater than aortic pressure during systole, but much less than the aortic pressure during diastole, flow through the arteries that supply the subendocardial portion of the left ventricle occurs only during diastole. When the heart rate increases, left ventricular coronary flow is reduced because diastole is shortened to a much greater degree than systole. Thus, left ventricular subendocardium is vulnerable to ischemia and is the most common site of myocardial infarction. However, in the right ventricle and atrium, aortic pressure is greater than right ventricular and atrial pressures during systole as well as in diastole. Coronary flow in these parts of the heart continues throughout the cardiac cycle.

The heart extracts large amounts of oxygen from the blood and, as compared to any other organ, yields the greatest arteriovenous oxygen difference. Oxygen consumption increases with heart rate, contractile state, intraventricular pressure or volume, myocardial wall tension, and total muscle mass. Oxygen consumption by the heart cannot be significantly increased by further extraction of oxygen from the blood but only by increasing coronary flow. The most important factors that affect the lumen size of the coronary arteriole are local metabolites. A drop in the oxygen tension or pH, or a rise in carbon dioxide, potassium, lactate, adenosine, or nitric oxide, is thought to bring about vasodilatation. Although the coronary arterioles also receive innervation from sympathetic vasoconstrictive fibers, this autonomic regulation is of minor importance. This fact reflects the necessity of maintaining coronary circulation in the face of adrenergically stimulated increases in cardiac output with concomitant systemic vasoconstriction.

The lumen of a coronary artery may become occluded by atherosclerosis and, less commonly, by vasospasm. The former is the major cause of myocardial infarction. **Angina pectoris** is paroxysmal chest pain caused by a transient ischemia of the myocardium (without infarction) due to coronary artery atherosclerosis. **Variant angina** (i.e., nonexercise related) is due to coronary artery vasospasm.

Questions

1. You are listening to the heart of a patient with suspected cor pulmonale. Which of the following would you be most likely to hear?
   A. Increased S1
   B. Increased S2
   C. Systolic murmur
   D. S3
   E. S4

   hypertension results in rapidly closing valves so S2 in intensified.

2. Which of the following does **not** fit the clinical picture of heart failure resulting from aortic atherosclerosis?
   A. Ventricular hypertrophy
   B. Decreased pulse pressure
   C. Decreased stroke volume
   D. Pulmonary edema
   E. Congestive heart failure
3. A 24-year-old motorcyclist has suffered severe cerebral trauma. As you watch the patient being intubated and hyperventilated, you realize these procedures are intended to
   A. Reduce the rate of intracerebral transcapillary flow
   B. Increase the partial pressure of oxygen to compensate for decreased cerebral flow due to increased intracranial pressure
   C. Decrease the partial pressure of carbon dioxide to reduce cerebral vasodilation and edema formation
   D. Reduce intracerebral capillary pressure (P_{cap})
   E. Accomplish all of the above
4. All of the following statements are true for cardiac and skeletal muscle fibers **except**
   A. The cardiac AP has a plateau and consequently a longer refractory period
   B. Both cardiac and skeletal muscle exhibit graded contraction
   C. Cardiac muscle fibers are stimulated by excitation from neighboring muscle cells, whereas skeletal muscle fibers are stimulated directly by motor neurons

D. Excitation-contraction coupling is dependent on extracellular calcium in myocardial cells but not in skeletal muscle cells
E. Both cardiac and skeletal muscle undergo contraction in response to increased intracellular calcium

5. Regarding the QRS complex, which of the following is **false?**
   A. It always precedes ventricular contraction
   B. It represents ventricular contraction
   C. It is caused mainly by current flow through fast Na^+ channels
   D. It may be prolonged in disturbances of ventricular conduction pathways
   E. It represents ventricular depolarization
6. Which of the following is the ECG manifestation of SA node depolarization?
   A. The T wave
   B. The P wave
   C. The QRS complex
   D. The U wave
   E. It is not manifested on ECG
7. The event in which the ventricles beat independent of and at a slower rate than the atria is known as
   A. First-degree heart block
   B. Second-degree heart block
   C. Third-degree heart block
   D. Fourth-degree heart block
   E. Fifth-degree heart block
8. The phase of the fast-response action potential that is associated with the voltage-gated calcium current is
   A. Phase 0
   B. Phase 1
   C. Phase 2
   D. Phase 3
   E. Phase 4
9. Phase 0 of the slow-response action potential is mediated by
   A. Calcium
   B. Sodium
   C. Potassium
   D. Chloride
   E. Hydrogen
10. At which point on the pressure-volume loop below is the semilunar valve closing?
    A. A
    B. B
    C. C
    D. D

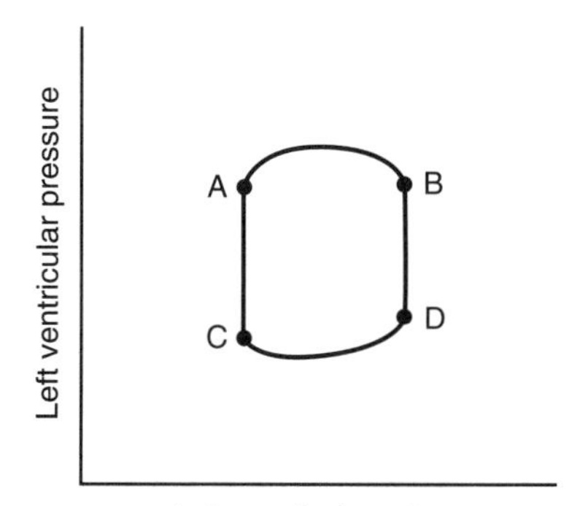

11. Regarding the pressure-volume loop above, during which phase does the greatest rate of pressure generation occur in the ventricles?
    A. Isovolumic relaxation
    B. Early diastolic filling
    C. Late diastolic filling
    D. Isovolumic contraction
    E. Systolic ejection
12. Which heart sound is heard at the end of systole?
    A. S1
    B. S2
    C. S3
    D. S4
    E. S5
13. A normal systolic-diastolic blood pressure of 25/8 is most likely to be measured in the
    A. Left ventricle
    B. Right ventricle
    C. Right atrium
    D. Aorta
    E. Pulmonary artery
14. The QRS is timed closest to
    A. Mitral closure
    B. Aortic opening
    C. Aortic closure
    D. Mitral opening
15. With aortic stenosis, pulse pressure
    A. Increases
    B. Decreases
    C. Does not change because both systolic and diastolic pressure increase
    D. Does not change because both systolic and diastolic pressure decrease
    E. Does not change with either systolic or diastolic pressure changes
16. With aortic insufficiency, pulse pressure
    A. Increases
    B. Decreases
    C. Does not change because both systolic and diastolic pressure increase
    D. Does not change because both systolic and diastolic pressure decrease
    E. Does not change with either systolic or diastolic pressure changes
17. What is the cerebral blood flow (total blood flow through the brain) when the brain consumes oxygen at a rate of 45 ml per minute, the carotid arterial oxygen content is 190 ml per liter, and the internal jugular (venous) oxygen content is 130 ml per liter?
    A. 0.50 liter per minute
    B. 0.75 liter per minute
    C. 1.00 liter per minute
    D. 1.25 liters per minute
    E. 1.50 liters per minute
18. Stroke volume increases with all of the following **except**
    A. Digitalis treatment of congestive heart failure
    B. Increased venomotor tone
    C. Administration of a β_2-blocker
    D. Administration of a β_1-agonist
    E. Administration of an arteriolar vasodilator
19. The increase in cardiac output during exercise is primarily due to which one of the following mechanisms?

A. Increased heart rate
B. The Frank-Starling mechanism
C. Decreased total peripheral resistance
D. Decreased preload
E. Increased afterload

20. Which of the following is **not** observed as a result of heart failure?
A. Increased preload
B. Increased afterload
C. Ventricular dilation
D. Ventricular hypertrophy
E. All of the above are observed

21. Which statement concerning cardiac tissues is **false?**
A. The myocardium normally has the slowest rate of spontaneous discharge
B. The SA node normally has the highest rate of spontaneous discharge
C. The cardiac impulse normally travels from the SA node to the AV node to the His-Purkinje fibers, and then to the myocardium
D. As the cardiac impulse travels through the heart along its normal conduction pathway, the conduction velocity progressively increases
E. V_{max} (the upstroke velocity) is directly proportional to the conduction velocity

22. Regarding the rate of blood flow, which of the following statements is **false?**
A. It is inversely proportional to vessel length
B. It increases with increased pressure difference
C. It is inversely proportional to blood viscosity
D. It is inversely proportional to vessel radius
E. All of the above are true

23. Which of the following events in mitral valve stenosis is **false?**
A. Atrial contraction has a more significant role in ventricular filling than it does in the normal heart
B. Pulmonary capillary hydrostatic pressure may exceed plasma oncotic pressure throughout the length of pulmonary capillaries, resulting in the formation of pulmonary edema
C. A laterally oriented pressure transducer placed just distal to a stenotic mitral valve records lower pressure (transmural pressure) than a similarly placed transducer in a normal heart
D. A longitudinally oriented pressure transducer placed in the left atrium of a heart with mitral valve stenosis records increased pressure as compared to a normal heart
E. All of the above are true statements

24. Which of the following statements regarding the heart is **false?**
A. Decreased extracellular K^+ may produce depolarization and loss of myocardial cell excitability
B. Decreased extracellular Ca^{2+} may decrease contractile force and result in arrest in diastole
C. Increased extracellular Ca^{2+} may increase contractile force and result in arrest in systole
D. Decreased extracellular Na^+ may increase the force of contraction
E. Depolarization (V_M more positive) may decrease membrane excitability and conduction velocity (predisposing to blockage of impulse transmission and reentrant arrhythmia) and may also potentiate abnormal automaticity

25. Which of the following is **not** a measure of stroke volume?
A. Left ventricular end-diastolic volume minus left ventricular end-systolic volume

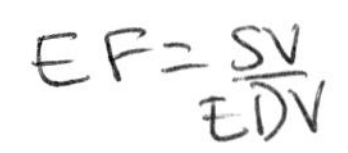

B. Ejection fraction times left ventricular end-diastolic volume

C. Ejection fraction times cardiac output
D. Cardiac output/heart rate

Questions 26–28. Match each term in the left column with the appropriately labeled variable in the right column.

26. Contractile state	A. Left ventricular end-diastolic volume
27. Preload	B. Left ventricular max dP/dt
28. Afterload	C. Aortic systolic pressure
	D. Aortic diastolic pressure
	E. Stroke volume

Use the following information to answer Questions 29 and 30:

Mean aortic pressure = 101 mm Hg
Mean right atrial pressure = 1 mm Hg
Mean pulmonary artery pressure = 10 mm Hg
Mean left atrial pressure = 5 mm Hg
Cardiac output = 5 liters per minute

29. The systemic vascular resistance (total peripheral resistance) is
    A. 10 units
    B. 20 units
    C. 100 units
    D. 200 units
    E. 500 units
30. Where a unit of resistance is mm Hg minutes per liter, the pulmonary vascular resistance is
    A. 1 unit
    B. 2 units
    C. 4 units
    D. 8 units
    E. 10 units
31. A medical student draws a blood sample from a patient's arm, taking a long time to fill the syringe. The venous pressure is 5 mm Hg and the pressure inside the syringe is 1 mm Hg. Which of the following procedures will **not** increase the rate of flow into the syringe?
    A. Increase the radius of the bore of the needle
    B. Increase the length of the needle
    C. Lower the patient's arm (increase the hydrostatic pressure)
    D. Pull on the plunger with more force
    E. Lower the patient's blood viscosity
32. Which of the following does **not** lead to increased interstitial fluid volume (edema)?
    A. Nephrosis leading to proteinuria
    B. A large burn resulting in loss of capillary membrane selective permeability
    C. Arteriolar dilation
    D. Congestive heart failure
    E. An increase in the hydrostatic pressure of the interstitial space
33. Blood flow to skeletal muscle increases during exercise because of
    A. Increased β_2 stimulation in skeletal muscle
    B. Increased α_2 stimulation of veins

C. Increased β_1 stimulation of the heart
D. Production of local metabolites
E. All of the above are true

34. Standing up quickly is expected to produce
A. Decreased activity in the afferent division of cranial nerve IX
B. Decreased activity of the vasomotor center
C. Increased vagal stimulation of the heart
D. Increased activity in the cardioinhibitory center
E. Increased venous compliance

35. During moderate exercise, all of the following will increase **except**
A. Heart rate
B. Arteriovenous oxygen difference
C. Stroke volume
D. Mean arterial blood pressure
E. Systemic vascular resistance

36. When the carotid sinus is stimulated, which of the following events occur?
A. Decreased norepinephrine is released at the SA node
B. Increased acetylcholine is released at the SA node
C. Total peripheral resistance decreases
D. Arterial blood pressure decreases
E. All of the above occur

↑ PNS
↓ SNS

37. Regarding the autonomic nervous system, which of the following statements is/are true?
A. The vasomotor center in the medulla maintains a state of partial vasoconstriction
B. The cardioinhibitory center in the medulla initiates a tonic vagal discharge
C. α_1 receptors mediate arteriolar vasoconstriction
D. α_2 receptors mediate venoconstriction
E. All of the above are true statements

Questions 38–40. Match each phrase in the left column with the appropriately labeled variable in the right column.

38. Greatest cardiac output	A. Heart
39. Greatest blood flow per gram tissue	B. Kidneys
40. Greatest oxygen consumption per gram tissue	C. Liver
	D. Brain
	E. Viscera

41. Which of the following is/are important in determining myocardial oxygen consumption rate?
A. Contractile state
B. Intraventricular pressure
C. Intraventricular volume
D. Heart rate
E. All of the above determine myocardial oxygen consumption rate

42. Which statement regarding the figure below is **incorrect?** The figure represents a vascular bed, the radii are as indicated, and the numbers indicate the relative resistance levels in different areas of the bed.
A. Resistance in section C is 16 times greater than in B
B. Flow in section B is 16 times greater than in C
C. The combined resistance through B and C is less than in B alone

D. Flow in section A is less than the combined flow through B and C
E. All of the above statements are correct

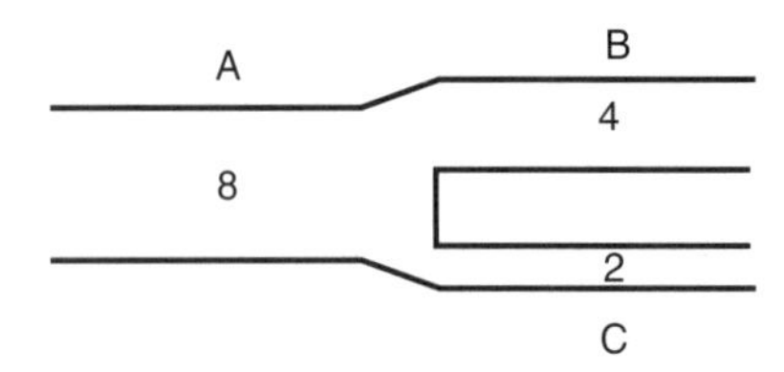

43. Which statement regarding the pressure-volume curves below is incorrect?
   A. Curve C represents the highest compliance
   B. Curve C represents the highest capacitance
   C. Venous compliance decreases with volume
   D. Aortic compliance decreases with volume
   E. Curve A is more likely to represent an elderly person who has a higher pulse pressure than is curve B

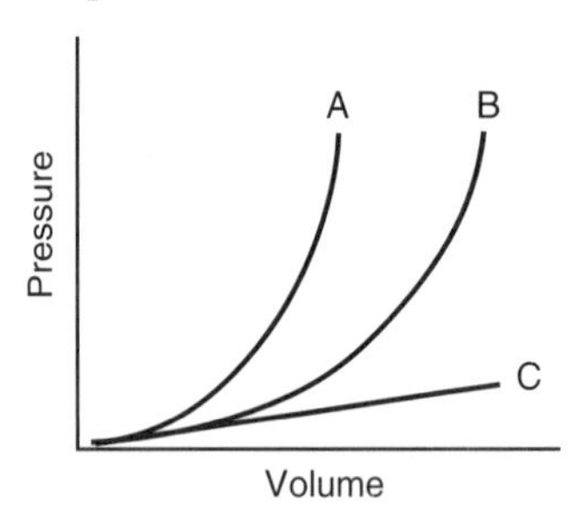

44. You are called to the emergency room to assist with a patient who suffered a massive hemorrhage in an industrial accident. Among the immediate expected reactions to this blood loss, which is **least likely?**
   A. Decreased activity in the afferent division of cranial nerve IX
   B. Increased activity of the vasomotor center
   C. Decreased vagal stimulation of the heart
   D. Decreased activity in the cardioinhibitory center
   E. Increased venous compliance
45. Regarding the heart, which of the following statements is **false?**
   A. The T wave corresponds with phase 3 of the fast-response fiber
   B. The QT interval decreases with heart rate
   C. ST segment elevation may indicate myocardial infarction
   D. The PR interval may be prolonged in disturbances of conduction through ventricular conduction pathways
   E. The duration of diastole decreases more than that of systole with increased heart rate
46. Which, if any, of the following statements is **true?**
   A. All arterioles respond to hypoxia with vasodilation
   B. The autonomic nervous system regulates flow to all organ systems
   C. Flow to all organ systems is based on metabolic need
   D. Autoregulation of flow always involves the relaxation of arterioles
   E. All of the above statements are false

Answers

1. B The second heart sound (S2) is caused by the closure of the semilunar valves, with the aortic sound usually louder and slightly before the pulmonic sound. Conditions that bring about more rapid closure, such as systemic or pulmonary hypertension, will increase the intensity of the second heart sound. A patient with cor pulmonale frequently has pulmonary hypertension and, consequently, an increased S2.

2. B Aortic atherosclerosis causes decreased aortic compliance. This results in *increased* pulse pressure because the increase in aortic volume with each heartbeat is not "absorbed" by a compliant aorta. Instead, the increased blood volume with each heartbeat results in increased systolic pressure and a widened (increased) pulse pressure. Because afterload increases with aortic atherosclerosis, the workload of the heart increases and hypertrophy may result. The noncompliant aorta increases resistance and, thus, reduces stroke volume. As a result, cardiac output may fall, leading to congestive heart failure and formation of pulmonary edema.

3. E Cerebral vessels vasodilate in response to hypoxia and hypercapnia. This augments intracerebral capillary pressure and transcapillary flow and results in increased cerebral edema formation. In the injured and swollen brain, cerebral edema is minimized by avoiding hypoxia and hypercapnia. Hyperventilation is therapeutic because it blows off carbon dioxide (reversing hypercapnia) and it increases the partial pressure of oxygen (reversing hypoxia). This results in cerebral vasoconstriction, which reduces P_{cap} and, thus, transcapillary flow.

4. B Only skeletal muscle exhibits graded contraction. Here, graded contraction occurs as a function of the number of stimulated muscle fibers. The heart, being a functional syncytium, fires in an all-or-none fashion (excitation spreads from fiber to fiber through gap junctions, a process not found in skeletal and smooth muscle). The skeletal muscle action potential does not exhibit a plateau. In skeletal muscle, depolarization (caused by Na^+ influx) results in Ca^{2+} release from the sarcoplasmic reticulum. In cardiac muscle, Ca^{2+} that enters during the plateau phase mediates additional release of Ca^{2+} from the sarcoplasmic reticulum. In both cardiac and skeletal muscle, the Ca^{2+} binds with troponin and the Ca^{2+}-troponin complex in turn binds to tropomyosin and facilitates the interaction between actin and myosin that produces contraction.

5. B The QRS complex is a recording of ventricular depolarization, which precedes contraction, and results mainly from current flow through fast Na^+ channels. The duration of the QRS complex may be prolonged in disturbances of ventricular conduction pathways.

6. E The initial discharge from the SA node (located in the right atrial wall) is too small to be seen by ECG.

7. C In first-degree heart block, all atrial impulses reach the ventricles, but are slowed, resulting in a long PR interval. In second-degree block, not all atrial impulses reach the ventricles and, thus, there are many P waves for each QRS complex. Third-degree block is a complete block, in which the ventricles beat independent of and usually at a slower rate than the atria. Fourth- and fifth-degree heart block do not exist.

8. C Phase 2, the plateau, is a state in which K^+ efflux is counterbalanced by a voltage-gated Ca^{2+} influx. The Ca^{2+} ions are responsible for excitation-contraction coupling.

9. A Phase 0 of the slow-response action potential is mediated mainly by Ca^{2+} ions, whereas phase 0 of the fast-response action potential is mediated mainly by Na^+ ions.

10. A The semilunar valves are the aortic and pulmonic valves. Semilunar valve closure marks the termination of ventricular systole. (See Fig. 2-9.)

11. D Isovolumic contraction begins with systolic contraction and proceeds from point D to point B in the pressure-volume loop. Isovolumic contraction is a period of rapid increase in pressure without change in volume and occurs when both valves (in each ventricle) are closed.

12. B The first heart sound (S1) is due to closure of the AV valves. The second heart sound (S2) occurs at the end of systole, when aortic pressure exceeds ventricular pressure and the semilunar valves are forced shut. A third sound (S3) may be heard during early diastole in children, in well-conditioned athletes (because of rapid ventricular filling), and in heart failure patients with reduced left ventricular compliance. In late diastole, a fourth heart sound (S4) may occur when the atrium contracts and propels blood into hypertrophied ventricles that have low compliance. There is no fifth heart sound.

13. E Ventricular systolic pressures equal the systolic pressures of the corresponding great arteries. The ventricular diastolic pressures, however, are significantly less than the corresponding diastolic pressures of the great arteries. In fact, while ventricular diastolic pressures fall to zero, aortic and pulmonary arterial pressures remain at approximately 80 and 8 mm Hg, respectively.

14. A The QRS wave results from ventricular depolarization, which immediately precedes ventricular contraction and the corresponding rise in ventricular pressure that forces the mitral valve shut. (See Figs. 2-4 and 2-5.)

15. B Pulse pressure is narrower (decreases) in aortic stenosis due to a slower rate of rise in pressure during ejection.

16. A Pulse pressure is wider (increases) in aortic insufficiency due to a faster rate of rise in pressure (due to increased preload) and a lower diastolic pressure (due to aortic regurgitation).

17. B Using Equation 2-3,

$$\text{CO} = \frac{45 \text{ ml O}_2\text{/min}}{190 \text{ ml O}_2\text{/liter} - 130 \text{ ml O}_2\text{/liter}}$$

$$= 0.75 \text{ liters blood/min}$$

18. C Digitalis treatment increases contractility and, hence, stroke volume. Increased venomotor tone increases preload and, therefore, stroke volume. Administration of a selective β_2-blocker will lead to vasoconstriction of all arteries that are endowed with β_2 receptors (e.g., skeletal muscle); this results in increased afterload and, thereby, decreased stroke volume. Administration of a β_1-agonist will increase heart rate and contractility, and, thus, stroke volume. Administration of an arteriolar vasodilator will decrease afterload and increase stroke volume.

19. A The increase in CO with exercise is primarily due to increased heart rate. Venous return (preload) increases with exercise due to venous pumping of blood as a result of movement of skeletal muscle. This accounts for a portion of the increase in stroke volume (100% contribution in cardiac transplant patients, in whom neural connections are severed). However, the increase in stroke volume during exercise is limited by a reduction in the time for diastolic filling (due to the increase in heart rate). The afterload reduction observed in vigorous exercise does not make a major contribution to the increase in CO.

20. E All of the compensatory responses listed in the question may occur in response to heart failure. Fluid retention due to activation of the renin-angiotensin-aldosterone axis increases preload. Increased sympathetic discharge has the positive effect of increasing contractility, but the negative effect of increasing afterload. Ventricular dilation and hypertrophy are also observed.

21. D Two points can be made regarding conduction velocity: Conduction velocity is directly proportional to V_{max}, and it varies considerably in different tissues. It is slowest in the SA node, speeds up in the atria, slows again in the AV node, is the fastest in the Purkinje fibers, and slows once more in the ventricles.

22. D The Poiseuille-Hagen formula (Equation 2-7) states that for laminar flow through a cylindrical tube, the rate of flow varies directly with the pressure difference and with the fourth power of vessel radius. In addition, flow varies inversely with vessel length and blood viscosity.

23. E Mitral valve stenosis presents a high-resistance orifice that impedes the normal flow of blood from left atrium to left ventricle. Clearly, the left atrium must then pump harder. Pressure builds up in the left atrium and backs up into the pulmonary circuit, resulting in pulmonary capillary hydrostatic pressure that exceeds plasma oncotic pressure, and, hence, the formation of pulmonary edema. The velocity of blood increases as it passes through the narrow orifice, as described by Bernoulli's principle (Equation 2-8). Potential energy (pressure) is converted to kinetic energy (which is not measured by a pressure transducer that measures lateral, transmural pressure).

24. A According to the Nernst equation (2-1), a decrease in extracellular K^+ hyperpolarizes the cell (makes V_M more negative), which will increase V_{max} (the upstroke velocity) and the amplitude of the action potential, thus increasing conduction velocity and excitability. On the other hand, depolarization (a more positive V_M) renders cardiac cells less excitable, predisposing them to slowing of conduction or to complete blockage, thus creating the conditions for reentrant arrhythmia. Depolarization may also increase automaticity (i.e., increased heart rate). Increased extracellular Ca^{2+} results in increased Ca^{2+} influx during phase 2, and thus increased contraction, according to the theory of excitation-contraction coupling. Decreased extracellular Na^+ reduces the rate of exchange of Na^+ for Ca^{2+} via the Na^+-Ca^{2+} pump, increasing intracellular Ca^{2+} and, thus, the force of contraction.

25. C Stroke volume is left ventricular end-diastolic volume minus left ventricular end-systolic volume. Ejection fraction is equal to stroke volume divided by ventricular end-diastolic volume. Cardiac output equals the product of stroke volume and heart rate.

26. B Contractility represents the performance of the heart and can be defined as the rate of pressure change with respect to time.

27. A The volume of the left ventricle at the end of diastole reflects the degree of prestretching of the cardiac muscle and equals the preload.

28. D Afterload correlates well with diastolic pressure. Afterload is a function of TPR, and increased TPR reduces diastolic relaxation (reduces aortic compliance).

For questions 29 and 30, rearrange Equation 2-4 to yield: $R = \Delta P/Q$.

29. B Systemic vascular resistance = (101 mm Hg – 1 mm Hg)/5 liters/min = 20 units

30. A Pulmonary vascular resistance = (10 mm Hg– 5 mm Hg)/5 liters/min = 1 unit

31. B According to the Poiseuille-Hagen formula (Equation 2-7), flow $= \pi \Delta P r^4/8\eta L$. Flow increases with increased radius and pressure difference, but decreases with increased length and viscosity. Lowering the patient's arm raises the hydrostatic venous pressure according to Bernoulli's equation (2-8), which increases the pressure difference between the syringe and vein. Pulling the plunger with more force also increases this pressure difference.

32. E Transcapillary flow is governed by Starling's equation (2-13). The following events increase transcapillary flow: (1) proteinuria, because it lowers plasma oncotic pressure; (2) loss of the integrity of the capillary membrane; (3) arteriolar dilation, because it increases capillary hydrostatic pressure; and (4) congestive heart failure, because it increases venous pressure (and, thus, increases capillary hydrostatic pressure). An increase in the hydrostatic pressure of the interstitial space would oppose capillary hydrostatic pressure and therefore would not lead to increased interstitial fluid volume.

33. E Metabolites produced in working muscle lead to more vasodilation than sympathetic stimulation. Activation of the sympathetic nervous system with exercise has the following effects, all of which increase flow to skeletal muscle: (1) Increased β_2 stimulation in skeletal muscle causes vasodilation; (2) increased α_2 stimulation of veins causes venoconstriction, increasing preload and CO, and, thus, increasing delivery to skeletal muscle; and (3) increased β_1 stimulation of the heart increases CO.

34. A A sudden fall in MAP results in decreased stimulation of the baroreceptors and decreased afferent neural activity in cranial nerves IX and X. The vasomotor center is stimulated, resulting in increased sympathetic stimulation, that is, tachycardia and increased contractility, and the cardioinhibitory center is inactivated, resulting in decreased vagal activity to the heart. Venous compliance decreases with increased sympathetic activity, thus increasing preload.

35. E During exercise, sympathetic nervous activity increases and parasympathetic activity decreases. β_1 stimulation of the heart increases heart rate and contractility. α_2-stimulated constriction of veins increases preload and stroke volume to a moderate degree (presystolic filling is limited at high heart rates because diastole shortens in duration). MAP increases (systolic more than diastolic). Even though sympathetically mediated vasoconstriction decreases flow to the splanchnic regions, inactive muscle, and skin, overall systemic vascular resistance falls as the production of vasoactive metabolites by working muscles opens local vascular beds. Note also that skin arterioles dilate later when heat dissipation is necessary.

36. E The carotid sinus is a baroreceptor that responds to increased pressure by (1) reducing sympathetic output, leading to decreased norepinephrine release at the SA node, which decreases total peripheral resistance and arterial blood pressure, and (2) increasing parasympathetic output, leading to increased acetylcholine release at the SA node, which slows heart rate by reducing the rate of spontaneous depolarization.

37. E The vasomotor center in the medulla emits a tonic sympathetic discharge that maintains a state of partial vasoconstriction. The cardioinhibitory center (or vagal center) in the medulla initiates a tonic vagal discharge that slows heart rate. As indicated in Table 2-1, α_1-adrenergic receptors mediate arteriolar vasoconstriction and α_2-adrenergic receptors mediate venoconstriction.

38. C The liver, because it receives both the hepatic and the portal circulations, receives the largest percentage of cardiac output.

39. B The kidneys have a very large flow because of their role in maintaining homeostasis. They also utilize a large amount of oxygen in the transport of ions and, as such, are vulnerable to ischemia. (Note that the carotid bodies receive the largest blood flow per gram of tissue of any organ in the body, a fact consistent with their primary role of monitoring the state of arterial oxygenation.)

40. A The heart extracts the most oxygen per gram of tissue and also exhibits the widest arteriovenous oxygen difference.

41. E Any factor that increases the work of the heart increases its oxygen consumption rate. These factors include heart rate, contractile state, intraventricular pressure, and intraventricular volume.

42. D Since resistance is inversely proportional to the fourth power of the radius, the resistance in section C is 16 times greater than in B and, hence, flow is 16 times greater in B than in C. Vessels B and C are in parallel, and therefore their combined resistance is less than either alone (see Equation 2-12). However, even though the combined resistance of B and C is less than that of A, this does not mean that flow through B and C is greater than that of A. Flow through B and C is equal to that in A, as described by the continuity equation (2-5).

43. C Curve C represents veins, curve B represents the aorta of a young person, and curve A the aorta of an elderly person. Compliance is the capacity of a vessel (or the heart) to increase in volume with increased pressure. High-capacitance vessels (e.g., veins) accommodate large changes in volume with little change in pressure, and are therefore highly compliant as well. The compliance of curve C (representing veins) does not change appreciably with volume. Aortic compliance normally decreases with age. This increases systolic pressure because the aorta does not expand as well to accept increased volume; hence, systolic pressure rises faster. Pulse pressure, the difference between systolic and diastolic pressure, is increased.

44. E A sudden fall in MAP results in decreased stimulation of the baroreceptors and decreased afferent neural activity in cranial nerves IX and X. The vasomotor center is stimulated (resulting in increased sympathetic stimulation, i.e., tachycardia and increased contractility). The cardioinhibitory center is inactivated, resulting in decreased vagal activity to the heart. Venous compliance decreases with increased sympathetic activity, which increases preload.

45. D The T wave is a recording of ventricular repolarization, which occurs in phase 3. The duration of the QT interval, during which ventricular contraction (systole) occurs, varies inversely with heart rate. Note that the duration of diastole decreases more than systole with increased heart rate. The ST segment, which is the isoelectric line, may be elevated in myocardial infarction, as damaged cells are depolarized. The PR interval may be prolonged in disturbances of conduction through the AV node, but not in disturbances of conduction through ventricular conduction pathways.

46. E All the statements are false. Pulmonary arterial smooth muscle is distinguished from all other vascular smooth muscle in that it responds to hypoxia with constriction and to high oxygen content with dilation. In the cerebral circulation, the autonomic nervous system contributes to the maintenance of MAP such that flow to the brain is constant, but does not significantly regulate cerebral vasculature (nor does the autonomic nervous system regulate the coronary vasculature). The kidneys and lungs receive much more blood than is necessary to meet their metabolic needs. These large volumes are filtered in the kidneys and utilized for gaseous exchange in the lungs. Autoregulation in the lungs is accomplished through passive recruitment and distention of capillary beds.

Bibliography

Berne, R. M., and Levy, M. N. *Cardiovascular Physiology* (6th ed.). St. Louis: Mosby–Year Book, 1992.

Cardiovascular Physiology Syllabus. Stanford, CA: Stanford University School of Medicine, 1993.

Wingard, L. B., Brody, T. M., Larner, J., and Schwartz, A. *Human Pharmacology.* St. Louis: Mosby, 1991.

3

Renal Physiology

Charles H. Tadlock

The composition and constancy of the body fluids are regulated primarily by kidney function. Fluctuations in fluid balance, electrolyte concentrations, acid-base status, and blood pressure affect renal function in maintaining homeostasis. Losses from skin, lungs, and intestine are important (Table 3-1); however, the responsibility for adjusting solute and water excretion is borne by the kidney. By the process of **filtration, secretion,** and **reabsorption,** the kidney directly regulates the quantity and composition of the urine (Table 3-2) and thereby indirectly regulates that of the other body fluid compartments. These processes result in the reabsorption of such essential substances as glucose, amino acids, bicarbonate, sodium, potassium, and chloride, and the elimination of urea, uric acid, creatinine, phosphates, sulfates, and hydrogen ions (Table 3-3). The mechanisms that enable the kidney to respond to daily alterations in fluid intake and diet, and to excrete waste products in amounts that precisely balance the quantities acquired by ingestion and metabolic transformation, are the subject matter of renal physiology.

Endocrine Functions of the Kidney

The kidneys function not only as execretory organs but also as endocrine organs. Functional kidney tissue is necessary for the activation of erythropoietin from its protein precursor in the blood and for the conversion of circulating 25-hydroxycholecalciferol to 1,25-dihydroxycholecalciferol **(vitamin D).** Furthermore, the hormone **renin** is manufactured and released by modified smooth muscle cells in the walls of the afferent arterioles. Acting via the **renin-angiotensin-aldosterone** system, renin affects virtually all aspects of water and electrolyte balance. **Prostaglandins** are also produced by the kidneys; their exact role(s) have yet to be clarified.

Functional Anatomy of the Kidney

Macroscopic Anatomy of the Kidney

The kidney is divided into the outer **cortex** and inner **medulla** (Figs. 3-1 and 3-2). Extensions of renal cortex or columns divide the medulla into distinct pyramidal-shaped regions, the **renal pyramids.** The apex of each of the renal pyramids is the **papilla** and opens into a minor **calyx.** The minor calyces join together to form the funnel-shaped renal pelvis, which is the dilated proximal portion of the ureter into which the urine flows.

Table 3-1. Average daily fluid balance[a]

	Intake (liters/day)		Output (liters/day)
Fluid	1–2	Urine	1–2
Food	0.7–1.0	Insensible losses[b]	0.5–1.0
Water of oxidation	0.25–0.3	Gastrointestinal[c]	0.1
Total	2–3 liters		2–3 liters

[a]For a sedentary average adult.

[b]Insensible losses refers to losses from the skin (perspiration), and the lungs and oropharynx (to humidify the respired gases).

[c]Water content of feces.

Table 3-2. Composition of urine

Characteristic	Range
Volume	400 ml/day–>4 liters/day
Osmolality	50–1200 mOsm/liter
pH	4.5–7.8
$[Na^+]$	0–300 mEq/liter
$[K^+]$	20–70 mEq/liter

Table 3-3. Filtration, excretion, and reabsorption of water, electrolytes, and solutes

Substance	Measure	Filtered*	Excreted	Reabsorbed	Filtered load reabsorbed (%)
Water	liters/day	180	1.5	178.5	99.2
Na^+	mEq/day	25,200	150	25,050	99.4
K^+	mEq/day	720	100	620	86.1
Ca^{2+}	mEq/day	540	10	530	98.2
HCO_3^-	mEq/day	4320	2	4,318	99.9+
Cl^-	mEq/day	18,000	150	17,850	99.2
Glucose	mmol/day	800	0.5	799.5	99.9+
Urea	g/day	56	28	28	50.0

*The filtered amount of any substance is calculated by multiplying the concentration of that substance in the ultrafiltrate by the glomerular filtration rate, (e.g., the filtered load of Na^+ is calculated as $[Na^+]_{ultrafiltrate}$(140 mEq/liter) × glomerular filtration rate (180 liters/day) = 25,200 mEq/day).

From B. A. Stanton and B. M. Koeppen, Elements of Renal Function. In R. M. Berne and M. N. Levy (eds.), *Principles of Physiology*. St. Louis: Mosby–Year Book, 1990.

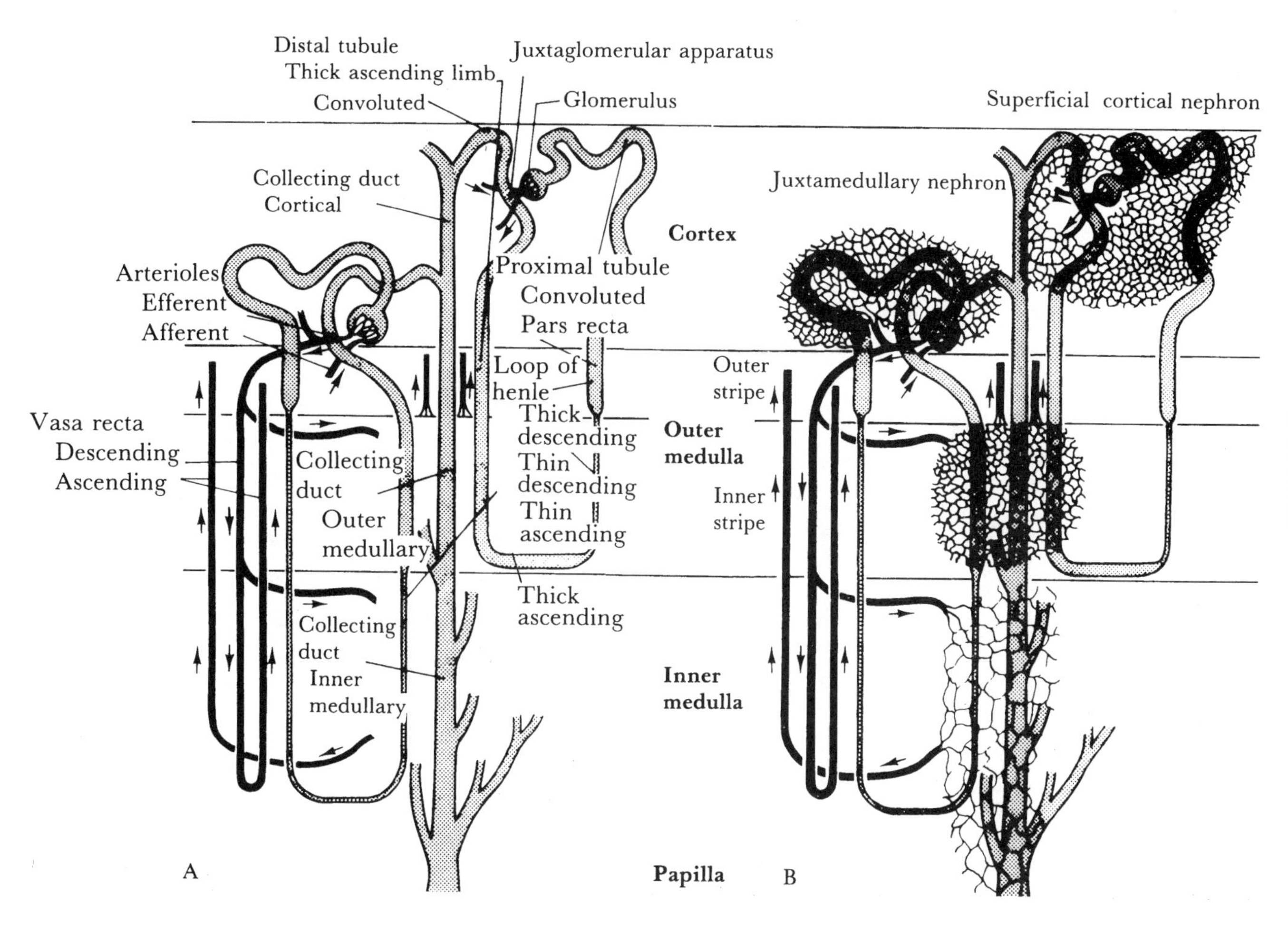

Fig. 3-1. A. Superficial cortical and juxtamedullary nephrons and their vasculature. The glomerulus plus the surrounding Bowman's capsule are known as the **renal corpuscle**. There is some overlapping nomenclature; for example, the loop of Henle consists of the proximal straight tubule, the descending and ascending thin limbs, and the ascending thick limb, even though the first and last parts are also considered to belong to the proximal and distal tubules, respectively. The beginning of the proximal tubule — the so-called urinary pole — lies opposite the vascular pole, where the afferent and efferent arterioles enter and leave the glomerulus. The ascending thick limb of the distal tubule is always associated with the vascular pole belonging to the same nephron; the juxtaglomerular apparatus is located at the point of contact (see also Fig. 3-19). B. Capillary networks have been superimposed on the nephrons illustrated in *A.* (From H. Valtin. *Renal Function: Mechanisms Preserving Fluid and Solute Balance in Health* [2nd ed.]. Boston: Little, Brown, 1983. P. 4.)

Renal

Microscopic Anatomy of the Kidney

The functional and structural unit of the kidney is the **nephron.** Each human kidney consists of approximately one million nephrons, all of which ultimately contribute fluid to the collecting system. The **distal tubules** of the nephrons merge into the **collecting system,** forming sequentially larger collecting ducts until ultimately only 10 to 25 open at the level of the papilla to empty into each of the minor calyces.

The nephron is divided into several histologically and functionally distinct segments, each of which occupies a characteristic location in the cortex or medulla:

1. The glomerulus
2. Proximal convoluted tubule

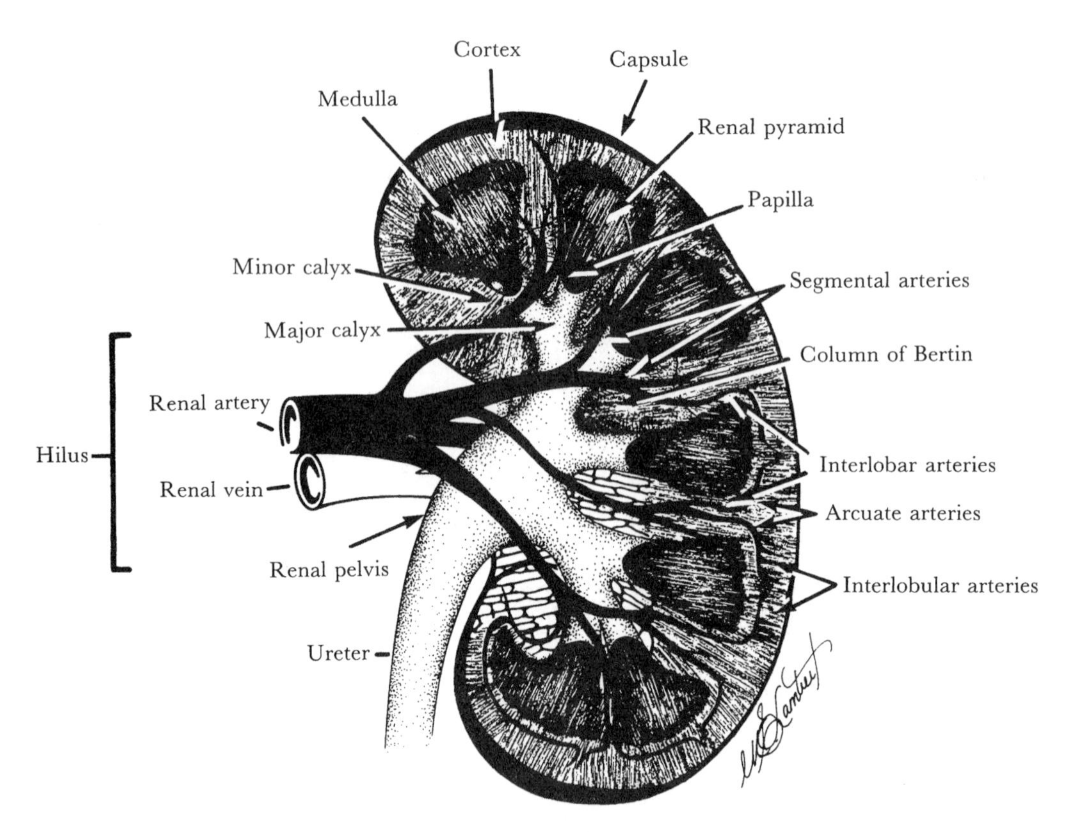

Fig. 3-2. Sagittal section of a human kidney showing the major gross anatomic features. The renal columns are extensions of cortical tissue between the medullary areas. (From J. B. West, *Best and Taylor's Physiological Basis of Medical Practice* [12th ed.]. P. 420. © 1990 by Williams & Wilkins, Baltimore.)

3. Loop of Henle
4. Distal convoluted tubule

Two distinct classes of nephrons are recognized on the basis of the location of their glomeruli in the cortex: outer cortical and juxtamedullary (see Fig. 3-1). **Outer cortical nephrons** are distinguished by

1. The location of their glomeruli in the outer portion of the cortex
2. A short loop of Henle extending only into the outer medulla
3. Efferent arterioles that divide to form the **peritubular capillary plexus**

Juxtamedullary nephrons are distinguished by

1. The location of their glomeruli in the cortex near the corticomedullary junction
2. A long loop of Henle extending deep into the medullary substance with thin descending and thin ascending limbs
3. Efferent arterioles that form the **vasa recta** in addition to contributing to the **peritubular capillary plexus**

The short-looped outer cortical nephrons are the overwhelming majority, comprising seven-eighths of the total, the remainder being juxtamedullary.

The **collecting system** consists of **cortical, medullary,** and **papillary collecting ducts.**

The formation of urine begins in the glomerulus, where the ultrafiltrate of plasma is extruded into **Bowman's space,** the proximal end of the nephron. After passing through and being modified by the various tubule segments, the fluid in its final form, the urine, passes into the calyceal system. No further modification of the urine is now possible. The urine then flows through the minor and major calyces, into the renal pelvis, ureter, bladder, and urethra, and out of the urethral orifice.

The Renal Circulation

The renal artery divides successively into the interlobar, arcuate, and interlobular arteries and finally into the afferent arterioles. The **afferent arteriole** enters the glomerulus and forms the **glomerular tuft of capillaries** and then reforms to exit the glomerulus as the **efferent arteriole.** The term **glomerulus,** in strict usage, refers only to the glomerular tuft of capillaries. Efferent arterioles of cortical nephrons then form a second capillary bed around the cortical tubules (i.e., the **peritubular capillary plexus**). Efferent arterioles exiting from juxtamedullary nephrons contribute to the peritubular plexus and also form an additional capillary bed, the descending **vasa recta,** which descends straight down into the medulla (see Fig. 3-1B). Both of these capillary beds drain into a common venous plexus. The thin limbs of the loop of Henle are surrounded by this plexus. The alignment of capillary beds in series and their arrangement about the tubular system make possible the reabsorption of the essential substrates, including water, that were filtered out in the glomerular capillary bed. The venous drainage system is named in exactly the same manner as the arterial system, with interlobular, arcuate, interlobal, and renal veins.

The Renal Nerves

The renal nerves are primarily composed of sympathetic efferents from the mesenteric plexus. A few afferent nerves and some cholinergic innervation by vagal fibers are also present but are of unknown significance. Stimulation of renal nerves will result in the following:

1. A marked decrease in renal blood flow as a result of vasoconstriction, primarily of the afferent and efferent arterioles
2. Increased **renin** secretion by the **juxtaglomerular apparatus** (β_2 receptor mediated)
3. Increased reabsorption of Na^+ and water by the renal tubules, an effect independent of either the renin-angiotensin-aldosterone system or prostaglandins

Body Fluid Compartments and Their Composition

Total Body Water

The percentage of the human body composed of water depends primarily on the proportion of fat present since fat cells contain very little water (Table 3-4). The average nonobese male is approximately 60 percent water by weight. The average nonobese female is approximately 50 percent water due to the slightly greater proportion of body fat present. Neonates are 70 percent water by weight.

The volume of the total body water (TBW) as well as the volumes of its various compartments can be determined experimentally by using the **indicator-dilution principle** and calculating the **volume of distribution** of various markers (Table 3-5) using the following equation:

Table 3-4. Percent water of various body fluid compartments

Extracellular fluid (interstitial)	99
Plasma	94
Whole blood	88
Cell (intracellular fluid)	75
Bone	25
Fat	20

Table 3-5. Substances and equations used to measure the size of the major body fluid compartments

Compartment	Substance	Equation
TBW	Antipyrine	Volume compartment = (amount marker – amount of marker lost)/[marker]
	Deuterium	
	Tritium	
ECF	Inulin	Same as above
	Raffinose	
	Sucrose	
	Mannitol	
	Thiocyanate	
	Radiochloride	
	Radiosodium	
	Radiobromide	
Plasma	Iodinated I-131 serum albumin	Same as above
	Evans blue, or T-1824	
	Chromicized Cr-51 erythrocytes	
Interstitial fluid	Not measured directly	Interstitial = ECF – plasma
ICF	Not measured directly	ICF = TBW – ECF

Adapted from H. Valtin. *Renal Function: Mechanisms Preserving Fluid and Solute Balance in Health* (2nd ed.). Boston: Little, Brown, 1983. P. 26.

$$\text{Volume of compartment} = \frac{(\text{amount of marker given}) - (\text{amount of marker lost})}{(\text{concentration of the marker in the compartment})}$$

Body Fluid Compartments

The body fluid is divided into two major compartments, **extracellular fluid (ECF)** and **intracellular fluid (ICF).** The ECF can be further subdivided into four separate subcompartments (Fig. 3-3). The percentage of the TBW found in each compartment and subcompartment is as follows:

ICF = 55% of TBW

ECF = 45% of TBW

  Interstitial = 20%

  Bone and connective tissue water = 15%

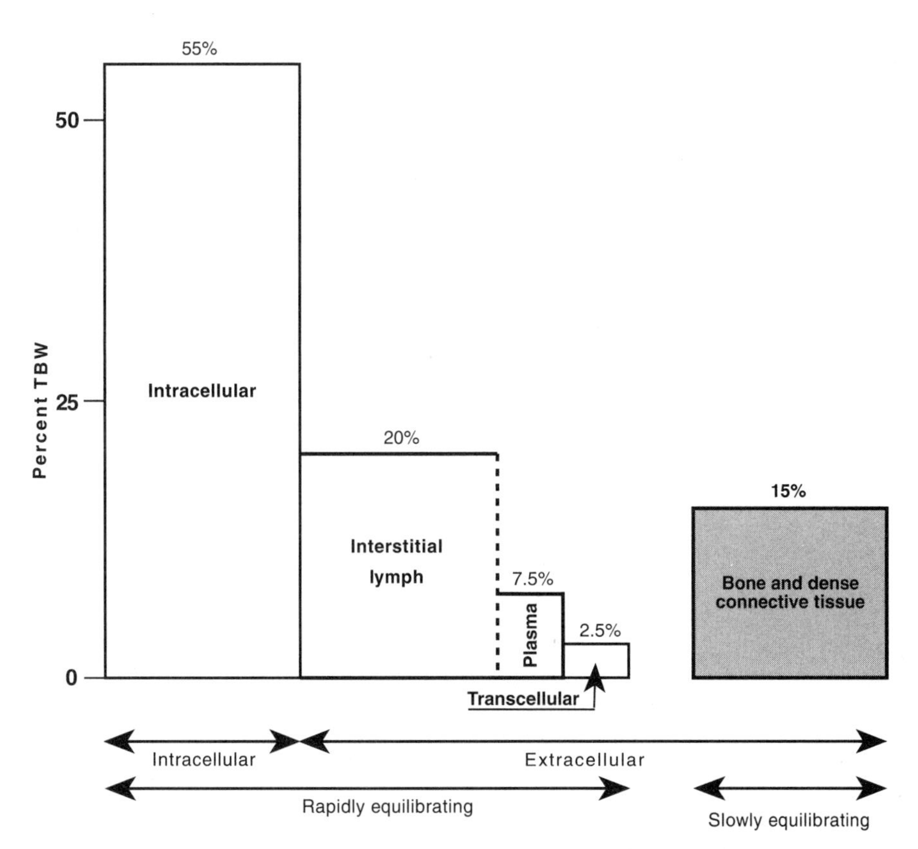

Fig. 3-3. Distribution of body water as a percentage of the total body water (%TBW). Intracellular water constitutes the largest portion of the body water. The balance of the body water is defined as the extracellular fluid. The plasma is separated from the interstitial-lymph compartments by the capillary membrane; the dashed line is meant to convey that the capillary membrane is highly permeable to water and solutes, with the important exception of the plasma proteins. The shadowed portions of the dense connective tissue and bone compartments appear not to be accessible to the commonly employed saccharide marker molecules.
The transcellular fluid is exemplified by the fluid contained in the gastrointestinal and urinary tracts.

Plasma water = 7.5%

Transcellular = 2.5% (i.e., water located in the viscera such as the intestine or gall bladder)

The bone and connective tissue water compartment is frequently neglected in discussions of the ratio of the ECF-ICF since it is difficult to measure and requires equilibration times greater than those of other compartments. As a consequence, the ICF is often said to represent two-thirds of the TBW and the ECF one-third.

Composition of the Fluid Compartments

Figure 3-4 illustrates the compositions of the extracellular and intracellular fluids. Table 3-4 lists the percentage of water found in various compartments. The ICF, for instance, is only 75 percent water. A working knowledge of the relative proportions of Na^+, K^+, Cl^-, and HCO_3^- in one compartment compared with those in another is vital to understanding renal physiology (Table 3-6).

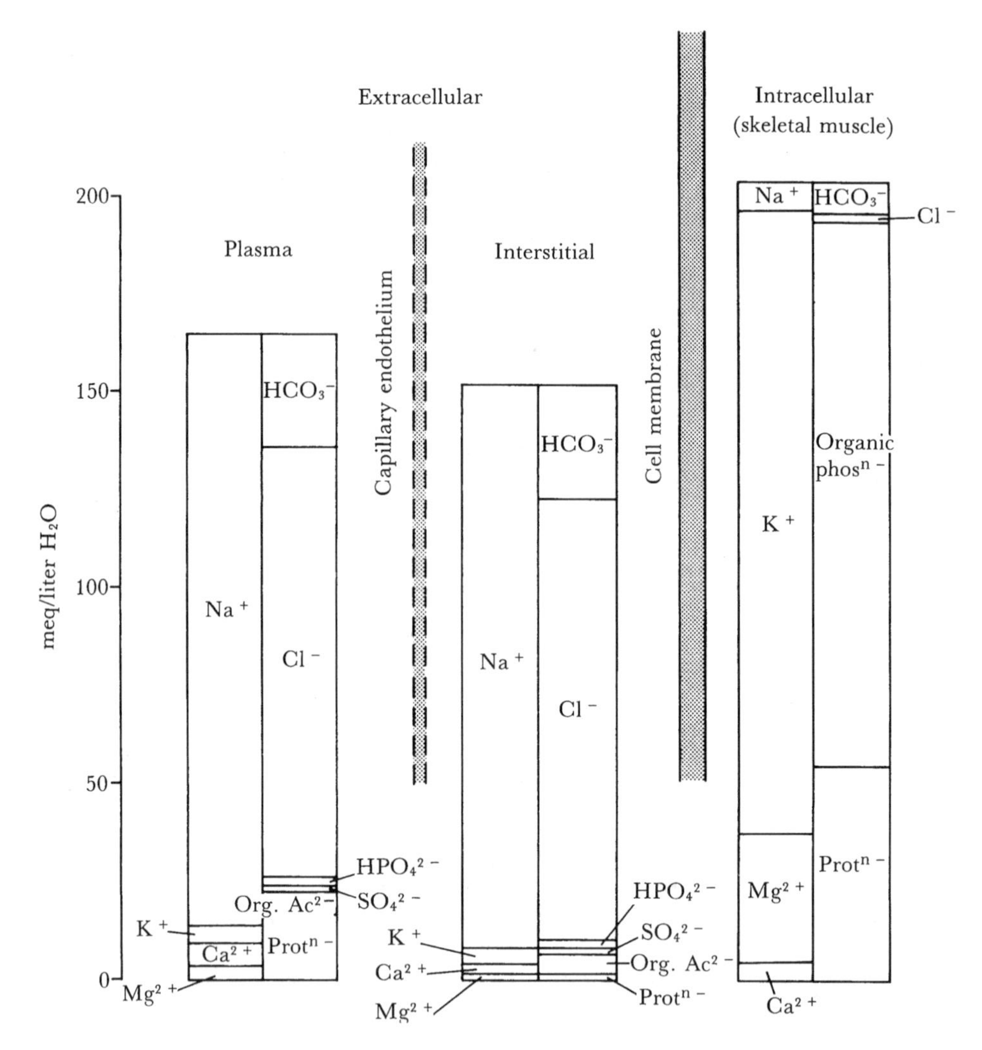

Fig. 3-4. Solute composition of the major body fluid compartments. The concentrations are expressed as chemical equivalents to emphasize that the compartments are made up primarily of electrolytes. Intracellular values are estimates only. (Modified from J. L. Gamble. *Chemical Anatomy, Physiology and Pathology of Extracellular Fluid* [6th ed.]. Cambridge, MA: Harvard University Press, 1954.)

Table 3-6. Concentrations of important ions in intracellular and extracellular fluids

Ion	ECF (mEq/kg H_2O)	ICF (mEq/kg H_2O)
K^+	4–5	150–160
Na^+	144–150	5–10
Cl^-	110–114	5–10
HCO_3^-	24	15
Plasma pH	7.4 or 40×10^{-9} mEq H^+/kg H_2O (the ICF is more acidic)	
Plasma protein	7g/dl	

Principles of Water and Electrolyte Balance

Effective and Ineffective Osmoles

Maffly's principle states that water goes where the osmoles go, that is, it flows from a region of low osmolality to a region of higher osmolality. Because sodium is the main extracellular osmole, this means that water goes where the sodium goes. Abnormalities in Na^+ balance are reflected, therefore, in the amount and distribution of extracellular body water (e.g., edema).

A corollary to Maffly's principle is that the movement of water across membranes is also governed by the distribution of osmoles. Only solutes that do not freely cross the membrane will effect the movement of water across that membrane. Such solutes are called **effective osmoles** because the addition of an effective osmole to one side of a membrane will osmotically obligate water to flow to that side of the membrane until equilibrium is established. Sodium is an effective osmole because it remains in the ECF and does not cross into the ICF; therefore, if sodium is added to the ECF it draws water out of the cells until equilibrium is attained:

$$[\text{Osmoles}]_{ECF} = [\text{osmoles}]_{ICF}$$

Other ECF-effective osmoles of importance include glucose, glycerol, and mannitol. Examples of ineffective osmoles include urea, alcohol, and any other osmole that moves rapidly across cell membranes and, therefore, does not osmotically obligate water to follow it. In uremia the osmolality of the body's fluids measured by the laboratory is increased by its increased content of urea; however, because urea is an ineffective osmole no change in the ECF-ICF ratio occurs.

Determinants of Plasma Volume

The volume of the ECF is determined primarily by total body stores of sodium; however, the amount of the ECF in the plasma subcompartment also depends on the oncotic and hydrostatic pressures. The relationship between hydrostatic pressure, oncotic pressure, and fluid flow across a capillary membrane is given in **Starling's equation** (see Chap. 2).

The major solutes in the plasma are Na^+, HCO_3^-, Cl^-, glucose, and urea. Because Na^+ is the major cation in the ECF, the total concentration of electrolytes can be estimated by doubling the Na^+ concentration and thereby accounting for the anions present in the ECF. One can then add the osmolalities of the two major nonelectrolytes and estimate the total osmolality. This is summarized in the following equation:

$$\text{Serum osmolality} = 2[Na^+] + [\text{glucose}] + [\text{urea}]$$

where $[Na^+]$ is the concentration of sodium ions in mEq per liter, [glucose] is the concentration of glucose molecules in mmole per liter, and [urea] is the concentration of urea in mmole per liter. Note that Na^+ is a univalent ion; therefore, 1 mEq per liter of Na^+ is equal to 1 mOsm per liter Na^+. Similarly, 1 mmole of glucose or urea is equal to 1 mOsm of glucose or urea. Thus, this equation estimates osmolality of the serum in mOsm per liter. For the purposes of this text, 1 liter of serum or plasma is equivalent to 1 kg water; therefore, these are used interchangeably when discussing concentration. Serum is actually plasma with the plasma proteins removed. Plasma is, in fact, approximately 91 percent water and 9 percent solute. This error cancels out because the activity of ions in plasma is also 91 percent, and it is activity rather than actual concentration that is measured clinically.

Glucose and urea are measured in mg per liter; a more useful form of the serum osmolality equation takes this into account. Given that the molecular weight of glucose is 180 and that of urea is 28, the following equation can be derived:

$$\text{Serum osmolality} = 2\,[\text{Na}^+] + \text{glucose}\,\frac{\text{mg/dl}}{18} + \text{urea}\,\frac{\text{mg/dl}}{2.8}$$

Thus, the normal serum osmolality can be calculated. Using normal values for $[Na^+]$ of 140 mEq per liter, for [glucose] of 90 mg/dl, and for [urea] of 14 mg/dl

$$\begin{aligned}\text{Serum osmolality} &= 2\,(140) + \frac{90\text{mg/dl}}{18} + \frac{14\text{mg/dl}}{2.8}\\ &= 2(140) + 5 + 5\\ &= 290\ \text{mOsm/kg H}_2\text{O}\end{aligned}$$

Osmolality of the body fluids is 290 mOsm/kg water everywhere except in the plasma (and, of course, the transcellular water), in which, due to the presence of plasma proteins, primarily albumin, it is 291 mOsm/kg water.

One milliosmole per kilogram water net **oncotic** pressure (i.e., osmotic pressure due to proteins and other large molecules restricted to the plasma; see also the Gibbs-Donnan effect) is equivalent to roughly 20 mm Hg of hydrostatic pressure. This oncotic pressure just balances the net capillary hydrostatic force of about 20 mm Hg (see Fig. 3-7).

Distribution of Ions in Body Compartments

If the concentrations of a singly charged ion on either side of a semipermeable membrane are known, then the **Nernst equation** can be used to determine the electric potential generated by the tendency of ions to move along their concentration gradients.

$$\text{Potential (mV)} = -61.5 \log [X]_i/[X]_o \text{ for univalent ions at } 37°\text{C}$$

where $[X]_i$ is the concentration of the ion in the cell and $[X]_o$ is the concentration of the ion outside of the cell.

The Gibbs-Donnan Equilibrium

The distribution of ions and the resulting osmotic pressure could not, however, be entirely predicted by a simple extension of the principles of the Nernst equation. It could not, for example, explain the unequal distribution of ions on two sides of a semipermeable membrane (e.g., capillary wall). This discrepancy resulted from the Gibbs-Donnan equilibrium. The Gibbs-Donnan equilibrium occurs because proteins exist in the plasma in higher concentration than in the extracellular fluid, because the capillary endothelium is relatively impermeable to proteins, and because proteins have multiple negative charges per molecule. The proteins therefore exert an appreciable effect electrostatically on other more permeable ions, changing the equilibrium concentrations of those ions by about 5 percent. The osmotic forces exerted by the proteins themselves and by their electrostatic effects on osmotically active ions are referred to as **oncotic pressure.** In the following illustration of the principles that underlie the Gibbs-Donnan equilibrium, it is assumed that proteins have only a single negative charge. This will simplify the explanation and will change only the quantitative and not the qualitative conclusions.

Given two compartments separated by a membrane permeable to small ions but impermeable to large protein molecules, the effect of the Gibbs-Donnan relation would be as follows:

Initial conditions		Equilibrium conditions	
A	B	A	B
5 Na^+	10 Na^+	9 Na^+	6 Na^+
5 $protein^-$	10 Cl^-	4 Cl^-	6 Cl^-
		5 $protein^-$	
10 particles	20 particles	18 particles	12 particles

Chloride moves down its concentration gradient (RT log $[Cl^-]_A/[Cl^-]_B$) and against its electric potential gradient (FE) until the two forces exactly cancel; that is, equilibrium for chloride is reached. The work performed is

$$W = RT \log \frac{[Cl^-]_A}{[Cl^-]_B} + FE$$

where W = work, R = gas constant, T = absolute temperature (degrees kelvin), F = Faraday's constant, E = electric potential difference between compartments, $[Cl^-]_A$ = concentration of chloride in compartment A, and $[Cl^-]_B$ = concentration of chloride in compartment B. In other words, chloride (and similarly any other permeable anions) would move along its concentration gradient into the protein compartment until the negative potential created by the accumulating excess negative charges exactly balanced the propensity of the chloride and other anions to obtain equal concentrations in both compartments.

The work needed to move Na^+ against its final concentration gradient (at equilibrium Na^+ will be in higher concentration in the compartment with the protein) but with the electrical gradient created by the anions would be:

$$W = RT \log \frac{[Na^+]_A}{[Na^+]_B} - FE$$

Since, at equilibrium, the sum of two equations must be equal to zero, that is, the total work done by the system must be zero, the two equations simplify to

$$[Na^+]_A \times [Cl^-]_A = [Na^+]_B \times [Cl^-]_B$$

Note that changing the initial distribution of permeable ions will not change the equilibrium obtained.

The following rules can now be used to predict qualitatively the net effect of the Gibbs-Donnan relation:

1. The product of the concentrations of diffusible ions in one compartment should equal the product of the same diffusible ions in the other compartment (in this example, $9 \times 4 = 36$ in compartment A and $6 \times 6 = 36$ in compartment B).
2. Within a given compartment the total cationic charge must equal the total anionic charge (9 of each in compartment A and 6 of each in compartment B; a small potential difference does exist across the membrane boundaries but is negligible except for its effects on ion distribution).

3. The concentration of diffusible cations will be greater in the compartment with the proteins. Similarly, the concentrations of *diffusible* anions will be less in that compartment.
4. The total net osmotic pressure will be greater in the compartment with the proteins than in the other (if the ions are univalent). For proteins this increased pressure is due not only to the osmotic force of the proteins themselves but also to their effect on the diffusible ions since, even neglecting the proteins, in our example there are 18 particles in compartment A compared to 12 in B. The total osmotic force exerted by the proteins themselves and by the Gibbs-Donnan effect is called the oncotic pressure.

The assumption that the proteins had only a single negative charge (when in fact they have multiple negative charges) has led to an overestimate of the ratio of protein-Na^+ and hence the osmolality due to protein. In fact the total contribution of the oncotic pressure of the protein is only 1 mOsm/kg H_2O. As discussed previously, 1 mOsm/kg H_2O equals 20 mm Hg or 0.2 cm H_2O hydrostatic pressure.

Fluid Balance

At equilibrium, **input = output;** except in complete renal shutdown, what goes in must come out. In a stable chronic condition, the body *must* be in balance. Acute changes in body weight may be assumed to be due to gain or loss of body water. The exact composition of fluid would depend on the nature of the insult (for example, excessive sweating would result in the loss of hypotonic fluid).

Physiology of the Glomerulus

The Glomerulus

The mesangial cells, the endothelial cells of the glomerular capillaries and the epithelia of Bowman's capsule, together with their fused basement membranes, constitute the **renal corpuscle.** The glomerular tuft of capillaries (the glomerulus proper) is encapsulated by the renal corpuscle (Fig. 3-5). The terms *renal corpuscle* and *glomerulus* are hereafter used interchangeably, as they frequently are, but strictly speaking they are not the same thing. The function of the renal corpuscle is to form an ultrafiltrate of plasma. Plasma water and small solutes move out of the glomerular capillaries propelled by hydrostatic pressure through fenestrae in the endothelial cells that form the capillary wall, filter through the basement membrane, and then pass through the slit pores between the podocyte endfeet to enter Bowman's space as an ultrafiltrate of plasma (Fig. 3-6). **Mesangial cells** provide structural support to the glomerular capillaries and also serve to clear the filtrate of any macromolecules that may have leaked across the filtration barrier. In addition, by virtue of the contractile elements they contain, it is thought that they may also serve to regulate intraglomerular blood flow.

Bowman's capsule is the most proximal portion of the nephron (see Fig. 3-5). It consists of parietal epithelial cells forming the outer spherical capsule and visceral epithelial cells whose foot processes interdigitate, surrounding the glomerular tuft of capillaries and forming an inner spherical capsule. The space between these two capsules is **Bowman's space.** Bowman's space is continuous with the lumen of the proximal convoluted tubule at the urinary pole of the capsule.

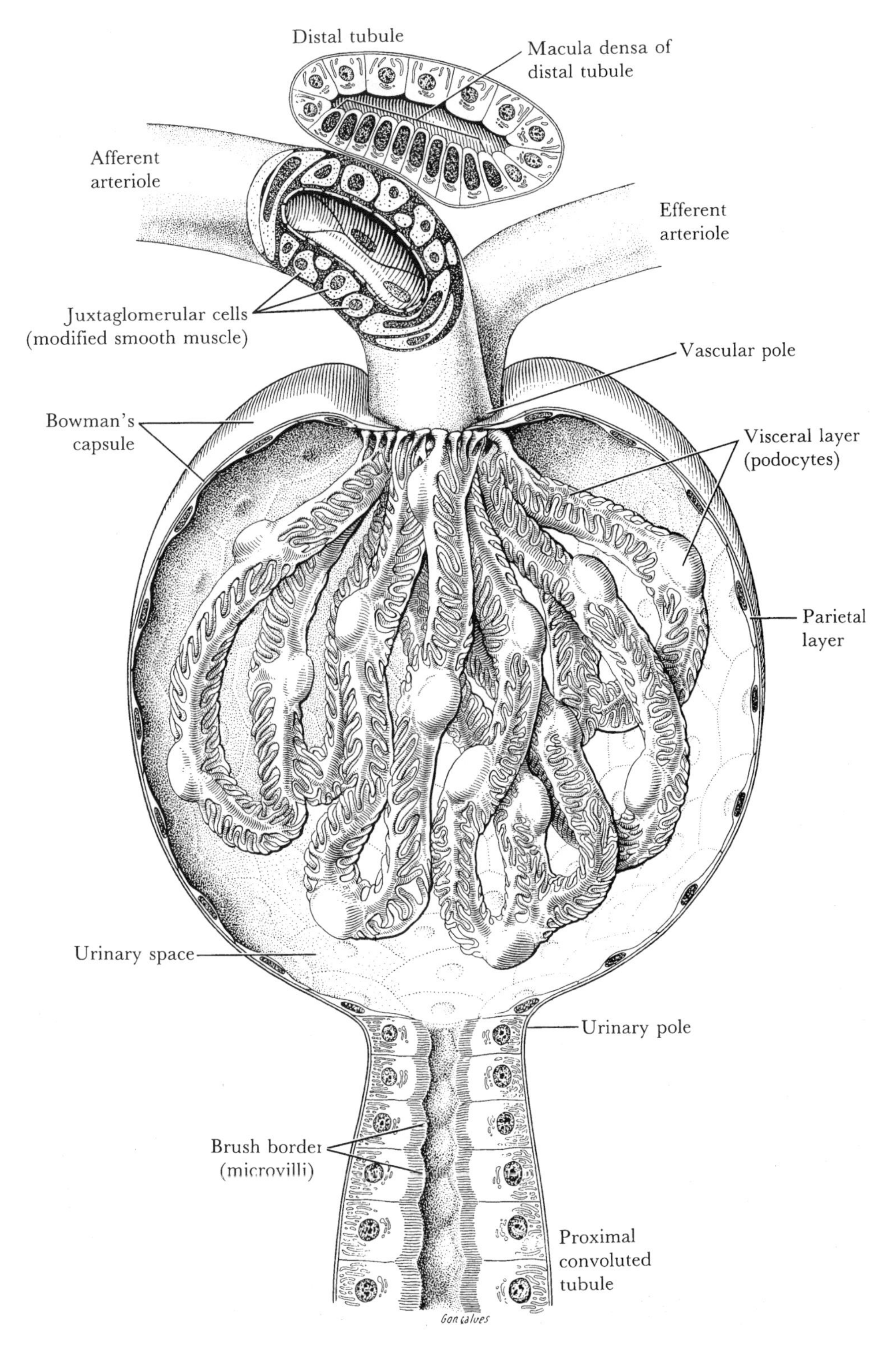

Fig. 3-5. The renal corpuscle. The upper part shows the vascular pole, with afferent and efferent arterioles and the macula densa. Note the juxtaglomerular cells in the wall of the afferent arteriole. Podocytes cover glomerular capillaries. Their nuclei protrude on the cell surface while their processes line the outer surface of the capillaries. Note the flattened cells of the parietal layer of Bowman's capsule. The lower part of the drawing shows the urinary pole and the proximal convoluted tubule. (From L. C. Junqueira and J. Carneiro. *Basic Histology* [7th ed.]. P. 374. © 1992 by Lange Medical Publications, Los Altos, CA.)

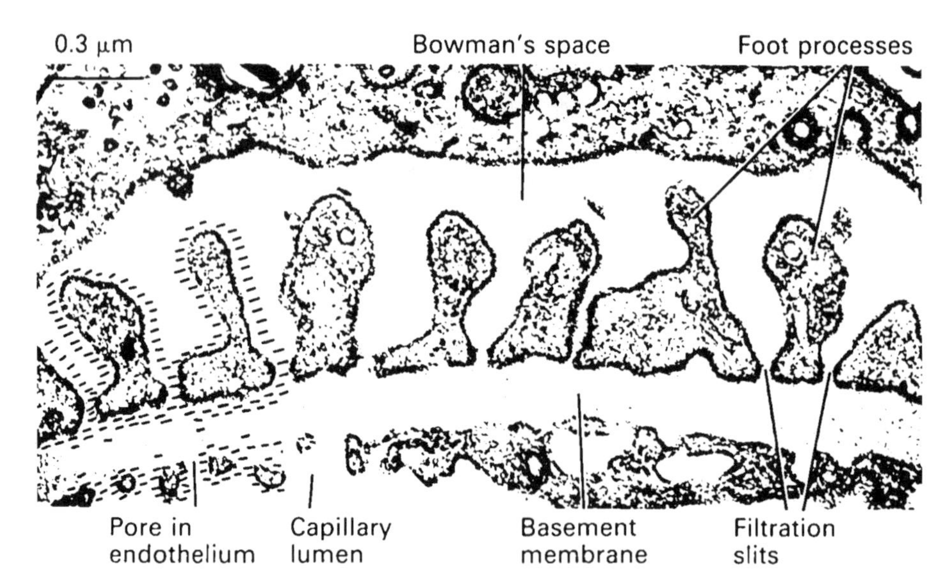

Fig. 3-6. Electron micrograph of a portion of the wall of a glomerular capillary, showing pores in the extremely attenuated endothelium. On the outer surface of the basal (basement) lamina are the foot processes of the podocytes, with the narrow filtration slits between them. The portion on the left indicates the distribution of charges in the glomerular capillary wall. (From H. Valtin. *Renal Function: Mechanisms Preserving Fluid and Solute Balance in Health* [2nd ed.]. Boston: Little, Brown, 1983. P. 47.)

The Glomerular Filtration Barrier

The components of the glomerular filtration barrier (i.e., between the capillary lumen and Bowman's space) (Fig. 3-6) are

1. The capillary endothelial fenestrae, which are 500 to 1000 angstroms (Å) in diameter
2. The fused basement membranes of the endothelial cells and the visceral epithelial podocytes
3. The podocytes with their foot processes separated by slit pores (250 Å wide), with diaphragms further limiting the actual pore size

The glomerular filtration barrier is **charge selective, size selective,** and, to an extent, **shape** (or deformability) **selective.** Because glomerular endothelium lacks the diaphragms that normally decrease the apertures of the fenestrae of other capillaries, they do not impose a barrier to the movement of proteins **(colloids)** although they do prevent the movement of erythrocytes. The basement membrane, on the other hand, is a continuous filamentous layer consisting of type IV and V collagen and negatively charged sialoglycoproteins. It is probably the size- and charge-selective component of the barrier. Finally, there are the visceral epithelial cell podocytes with their foot processes and slit pores whose exact function is unknown but which may also contribute to the filtration barrier. Molecules greater than 50 to 100 Å in diameter are unable to pass the filtration barrier. Molecules such as albumin that are somewhat smaller than this but which contain numerous negative charges are removed by a combination of charge- and size-selective filtration. Molecules that are unable to deform sufficiently are also excluded. (Perhaps they cannot squeeze through the right holes.) Overall, glomerular capillaries are significantly less permeable to proteins than are their systemic counterparts.

For many smaller substances protein binding limits filtration. Only that portion of a substance or drug that is unbound to protein is subject to filtration.

Glomerular Filtration Rate

The glomerular filtration rate (GFR) is the total volume per unit time of plasma ultrafiltrate leaving the capillaries and entering Bowman's space. In a healthy male adult, GFR is roughly 180 liters per day or approximately 120 ml per minute.

The renal blood flow is approximately 25 percent of the total cardiac output or 1200 ml whole blood per minute. Of all the organs only the liver receives a greater proportion of the cardiac output (roughly 30%). Because only the plasma (not the red cells) is available for filtration, 660 ml of plasma per minute (the **renal plasma flow [RPF]**) is available for modification during any given moment (55% of 1200). Roughly 120 ml per minute of the 660 ml per minute of plasma flow is filtered into Bowman's space as ultrafiltrate. From there it passes into the renal tubule system. Only about 1 percent of the filtered load of fluid eventually passes through the tubules and collecting system to emerge as urine. The average urine output or flow per minute ($\dot{V}$) is 1.2 ml per minute, that is, 1 percent of the 120 ml per minute filtered.

Determinants of the GFR

The four determinants of the magnitude of the GFR are

1. **Ultrafiltration coefficient** or constant, $\mathbf{K_f}$, which is simply the capillary permeability per meter squared multiplied by the surface area available for filtration
2. **Net oncotic pressure,** $\bar{\pi} = (\pi_c - \pi_t)$, in the glomerular capillary, in which π_c = glomerular capillary oncotic pressure and π_t = the oncotic pressure in Bowman's space; because there is normally no protein in Bowman's space, $\pi_t = 0$ and therefore $\bar{\pi} = \pi_c$
3. **Net hydraulic pressure** driving filtrate from the capillaries into Bowman's space, $\bar{\mathbf{P}} = (P_c - P_t)$, in which P_c = the hydraulic pressure in the capillary and P_t = the hydraulic pressure in Bowman's space
4. Initial or afferent glomerular capillary plasma flow, $\mathbf{Q_A}$

Note that the first three determinants of the GFR are related to each other by Starling's equation (see Chap. 2).

Glomerular and Systemic Capillaries

Glomerular capillaries differ from systemic capillaries in several important respects. In systemic capillary beds, such as those of the skeletal muscles, the hydraulic pressure driving filtration falls steadily as filtrate is forced out of the capillaries (Fig. 3-7). When the net hydraulic pressure just equals the net oncotic pressure, which keeps fluid in the capillaries, **filtration pressure equilibrium** is reached and no more fluid can leave the capillaries. As the blood reaches the venous side of the capillary bed, the net hydraulic pressure decreases and reabsorption of fluid begins (Fig. 3-7). Almost all the fluid filtered out on the arterial side of the capillaries is reabsorbed on the venous side. The remainder is returned to the venous circulation by the lymphatics, as are the proteins that manage to leave the capillaries.

The situation in the glomerular capillaries is quite different. The mean hydraulic pressure driving filtration remains nearly constant along the length of the capillary (Fig. 3-8). Filtration is slowed as blood reaches the end of the capillary bed because the capillary oncotic pressure rises to oppose the hydraulic pressure, thus decreasing

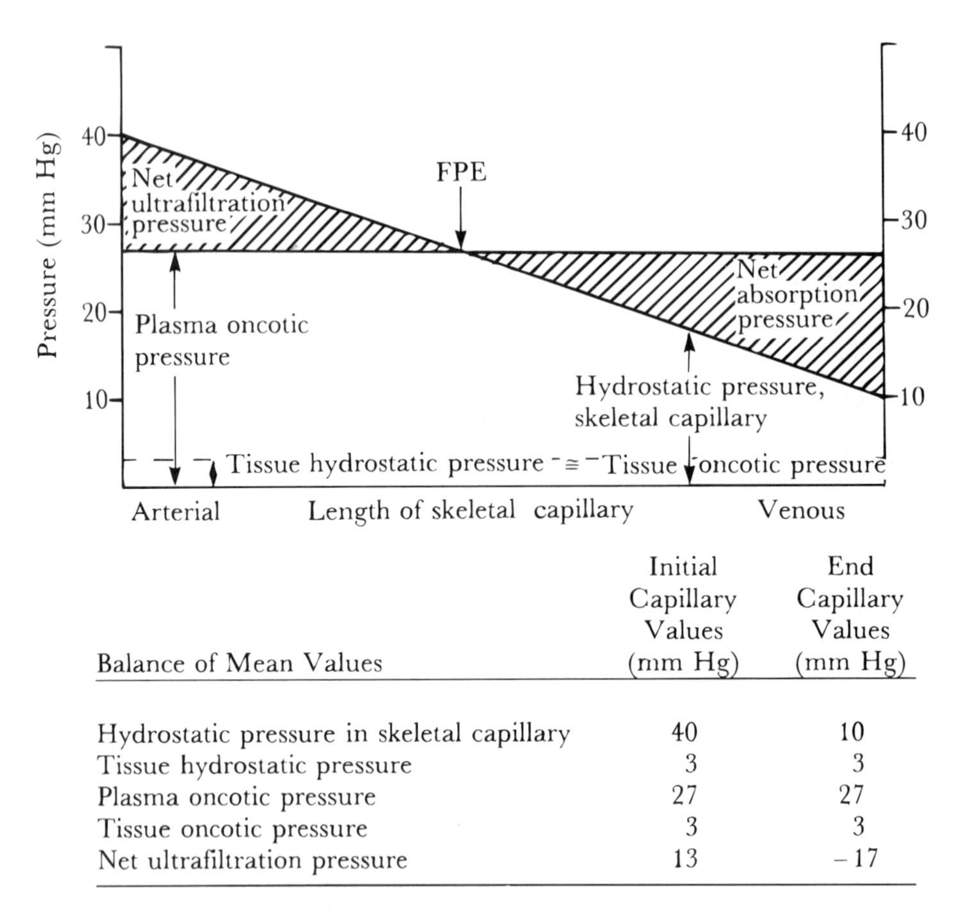

Balance of Mean Values	Initial Capillary Values (mm Hg)	End Capillary Values (mm Hg)
Hydrostatic pressure in skeletal capillary	40	10
Tissue hydrostatic pressure	3	3
Plasma oncotic pressure	27	27
Tissue oncotic pressure	3	3
Net ultrafiltration pressure	13	−17

Fig. 3-7. Starling's forces involved in skeletal capillaries. As shown, ultrafiltration pressure decreases mainly because hydrostatic pressure decreases. Hydrostatic pressure and plasma oncotic pressure are plotted against distance along the skeletal capillary. In this example, tissue hydrostatic pressure and tissue oncotic pressure fortuitously cancel. Initially, net hydrostatic pressure exceeds net oncotic pressure, driving fluid out of the capillaries. At some point along the capillaries, the net hydrostatic pressure equals the net oncotic pressure and filtration pressure equilibrium (FPE) occurs, with cessation of net fluid movement. Subsequently, net oncotic pressure exceeds net hydrostatic pressure and reabsorption begins. Any fluid not reabsorbed by the end of the permeable capillary bed is returned to the circulation by the lymphatic system. (Adapted from a lecture given by Channing R. Robertson, PhD, at Stanford Medical School, 1984.)

the net pressure driving filtration. The increase in capillary oncotic pressure is the result of the rapid loss of ultrafiltrate and consequent increase in protein concentration.

There is a large hydraulic pressure drop at the level of the efferent arteriole while the oncotic pressure remains nearly constant since no further filtration can occur. Thus, when the blood reaches the peritubular capillaries the oncotic pressure exceeds the hydraulic pressure and fluid is reabsorbed into the capillaries of the descending vasa recta or peritubular plexi.

Autoregulation

The kidneys are able to regulate the GFR under a wide range of conditions. This phenomenon is called **autoregulation** because there are intrinsic mechanisms in the kidney responsible for maintaining the nearly constant GFR. This is necessary since even small changes in the GFR can vastly alter the filtered load of solute and water.

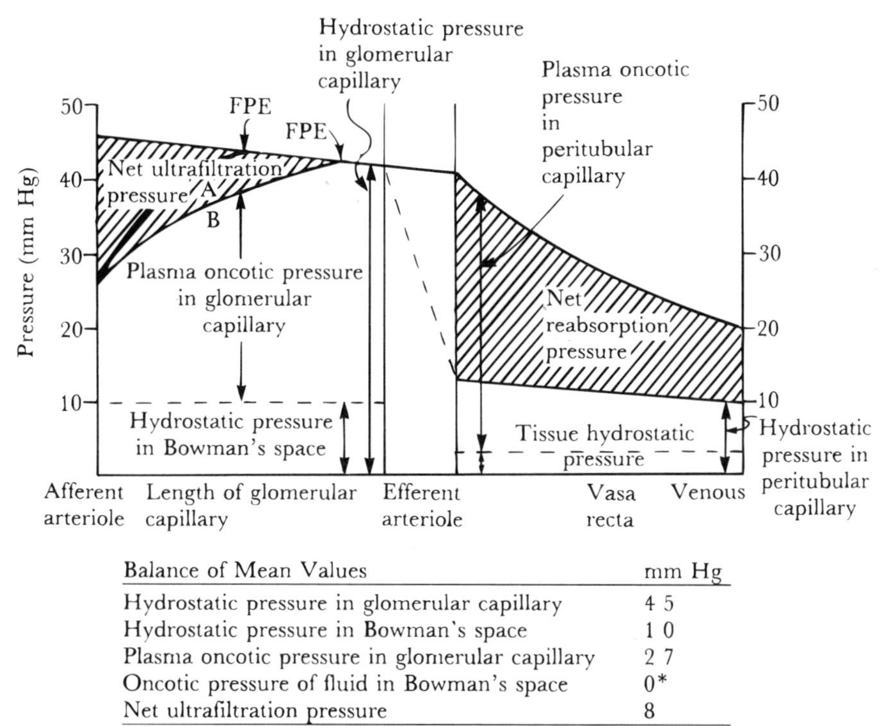

Balance of Mean Values	mm Hg
Hydrostatic pressure in glomerular capillary	45
Hydrostatic pressure in Bowman's space	10
Plasma oncotic pressure in glomerular capillary	27
Oncotic pressure of fluid in Bowman's space	0*
Net ultrafiltration pressure	8

* The concentration of protein in Bowman's space fluid is neglible. The estimated oncotic pressure is 0.3 mm Hg.

Fig. 3-8. Starling's forces involved in glomerular ultrafiltration (specific values vary with species). As shown, ultrafiltration pressure declines in glomerular capillaries, mainly because plasma oncotic pressure rises. This is in contrast to extrarenal capillaries, in which the decline in ultrafiltration pressure is due mainly to a decrease in intracapillary hydrostatic pressure (see Fig. 3-7). It is not yet known at what point in the capillary the sum of the hydrostatic pressure in Bowman's space and of plasma oncotic pressure exactly balances the hydrostatic pressure in the glomerular capillary. In humans, this point, called filtration pressure equilibrium (FPE), may not be reached in the glomerular capillary. In those species in which it is attained, the pattern for the rise in plasma oncotic pressure as a function of capillary length can vary. For example, an increase in the rate at which plasma enters the glomerular capillary will lead to a change in the pattern from curve A to curve B, and consequently to a rise in the mean net ultrafiltration pressure.

Hydrostatic pressure drops rapidly at the level of transition between the efferent arterioles and the vasa recta. Net oncotic pressure decreases much more slowly and therefore exceeds hydrostatic pressure in the vasa recta. This leads to net reabsorption of fluid. (From H. Valtin. *Renal Function: Mechanisms Preserving Fluid and Solute Balance in Health* [2nd ed.]. Boston: Little, Brown, 1983. P. 44.)

Over a range of arterial pressure from 80 to 200 mm Hg, both GFR and RPF remain quite constant (Fig. 3-9).

The kidneys are able to autoregulate glomerular capillary hydrostatic pressure, flow, and, indeed, filtration because there are precapillary sphincter muscles in both the afferent and efferent arterioles. Increasing or decreasing the size of the afferent arteriole allows more or less of the systemic pressure to be felt by the glomerulus, while changes in the diameter of the efferent arteriole can alter the filtration rate as well as help to govern flow (Fig. 3-10).

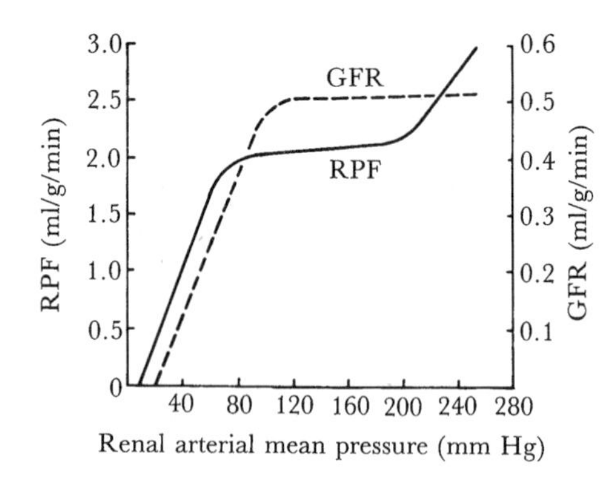

Fig. 3-9. Autoregulation in the kidney. GFR = glomerular filtration rate; RPF = renal plasma flow. (From R.E. Shipley and R. S. Study. Changes in Renal Blood Flow. *Am. J. Physiol.* 167:676, 1951.)

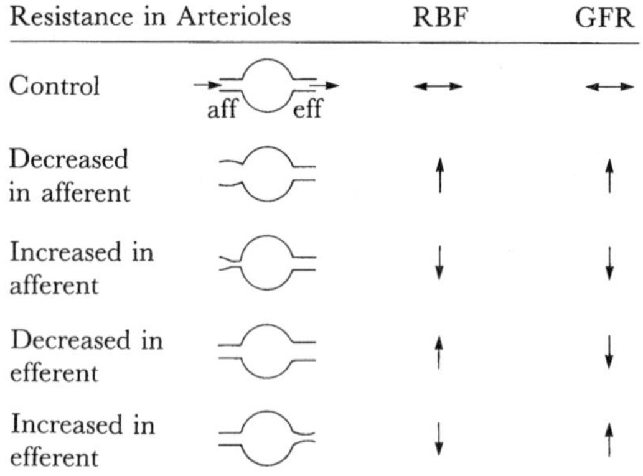

Resistance in Arterioles		RBF	GFR
Control	aff eff	↔	↔
Decreased in afferent		↑	↑
Increased in afferent		↓	↓
Decreased in efferent		↑	↓
Increased in efferent		↓	↑

Fig. 3-10. Changes in renal blood flow (RBF) and glomerular filtration rate (GFR) that will occur when resistance is altered in either the afferent or the efferent arterioles, provided that renal perfusion does not change. (From H. Valtin. *Renal Function: Mechanisms Preserving Fluid and Solute Balance in Health* [2nd ed.]. Boston: Little, Brown, 1983. P. 105.)

Calculation of Renal Clearance, Glomerular Filtration Rate, and Renal Blood Flow

Renal clearance is defined as that volume of plasma from which all of a given substance is removed per unit time in one pass through the kidneys (the units are, therefore, milliliters plasma per unit time). If a substance is neither secreted nor reabsorbed, then the clearance is equal to the GFR, and all of the plasma filtered has been cleared of the substance (e.g., inulin) (Table 3-7). The clearance would then be due to filtration of the substance into the glomerulus followed by its passage through the uriniferous tubule without further alteration (as in the case of **inulin**). Clearance can, however, be the result of any possible combination of filtration, reabsorption, and secretion. This relation is demonstrated by the following equation:

(Quantity excreted) = [(quantity filtered) + (quantity secreted)] − (quantity reabsorbed)

Table 3-7. Characteristics of a substance suitable for measuring the GFR by determining its clearance

Freely filtered (i.e., not bound to protein in plasma or sieved in the process of ultrafiltration)
Not reabsorbed or secreted by tubules
Not metabolized
Not stored in kidney
Not toxic
Has no effect on filtration rate
Preferably easy to measure in plasma and urine

From W. F. Ganong. *Review of Medical Physiology* (15th ed.). Los Altos, CA: Lange, 1991. P. 656.

If all of the substance is reabsorbed (e.g., glucose), then the clearance would be zero (none of the plasma is cleared of the substance), but the GFR would still be 120 ml per minute, as before. If all of the substance in the plasma is secreted, then the clearance would be equal to the RPF because all of the plasma volume passing through the kidneys would be cleared of the substance (e.g., **paraaminohippurate [PAH]).** In other words, the combined rates of filtration and secretion of the substance would equal the plasma flow rate.

Clearance can be calculated by the following equation:

$$C_x = \frac{(U_x)(V)}{(P_x)} = \frac{\text{amount excreted/min}}{\text{plasma concentration of x}}$$

in which C_x = clearance of x, $\dot{V}$ = urine flow in ml per minute, and P_x = plasma concentration of x.

Clearance could have been defined simply as the amount of substance X excreted per minute; however, this value would not in itself be an adequate indication of how efficiently or in what manner the kidneys were excreting the substance. Use of the equation for clearance creates an index of renal function that can be usefully compared to the calculated GFR.

If $\mathbf{C_x}$ **> GFR,** then there must be **net secretion.**
If $\mathbf{C_x}$ **< GFR,** then there must be **net reabsorption.**
If $\mathbf{C_x}$ **= GFR,** then probably there is *neither* net reabsorption nor net secretion (unless the two fortuitously canceled).

The renal clearance of **inulin** is often used to estimate GFR because it is neither secreted nor reabsorbed, and its elimination into the urine is, therefore, entirely due to filtration and independent of the plasma inulin concentration. Inulin is not bound to plasma proteins, its volume of distribution includes all of the ECF, and it is not metabolized. The renal clearance of **creatinine** is often used clinically in place of inulin because, unlike inulin, creatinine is a normal component of the blood. Creatinine, however, is secreted to a small extent in the proximal tubule (Fig. 3-11) and in disease states the extent of its secretion is unpredictable. The renal clearance of inulin is therefore a better estimate of GFR than is creatinine clearance.

At low concentrations of PAH, approximately 91 percent of the PAH entering the kidneys in the renal arteries is removed from the plasma by a combination of filtration into Bowman's space and active secretion into the proximal tubules. Excretion of PAH is thus the result of filtration plus net secretion. At plasma concentrations below the level at which the secretory transport maximum of PAH is reached, the clear-

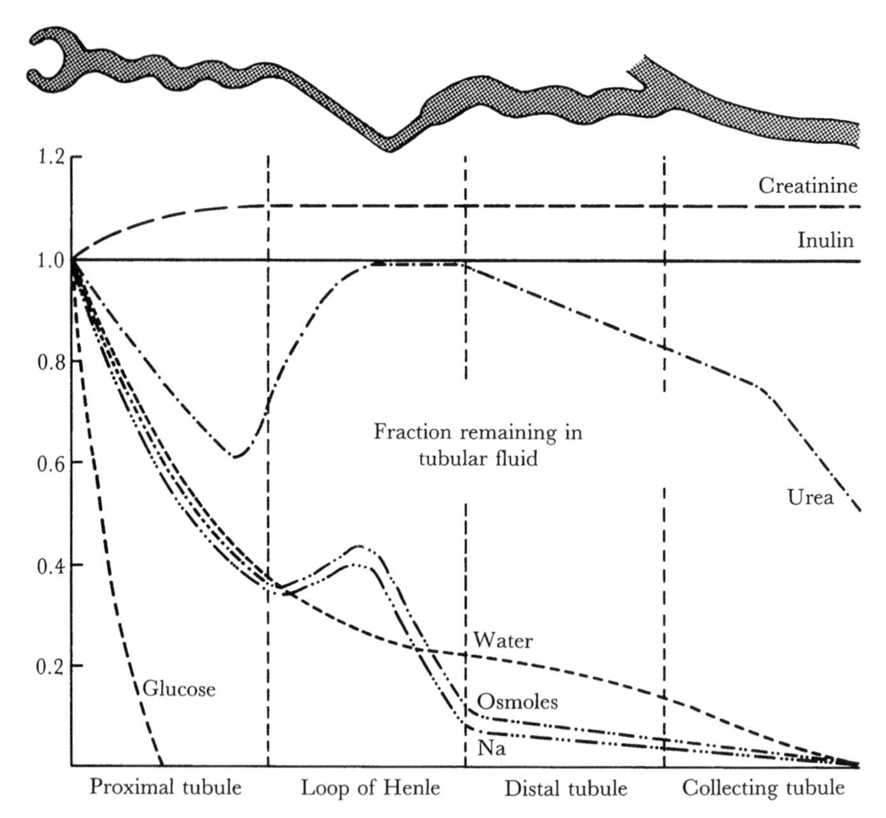

Fig. 3-11. Changes in the fraction of the filtered amount of substances remaining in the tubular fluid along the length of the nephron. (From L. P. Sullivan and J. J. Grantham. *Physiology of the Kidney* [2nd ed.]. Philadelphia: Lea and Febiger, 1982. P. 93.)

ance of PAH is used to estimate the RPF **(the effective renal plasma flow [ERPF]).** Given a urine flow of 1 ml per minute, and urine and plasma PAH concentrations of 6 mg/ml and 0.01 mg/ml, respectively, **RPF** can then be calculated using the formula

$$0.91 \text{ RPF} = \text{clearance of PAH} = \frac{(\text{U})(\text{V})}{(\text{P})} = \frac{(6 \text{ mg/ml})(1 \text{ ml/min})}{(0.01 \text{ mg/ml})} = 600 \text{ ml/min}$$

$$\text{RPF} = \frac{600 \text{ ml/min}}{0.91} = 659 \text{ ml/min}$$

Free Water Clearance

The kidneys can be remarkably miserly in their handling of water when such conservation is necessary, eliminating in the urine only that amount of water necessary to achieve excretion of waste products. Renal concentrating mechanisms can achieve urine concentrations of 1200 mOsm per liter. Conversely, in a state of water excess, urine can be excreted in copious amounts with a solute concentration of less than 100 mOsm per liter.

Free water is water free of solute. **Free water clearance (C_{H_2O})** is a calculation of the amount of distilled water that must be added to or removed from the urine to render the urine isosmotic with plasma. The concept of free water clearance is an attempt to quantify the states of water diuresis and conservation. It is calculated by subtracting the **clearance of osmoles, C_{Osm},** from the minute urine flow, $\dot{V}$.

$$C_{H_2O} = \dot{V} - \frac{U_{Osm}\dot{V}}{P_{Osm}}$$

If the C_{H_2O} **is greater than 0,** then water is being excreted in excess of solute (diuresis). If the C_{H_2O} **equals 0,** then the urine is being excreted isosmotically. If the C_{H_2O} **is less than 0,** then solute is being excreted in excess of H_2O (antidiuresis).

Negative free water clearance (T_{H_2O}) is the free water clearance with the sign reversed for convenience when discussing water-conserving states.

$$T_{H_2O} = -(C_{H_2O})$$

Free water clearance is not a true clearance, although it does contain a true clearance as a part of its definition.

Renal

Quantitation of Water Reabsorption

The ratio of tubular fluid (TF) to plasma (P) inulin concentrations is an indication of the degree of water reabsorption that has occurred. For instance, TF/P inulin is normally 3 for the proximal tubule, which means that the tubular fluid has three times the inulin concentration of the plasma. The fraction of water reabsorbed along the tubule can then be calculated by the following equation:

$$\text{Fraction of water reabsorbed} = 1 - \frac{1}{TF_{in} / P_{in}} = 2/3$$

Physiology of the Nephron and Collecting System: Beyond the Glomerulus

Reabsorption and Secretion in the Renal Tubules

Renal tubular reabsorption and secretion can occur as a result of **active transport, passive diffusion** (either **simple** or **facilitated**), or **secondary active transport.** In the renal tubules reabsorption and secretion of many substrates is coupled to the active transport of sodium via specific membrane carriers, glucose and amino acids being of particular importance. The rate of secretion or reabsorption by membrane carriers is directly proportional to the concentration of the substrate and to the affinity of the carrier for the substrate. Also characteristic of such transport systems is that each has a **maximal rate (T_{max})** at which it can transport solute above which the transport mechanism is said to be saturated (i.e., increases in solute delivery do not lead to any increase in transport since all the carriers are occupied). Transport maximums for the various solutes vary tremendously, from quite low to so high that their measurement is not practical. In some cases the transport may be bidirectional. The composition of the urine is determined by the balance that normally exists between filtration, secretion, and reabsorption (Table 3-8).

The Proximal Tubule

The proximal tubule is the main reabsorptive region of the nephron (Figs. 3-11 and 3-12). All reabsorption that occurs there is isosmotic (Fig. 3-13). The proximal tubule, therefore, does not affect the overall concentration of the TF, but it does alter the com-

Table 3-8. Renal handling of various substances

Renal handling
Primarily filtered
Inulin
Urea[a]
Creatinine[b]
Primarily secreted[c]
Hydrogen ions
Organic acids (e.g., PAH, penicillins)
Organic bases (e.g., choline)
Filtration and reabsorption
Water
Sodium
Chloride
Bicarbonate
Calcium
Magnesium
Phosphate
Glucose
Amino acids
Filtration, reabsorption, and secretion
K^{+}[d]
Uric acid
Filtration and renal tubular catabolism
Low-molecular-weight proteins (e.g., insulin, glucagon, PTH and ADH)

[a]Approximately half of the filtered load may be passively reabsorbed into the medullary interstitium.
[b]Approximately 20 percent is secreted in humans.
[c]Some passive reabsorption is also possible.
[d]Virtually all the potassium filtered is reabsorbed somewhere in the tubule. Net excretion is by passive secretion in the distal nephron.

position of the tubular fluid. The proximal tubule has two subdivisions, the **proximal convoluted tubule** and the **proximal straight tubule.**

The proximal tubule is characterized by a large reabsorptive surface formed by thousands of microvilli that project from the luminal surface of the proximal tubular cells and by extensive lateral intercellular channels that are important in the isosmotic reabsorption of NaCl and water (Fig. 3-14). The proximal tubule is the only region of the tubule in which the enzyme carbonic anhydrase may be found in the luminal membrane. Within the luminal membrane is a relatively weak H^+ ion secretory pump that is unable to acidify the urine to a significant extent (i.e., does not contribute to H^+ excretion) but serves instead to facilitate the reabsorption of $NaHCO_3$ in conjunction with carbonic anhydrase (Fig. 3-15). The basolateral membrane of the proximal tubule is well developed and contains the active Na^+-K^+ exchange pumps (Fig. 3-16). Numerous mitochondria line the basolateral membrane to provide energy for the active Na^+-K^+ pumps. Cell-to-cell junctions in the proximal tubule are permeable to water and electrolytes to a greater degree than those of other parts of the tubule. This results in significant backflux of fluid and solute, which may be important in the regulation of solute and water reabsorption (see Fig. 3-14).

The functions of the proximal tubule are as follows:

1. Reabsorption of two-thirds of the filtered Na^+ and water, predominantly by active transport of Na^+ at the basolateral membrane with water following passively
2. Passive reabsorption of chloride and other electrolytes along their electrochemical gradient

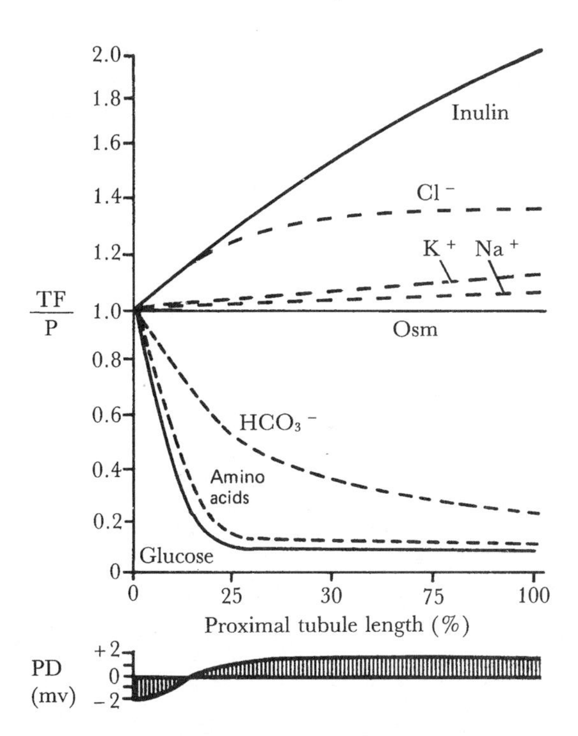

Fig. 3-12. Reabsorption of various solutes in the proximal tubule in relation to the potential difference (PD) along the tubule. TF/P = tubular fluid to plasma concentration ratio. (From W. F. Ganong. *Review of Medical Physiology* [15th ed.]. P. 661. © 1991 by Lange Medical Publications, Los Altos, CA.)

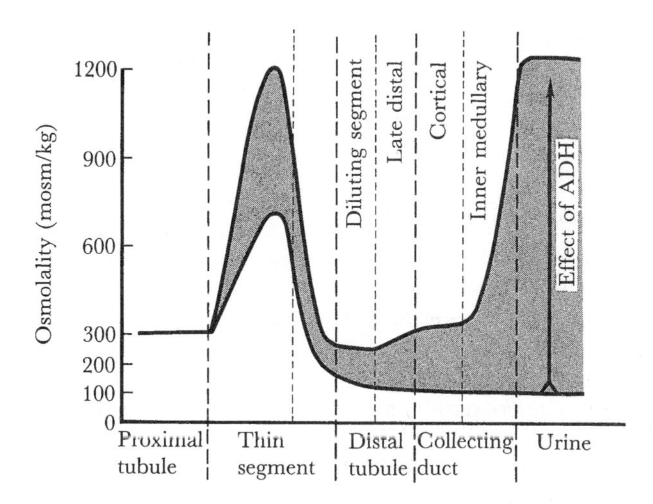

Fig. 3-13. Changes in osmolality of the tubular fluid as it passes through the tubular system. (From A. C. Guyton. *Textbook of Medical Physiology* [8th ed.]. Philadelphia: Saunders, 1991. P. 312.)

3. Reabsorption of all the filtered glucose and amino acids by cotransport with Na^+
4. Preferential reabsorption of $NaHCO_3$ (carbonic anhydrase dependent)
5. Organic acid secretion (e.g., diuretics, salicylates, antibiotics, *p*-aminohippurate) by an anionic pump located in the basolateral membrane
6. Organic base secretion (e.g., procainamide, choline) via a cationic pump located in the basolateral membrane

7. Vitamin reabsorption
8. Pinocytosis and breakdown of any protein remaining in the tubular fluid
9. Reabsorption of organic anions by cotransport with Na^+ (e.g., phosphate, sulfate, and lactate)
10. Active K^+ reabsorption
11. Secretion of ammonia

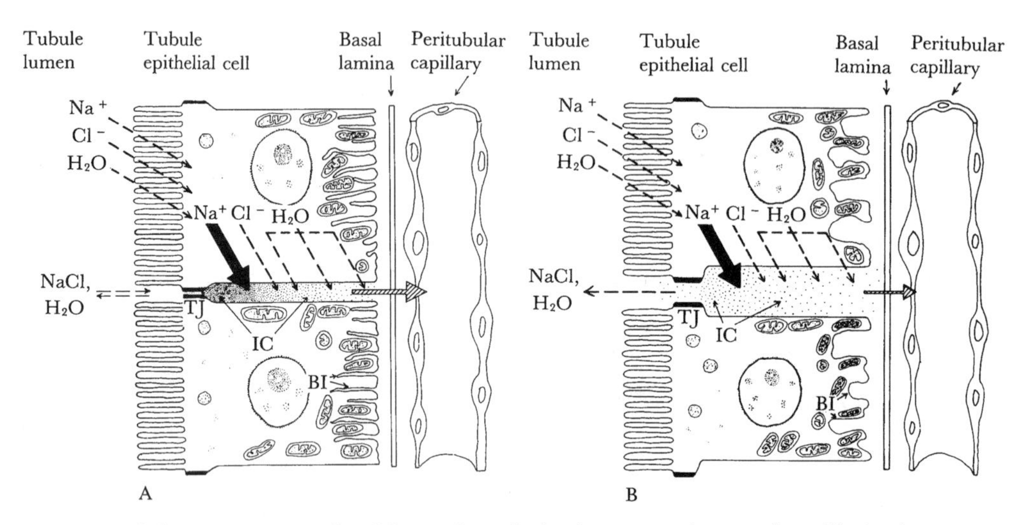

Fig. 3-14. Diagrammatic summary of the movement of Na^+ from the tubular lumen to the renal capillaries in locations where it is actively transported. A. Situation in hydropenic state. B. Situation when capillary uptake is reduced by increasing capillary pressure or decreasing plasma protein concentration. This causes widening of the intracellular junctions with backflow of solute and fluid into the tubule. Heavy solid arrow indicates active transport; dashed arrows indicate passive movement; hatched arrow indicates movement into renal capillaries as a function of Starling's forces across these capillaries. TJ = tight junction; IC = lateral intercellular space; BI = basilar infoldings. (From P. F. Mercer, D. A. Maddox, and B. M. Brenner. Current concepts of sodium chloride and water transport by the mammalian nephron. *West. J. Med.* 120:33, 1974.)

Fig. 3-15. A. Reabsorption of filtered HCO_3^- via H^+ secretion in the proximal tubule. The H^+ is secreted via the Na^+-H^+ exchange process in the proximal tubule and via an active process in the distal nephron; the HCO_3^- exits the cell to enter the peritubular space and be absorbed via a carrier-mediated process in conjunction with Na^+. The secreted H^+ reacts with HCO_3^- in the tubular fluid to form H_2CO_3, which then dissociates into carbon dioxide and water; that is, an HCO_3^- ion is lost from the tubular fluid. However, since the H^+ secretion process adds a HCO_3^- ion to the peritubular fluid and peritubular capillaries, the net result is HCO_3^- reabsorption. The proximal tubule is illustrated here, since approximately 90 percent of the filtered HCO_3^- is reabsorbed in the proximal tubule. However, similar reactions occur in more distal portions of the nephron (except for the absence of carbonic anhydrase in the luminal surface) and are responsible for the reabsorption of most of the remaining HCO_3^- in the tubular fluid. An active Na^+-K^+ exchange pump in the basolateral membrane actively pumps Na^+ from the cells into the peritubular fluid. This supplies the gradient for Na^+ reabsorption from the tubular fluid. B. Generation of a new HCO_3^- via H^+ secretion. The HCO_3^- generated in the H^+ secretion process represents a new HCO_3^- if the secreted H^+ ion reacts with NH_3 synthesized by the epithelial cells to form NH_4^+, or reacts with HPO_4^{-2} and other titratable acids in the tubular fluid to form $H_2PO_4^-$ and to acidify the remaining titratable buffers. Although the distal nephron is illustrated here, similar reactions can occur in the proximal tubule (except that proximal tubular H^+ secretion is primarily coupled to Na^+ reabsorption and is gradient limited). Most of the H^+ secreted in the proximal tubule reacts with HCO_3^- in the tubular fluid and therefore accomplishes the reabsorption of filtered HCO_3^- rather than the generation of new HCO_3^-. In the distal nephron HCO_3^- exits the cell in a carrier-mediated process in exchange for chloride. The distal tubular H^+ secretory pump can create a 1000-fold [H^+] gradient enabling it to acidify the urine. (Redrawn from J. B. West. *Best and Taylor's Physiological Basis of Medical Practice* [12th ed.]. Pp. 493-494. © 1990 by Williams & Wilkins, Baltimore.

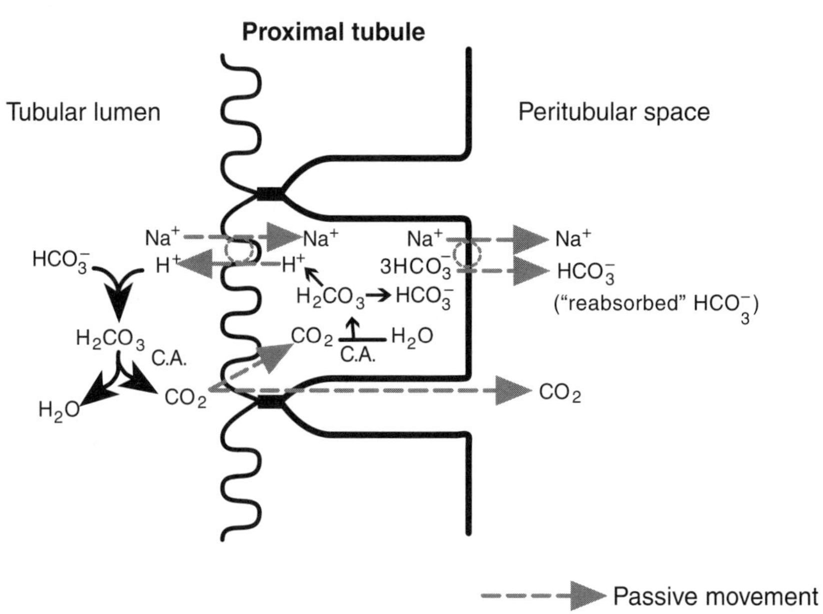

Proximal tubule
Tubular lumen
Peritubular space
Na+
H+
HCO3−
H2CO3
C.A.
H2O
CO2
Na+
H+
H2CO3 → HCO3−
CO2
C.A.
H2O
Na+
3HCO3−
Na+
HCO3−
("reabsorbed" HCO3−)
CO2
Passive movement

A

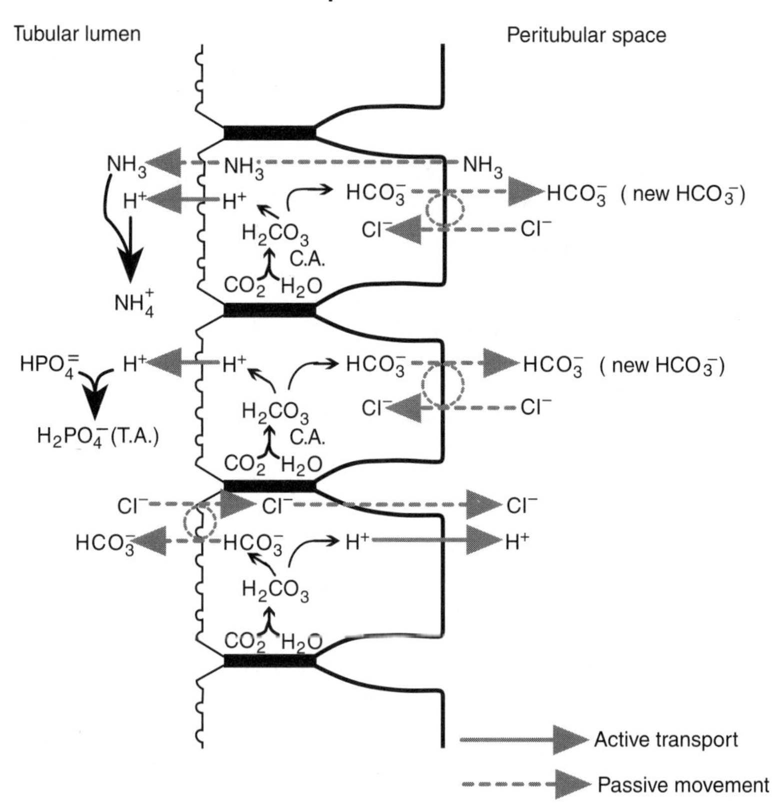

Distal nephron
Tubular lumen
Peritubular space
NH3
H+
NH4+
NH3
H+
HCO3−
H2CO3
C.A.
CO2
H2O
NH3
HCO3− (new HCO3−)
Cl−
Cl−
HPO4=
H+
H2PO4− (T.A.)
H+
HCO3−
H2CO3
C.A.
CO2
H2O
HCO3− (new HCO3−)
Cl−
Cl−
Cl−
HCO3−
Cl−
HCO3−
H+
H2CO3
CO2
H2O
Cl−
H+
Active transport
Passive movement

B

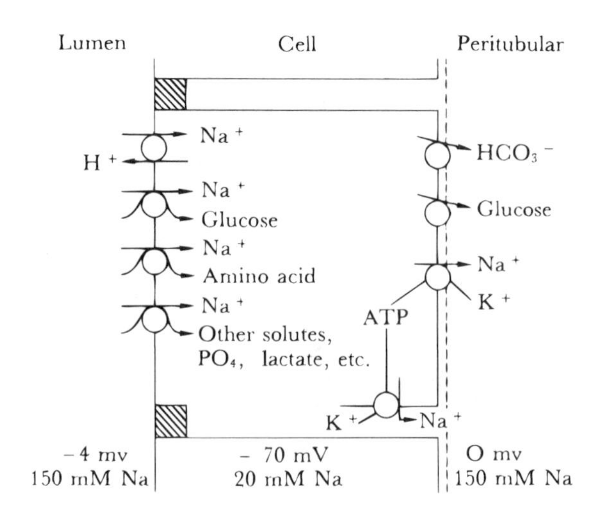

Fig. 3-16. Interaction of the transport of Na^+ and other solutes across early proximal tubule. (From B. M. Burg. Renal Handling of Sodium, Chloride, Water, Amino Acid and Glucose. In B. M. Brenner and F. C. Rector. *The Kidney* [2nd ed.]. Philadelphia: Saunders, 1981. P. 335.)

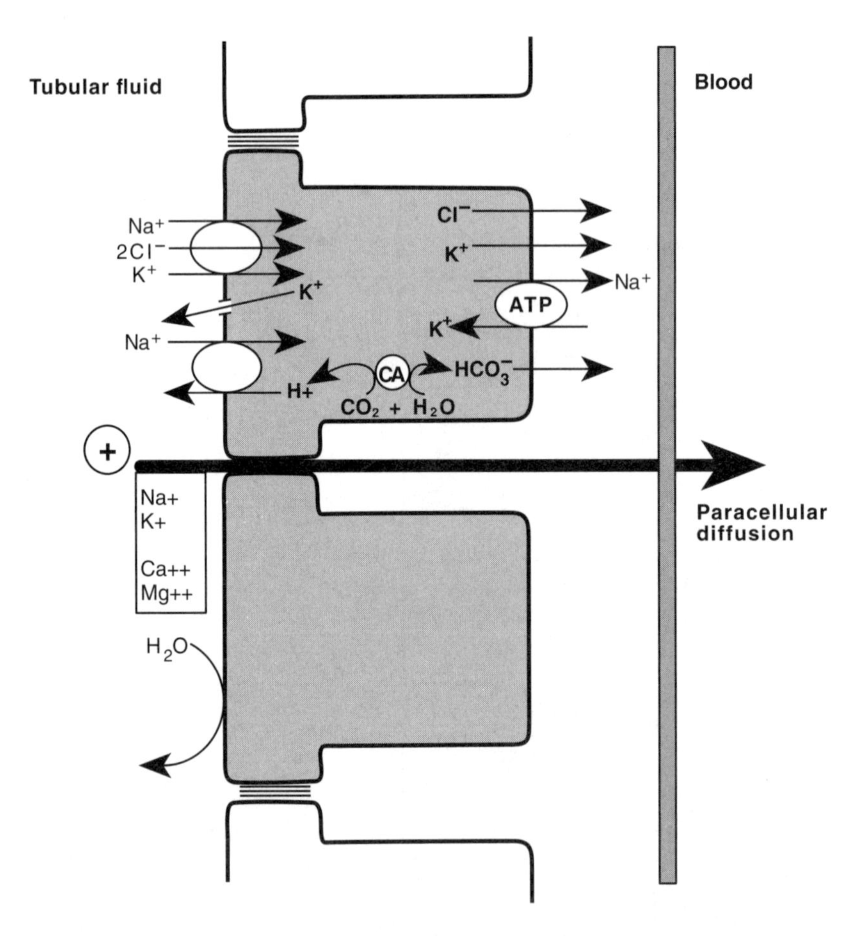

Fig. 3-17. The Na^+-K^+-$2Cl^-$ pump in the thick ascending limb of the loop of Henle. The lumen-positive transepithelial voltage results from the unique location of transport proteins in the apical and basolateral membrane and plays a major role in driving passive paracellular reabsorption of cations such as calcium. (Redrawn from R. M. Berne and M. N. Levy. *Physiology* [3rd ed.]. St. Louis: Mosby–Year Book, 1993. P. 746.)

The TF leaving this segment of the nephron is thus characterized by

1. A TF osmolality equal to that of plasma
2. Flow equal to one-third of the GFR
3. Absence of glucose, amino acids, and proteins
4. A somewhat higher concentration of chloride. Because normally almost all of the HCO_3^- is reabsorbed in the proximal tubule, some anion must remain behind in greater concentration to preserve electroneutrality. Therefore, even though there is net reabsorption of chloride, its relative concentration nonetheless increases in the proximal tubule (see Fig. 3-12).

The Loop of Henle

Anatomically, the loop of Henle consists of the **pars recta** of the proximal tubule, the **thin descending limb,** the **thin ascending limb** (present only in long-looped nephrons), and the **thick ascending limb.** Because it is an anatomic and not a physiologic entity, it is not surprising that its boundaries overlap those of other portions of the nephron. The thick ascending limb, for instance, may be considered a part of the distal tubule; its function is discussed separately. The physiologic significance of the loop of Henle is discussed in conjunction with the renal concentrating and diluting mechanism.

The Thick Ascending Limb of the Loop of Henle

The thick ascending limb is the diluting segment of the nephron because solute is actively pumped out of the thick ascending limb, leaving behind dilute tubular fluid. In the absence of antidiuretic hormone, the urine will be dilute as well, because the fluid diluted in this segment of the nephron, by removal of NaCl, is diluted even further in the distal tubule and collecting system in which additional active Na^+ pumping occurs. The thick ascending limb is extremely impermeable to water (which makes sense physiologically because otherwise water would be osmotically obligated to follow the solute).

The luminal membrane of the thick ascending limb has a carrier that requires four sites to be occupied in order to carry solute from the lumen to the cell interior (Fig. 3-17). The **stoichiometry of the carrier** is **$1Na^+ : 1K^+ : 2Cl^-$**. Regardless of the exact stoichiometry, the important point is that active pumping of both sodium and chloride occurs in this segment of the nephron. It is the only segment in which active Cl^- pumping normally occurs. A positive intraluminal potential difference provides the driving force for the reabsorption of all cations but particularly divalent cations (Mg^{2+} and Ca^{2+}) from the thick ascending limb. Medullary portions of the thick ascending limb contribute to the medullary hypertonicity by pumping solute into the interstitium and, therefore, contributing to the dilution of urine as well as to its concentration under the appropriate circumstances.

Distal Convoluted Tubule

The distal convoluted tubule is also permeable to water. Sodium is actively reabsorbed in this segment with Cl^- following passively. The reabsorption of Na^+ and the active secretion of H^+ and K^+ are under the influence of the mineralocorticoid hormone **aldosterone.** Both the distal convoluted tubule and collecting duct are subject to the actions of aldosterone. The excretion of K^+ is primarily dependent on the amount secreted in the distal nephron in response to aldosterone. The distal tubule has an extremely powerful H^+ secretory pump able to acidify urine to a pH of 4.5 (almost 3 pH units below the plasma pH).

Except for very distal portions, the distal convoluted tubule does not appear to respond to antidiuretic hormone.

The Collecting System

The major function of the collecting system is the reabsorption of water in response to antidiuretic hormone. In the presence of ADH, the collecting system becomes very permeable to water and urea. A very concentrated urine is then formed as water flows along the osmotic gradient out of the tubule and into the interstitium. In the absence of ADH, the collecting system is impermeable to water and dilute tubular fluid from the thick ascending limb and the distal tubule is excreted into the minor calyces. The collecting system, like the distal tubule, is subject to mineralocorticoid action with active reabsorption of Na^+ and secretion of K^+ and H^+.

The Renin-Angiotensin-Aldosterone System

The renin-angiotensin-aldosterone system regulates NaCl and water reabsorption by the kidneys (Table 3-9) and responds to sudden alterations in fluid status to help maintain homeostasis (e.g., hemorrhage, Fig. 3-18).

The **juxtaglomerular apparatus (JGA)** is the site of synthesis and release of the hormone **renin** (Fig. 3-19). The juxtaglomerular apparatus consists of four components:

1. Modified smooth muscle cells in the walls of the afferent arterioles
2. Modified smooth muscle cells in the walls of the efferent arterioles
3. Extraglomerular mesangium
4. The **macula densa** cells in a short region of the wall of the distal tubule

The juxtaglomerular apparatus is located where the distal tubule of the nephron passes between the afferent and efferent arterioles of that same nephron. The JGA is therefore immediately adjacent to its own glomerulus.

Renin is synthesized and secreted by the modified smooth muscle cells in the walls of the arterioles when a decrease in ECF volume is sensed by the kidney. A decreased

Table 3-9. Hormones that regulate NaCl and water reabsorption

Segment	Hormone	Effects on NaCl and water reabsorption
Proximal tubule		
	Angiotensin II	↑NaCl ↑H_2O
	Glucocorticoids	↑NaCl ↑H_2O
Thick ascending limb		
	Aldosterone	↑NaCl
	Vasopressin	↑NaCl
Distal tubule/collecting duct		
	Aldosterone	↑NaCl
	Atrial natriuretic peptide	↓NaCl ↓H_2O
	Prostaglandins	↓NaCl ↓H_2O
	Bradykinin	↓NaCl
	Vasopressin	↑NaCl ↑H_2O

From R. M. Berne and M. N. Levy. *Physiology* (3rd ed.). St. Louis: Mosby–Year Book, 1993. P. 751.

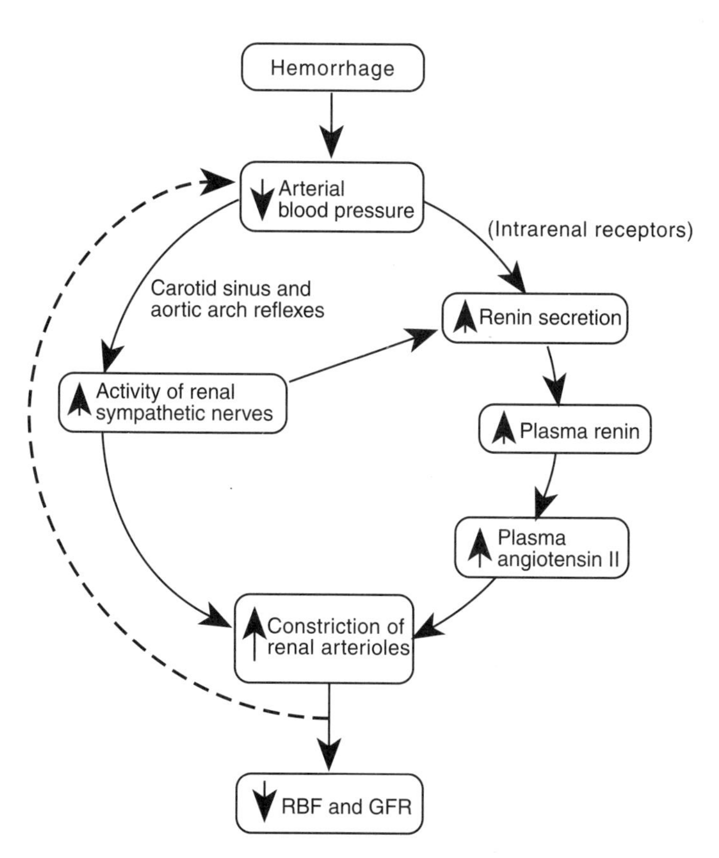

Fig. 3-18. Pathway by which hemorrhage activates renal sympathetic nerve activity and stimulates angiotensin II production. (Modified from A. J. Vander. *Renal Physiology* [2nd ed.]. New York: McGraw-Hill, 1980.)

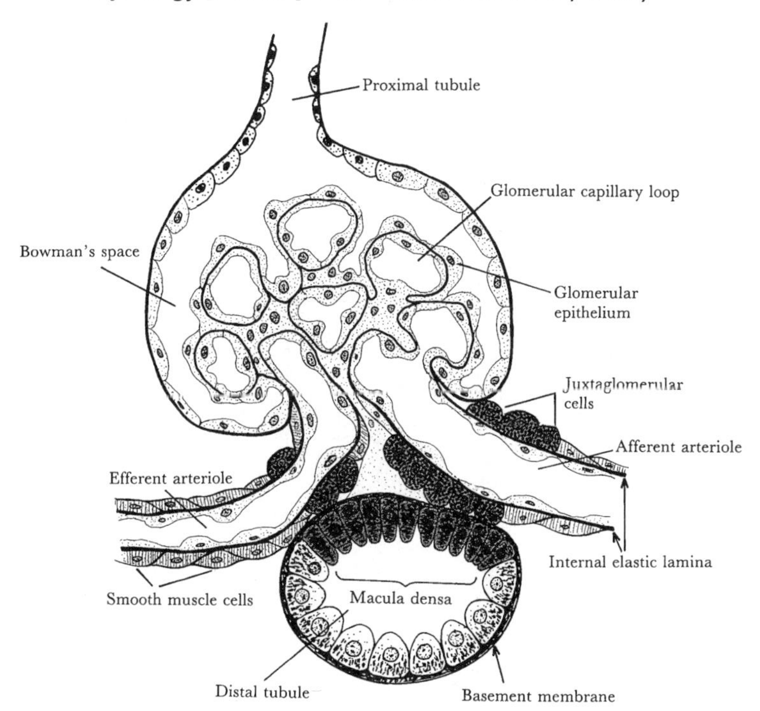

Fig. 3-19. Schematic diagram of the juxtaglomerular apparatus. (Modified from A. C. Guyton. *Textbook of Medical Physiology* [8th ed.]. Philadelphia: Saunders, 1991. P. 294.)

NaCl load reaching the macula densa cells in the TF or a decreased perfusion pressure detected by baroreceptors in the afferent arterioles results in increased renin secretion and release. Other causes of renin release include increased beta-adrenergic stimulation, decreased serum potassium, and a decrease in the arterial pressure detected by baroreceptors in the major blood vessels (affecting vasomotor centers in the central nervous system to increase sympathetic discharge).

These causes of renin release are interrelated. A decrease in the ECF volume, the perfusion pressure at the afferent arteriole, or the blood flow through the kidneys will all cause a decrease in the GFR (Fig. 3-20). This in turn will result in decreased flow in the tubule and more time for reabsorbing NaCl. The macula densa then senses the decreased load of NaCl and, in some unknown fashion, this activates renin release to correct the situation. Beta-adrenergic stimulation by its vasoconstrictor effect on the renal vasculature also results in a decreased GFR and has a direct effect on the release of renin. An acute increase in ECF volume, however, results in opposite actions (Fig. 3-21).

Renin causes circulating **angiotensinogen** (a decapeptide) to be cleaved resulting in **angiotensin I** (an octapeptide) (Fig. 3-22). Angiotensin I is then converted to **angiotensin II,** the most potent vasoconstrictor known, by a **converting enzyme** located primarily in the lung vasculature.

Fig. 3-20. Integrated responses to a decrease in effective circulating volume. These responses include increased renal sympathetic nerve activity, increased secretion of renin with resulting increased levels of angiotensin II and aldosterone, inhibition of atrial natriuretic peptide secretion by atrial myocytes, and stimulation of ADH secretion by the posterior pituitary. The nephron responds as follows (numbers refer to diagram): (1) Afferent and, to a lesser extent, efferent arterioles constrict lowering GFR. (2) Sodium reabsorption by the proximal tubule increases. (3) Sodium reabsorption by the distal nephron and collecting ducts is also enhanced. (Redrawn from R. M. Berne and M. N. Levy. *Physiology* [3rd ed.]. St. Louis: Mosby–Year Book, 1993. P. 779.)

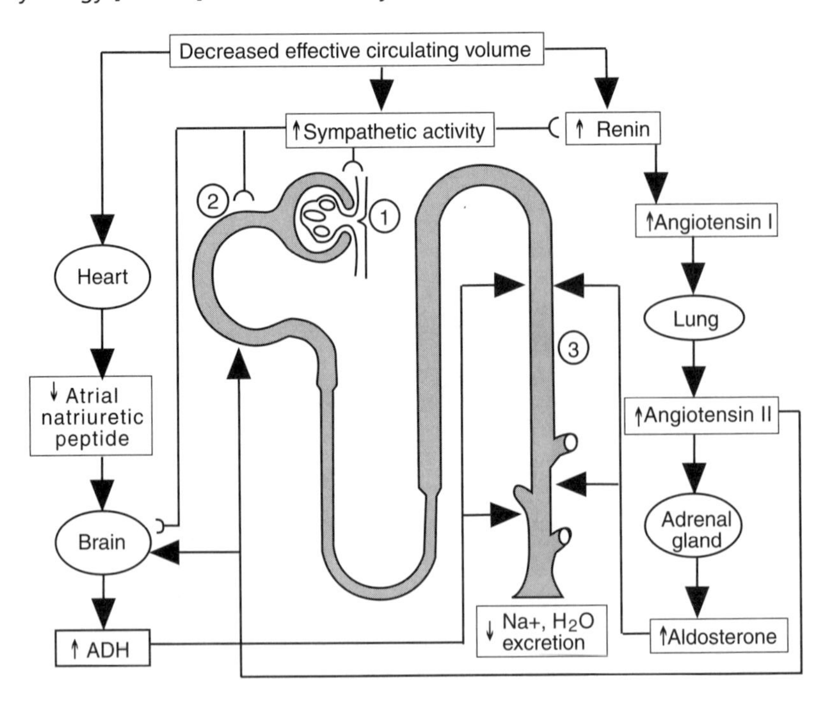

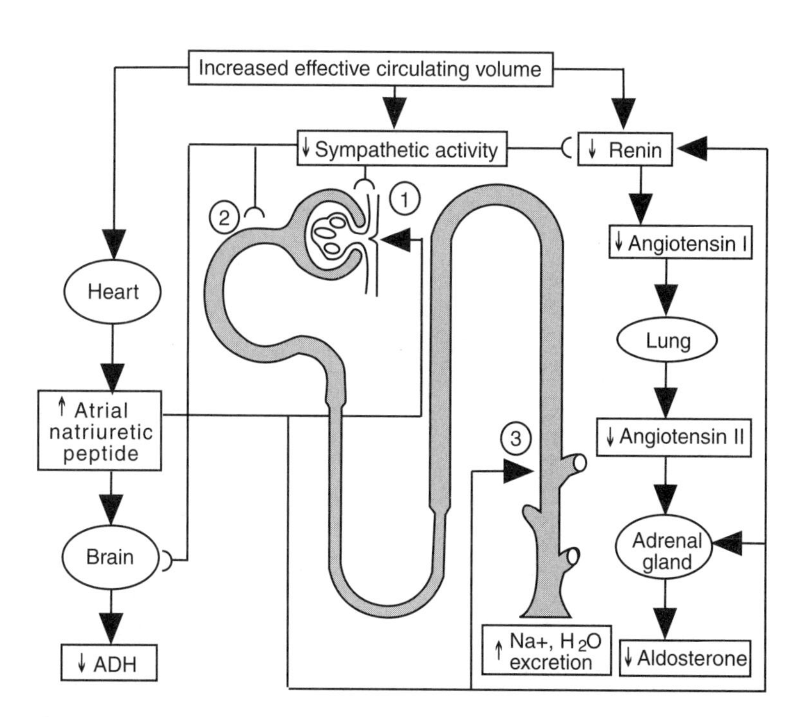

Fig. 3-21. Integrated response to expansion of the effective circulating volume. These include decreased activity of the renal sympathetic nerves, release of atrial natriuretic peptide from the atrial myocytes, inhibition of ADH secretion by the posterior pituitary, and decreased renin secretion causing decreased production of angiotensin II and thus decreased secretion of aldosterone by the adrenal cortex. The numbers that follow refer to those in the diagram: (1) Decreased sympathetic nerve activity results in increased GFR by lowering the vasomotor tone of the afferent and efferent arterioles. The afferent arteriole dilates more than the efferent, increasing the perfusion pressure within the glomerular capillaries and increasing the amount filtered (the filtered load of sodium increases in parallel). (2) Sodium reabsorption decreases in the proximal tubule. (3) Sodium reabsorption decreases in the distal nephron and collecting system. (From R. M. Berne and M. N. Levy. *Physiology* [3rd ed.]. St. Louis: Mosby–Year Book, 1993. P. 777.)

Angiotensin II has the following actions:

1. Potent vasoconstrictor effect on both arterial and venous beds
2. Increases the synthesis and release of **aldosterone**
3. Increases the release of **ADH** centrally
4. Increases blood pressure by a central mechanism
5. Increases thirst by a central mechanism
6. Feedback inhibition of the release of renin
7. Constricts both afferent and efferent arterioles but also increases levels of prostaglandins; the prostaglandins decrease the relative amount of afferent constriction and thereby act to maintain the GFR despite high levels of circulating angiotensin II

Aldosterone

The major stimulus for the release of aldosterone is increasing ECF levels of K^+. **Adrenocorticotropic hormone (ACTH)** acts in a permissive role only, making possible the synthesis of aldosterone by its trophic effect on the adrenal cortex but not directly controlling the degree of production. Angiotensin II acts as a major stimulus for the synthesis and release of aldosterone, as does one of its metabolites, angiotensin

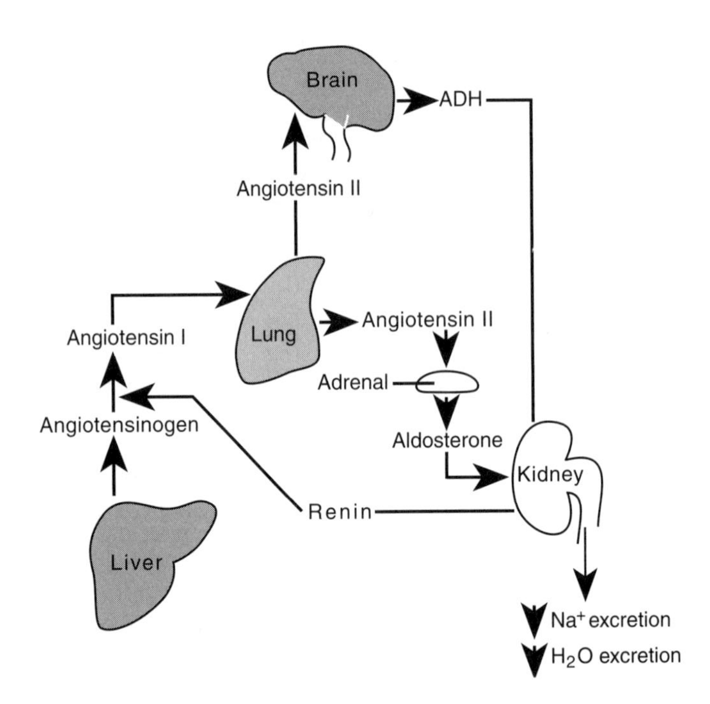

Fig. 3-22. Schematic diagram of the essential components of the renin-angiotensin-aldosterone system's response to a decrease in ECF volume or blood pressure, or an increase in sympathetic tone. (From R. M. Berne and M. N. Levy. *Physiology* [3rd ed.]. St. Louis: Mosby–Year Book, 1993. P. 773.)

III. Angiotensin III is not known to have any other important physiologic actions. The fourth and final regulatory mechanism known to control the secretion of aldosterone is the quantity of body sodium, the exact mechanism of which is unknown. Aldosterone is synthesized in the **zona glomerulosa** of the adrenal cortex and high levels of aldosterone have a negative feedback effect on the release of renin.

Aldosterone acts on the distal tubule and collecting ducts to cause the secretion of K^+, and to a lesser extent H^+, in exchange for Na^+. Aldosterone therefore leads to the conservation of Na^+ and the excretion of K^+ and H^+. High aldosterone levels lead to the reabsorption of virtually all the Na^+ remaining in the TF. Loss of Na^+ in the urine may decrease to only a few mEq/liter/day. Potassium loss in the urine, on the other hand, will increase severalfold. The retention of Na^+ will result in the rise of the sodium concentration by, at most, a few mEq per liter because water is also retained. If the levels of aldosterone are excessive, however, total body sodium and water will increase by up to 20 percent as the retained sodium osmotically obligates water to follow it into the body. The phenomenon known as **escape from aldosteronism** occurs when aldosterone levels have been high for more than a few days. Sodium and water reabsorption by the proximal tubule decreases as fluid levels in the body rise. Aldosterone continues to act on the distal and collecting tubules in an attempt to retain sodium and excrete K^+ and H^+ ions. However, the increased Na^+ load resulting from decreased reabsorption by the proximal tubule exceeds the distal and collecting tubules' ability to reabsorb Na^+, resulting in the return of Na^+ excretion toward normal levels. Potassium excretion continues at high levels. Sodium and water retention of up to 10 percent above normal and moderate to severe hypertension are frequent sequelae.

In **Conn's syndrome** an autonomously secreting tumor releases excessive amounts of aldosterone. As might be predicted, Conn's syndrome results in hypertension, hypokalemia, hypervolemia, and low plasma renin activity. When the serum K^+ level drops to half of normal, muscle weakness or paralysis, which can result in death secondary to hyperpolarization of nerve and muscle cells, occurs. The increased exchange of H^+ for Na^+ may lead to a mild alkalosis.

Total lack of aldosterone will cause plasma sodium concentrations to fall by 5 to 8 percent initially as is lost in the urine. The ECF becomes volume depleted as Na^+ levels fall. If left untreated, progressive hyperkalemia eventually leads to cardiac arrhythmias and death.

Renal Handling of Sodium, Potassium, and Chloride

Sodium, chloride, and potassium are freely filtered in the glomerulus. The proportions of these three ions reabsorbed in various parts of the nephron are approximately:

67 percent in the proximal tubule
25 percent in the loop of Henle
5 percent in the distal convoluted tubule
2 to 3 percent in the collecting tubule

While there is an average net reabsorption of greater than 99 percent of Na^+ and Cl^-, and 80 to 90 percent of K^+, the mechanisms involved in each case are quite distinct (Table 3-10).

The contributions of the distal convoluted tubule and the collecting tubules are particularly important despite the relatively small percentages reabsorbed there, because the final regulation of the composition of the urine takes place in these two segments. Thus, the fine control of excretion is governed by the most distal part of the nephron.

Table 3-10. Signals involved in the control of renal Na^+ and water excretion

Renal sympathetic nerves (↑ activity: ↓ Na^+ excretion)
- ↓ Glomerular filtration rate
- ↑ Renin secretion
- ↑ Proximal tubule Na^+ reabsorption

Renin-angiotensin-aldosterone (↑ secretion: ↓ Na^+ excretion)
- ↑ Angiotensin II levels stimulate proximal tubule Na^+ reabsorption
- ↑ Aldosterone levels stimulate collecting duct Na^+ reabsorption
- ↑ ADH secretion

Atrial natriuretic peptide (↑ secretion: ↑ Na^+ excretion)
- ↑ Glomerular filtration rate
- ↓ Renin secretion
- ↓ Aldosterone secretion
- ↓ Na^+ reabsorption by the collecting duct
- ↓ ADH secretion

ADH (↑ secretion: ↓ H_2O and Na^+ excretion)
- ↑ H_2O absorption by the collecting duct
- ↑ NaCl reabsorption by the thick ascending limb of Henle's loop
- ↑ Na^+ reabsorption by the collecting duct

From R. M. Berne and M. N. Levy. *Physiology* (3rd ed.). St. Louis: Mosby–Year Book, 1993. P. 772.

The primary active transport of Na^+ provides the driving force that is ultimately responsible for the reabsorption and secretion of many of the solutes transported by the kidneys. In the proximal tubules, roughly 70 percent of the Na^+ enters the proximal tubular cells passively by either facilitated or simple diffusion and is subsequently actively pumped out into the lateral intercellular space. Na^+-K^+ ATPase exchange pumps located in the basolateral membrane provide the energy for the active transportation of Na^+ out of the cells in exchange for K^+ from the peritubular fluid (Figs. 3-16 and 3-23). The resulting low concentration of Na^+ in the cells creates a concentration gradient that favors the passive uptake of Na^+ from the tubular fluid. The transportation of Na^+ into the cells can be coupled to that of other solutes by either **cotransport** (glucose, amino acids, HCO_3^-, lactate, sulfate, and phosphate) or **exchange** (H^+). Because this transportation of solute is coupled to the transport of sodium rather than directly to metabolism, it is referred to as **secondary active transport** (see Fig. 3-16).

Cotransport of Na^+ and glucose or amino acids results in a lumen-negative potential difference in the earliest portion of the proximal tubule because glucose and amino acids are electrically neutral (see Fig. 3-12). A positive intraluminal potential difference quickly supervenes, however, as a consequence of the preferential secretion of H^+ and reabsorption of bicarbonate in the early proximal tubule. Bicarbonate is less permeable to the proximal tubular membrane than is Cl^-. Because Cl^- is left behind as $NaHCO_3$ and water are reabsorbed, its concentration in the tubular fluid increases. As the Cl^- diffuses out of the tubular fluid into the cell along its concentration gradient, it creates a lumen-positive transepithelial potential difference that results in additional sodium reabsorption. Unlike the balance of sodium reabsorption, the 30 percent of Na^+ reabsorption attributable to this lumen-positive potential difference actually bypasses the proximal tubular cells and enters *between* cells (Fig. 3-23). The proximal tubule is so permeable to small ions that only a small potential difference is required to drive large amounts of Na^+ out of the tubule.

Of the sodium entering through the cell (70%), the following is a reasonable estimate of the contribution of each transport mechanism (Fig. 3-23):

1. Five percent by cotransport with glucose or amino acids using specific membrane carriers, or by cotransport with any one of several organic anions (sulfate, lactate, and phosphate)
2. Fifteen percent by a tightly coupled Na^+-H^+ exchange pump in the luminal membrane; because the H^+ so secreted titrates HCO_3^-, the net result is the reabsorption of $NaHCO_3$ without any net secretion of hydrogen ion

Fig. 3-23. Mechanisms for Na^+ reabsorption in the proximal tubule. A. Na^+-solute symport. Apical entry of Na^+ along its electrochemical gradient is coupled (as indicated by the dashed circle) to the transport of an organic solute (e.g., glucose, amino acids) or phosphate. The Na^+ is then extruded across the basolateral surfaces by the NA^+-K^+ ATPase (illustrated here across the lateral surface). Since glucose, the major solute cotransported with Na^+, is uncharged, this reabsorption of Na^+ generates a lumen-negative transepithelial potential difference (PD) of approximately –2 mV. The potential is limited because Cl^- accompanies Na^+ by diffusing across the "leaky" tight junctions. B. Na^+-H^+ exchange (antiport). The apical entry of Na^+ is coupled (as indicated by the dashed circle) to the secretion of H^+ into the tubular fluid. The Na^+ is then actively extruded across the basolateral surfaces by the Na^+-K^+ ATPase. If the secreted H^+ is derived from H_2CO_3, the net result of the Na^+-H^+ antiport is Na^+-HCO_3^- reabsorption (B1); if the secreted H^+ is derived from formic acid, the net result of the Na^+-H^+ antiport is Na^+-Cl^- reabsorption (B2). C. Cl^- driven Na^+ transport. The Cl^- concentration in the tubular fluid increases to about 132 mmole/per liter in segments 2 and 3, resulting in a concentration gradient for the diffusion of Cl^- across the "leaky" tight junctions into the peritubular fluid, where the Cl^- concentration is about 110 mmole/per liter. This reabsorption of Cl^- generates a lumen-positive transepithelial PD in segments 2 and 3, but only of about +2 mV because Na^+ accompanies Cl^- across the "leaky" tight junctions. (From J. B. West. *Best and Taylor's Physiological Basis of Medical Practice* [12th ed.]. P. 452. © 1990 by Williams & Wilkins, Baltimore.)

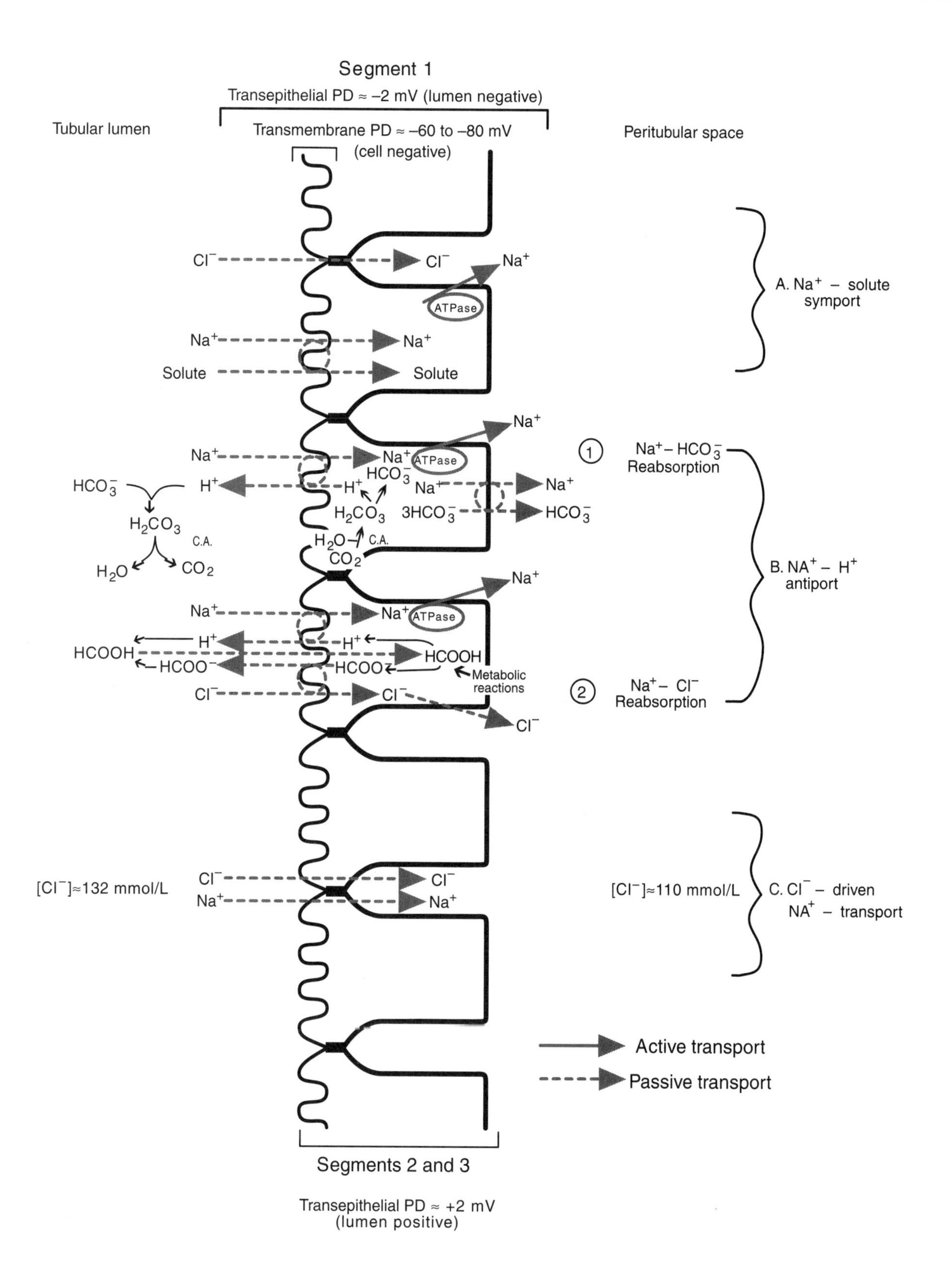

Segment 1
Transepithelial PD ≈ −2 mV (lumen negative)
Transmembrane PD ≈ −60 to −80 mV (cell negative)
Tubular lumen
Peritubular space
Cl⁻
Na⁺
ATPase
Solute
A. Na⁺ − solute symport
HCO₃⁻
H⁺
H₂CO₃
C.A.
H₂O
CO₂
3HCO₃⁻
1 Na⁺− HCO₃⁻ Reabsorption
B. NA⁺ − H⁺ antiport
HCOOH
HCOO⁻
Metabolic reactions
2 Na⁺− Cl⁻ Reabsorption
[Cl⁻]≈132 mmol/L
[Cl⁻]≈110 mmol/L
C. Cl⁻ − driven NA⁺ − transport
Active transport
Passive transport
Segments 2 and 3
Transepithelial PD ≈ +2 mV (lumen positive)

Renal

3. Fifty percent by simple diffusion with Cl^- (Cl^- is reabsorbed passively by the proximal tubular cells)

In the thick ascending limb of the loop of Henle, both Na^+ and Cl^- are actively reabsorbed. In the distal convoluted and collecting tubules, Na^+ is transported actively against both a large concentration gradient and a large electrical gradient in response to aldosterone. The Na^+ is reabsorbed in exchange for either H^+ or K^+. The collecting duct has a particularly powerful pump able to decrease the Na^+ concentration to almost zero. Chloride generally follows passively in the distal convoluted and collecting tubules, although in cases of severe Cl^- deprivation a Cl^- reabsorptive pump, normally masked by the quantitatively greater passive reabsorption of chloride, is demonstrable.

The magnitude of the net excretion of Na^+ and Cl^- depends, then, on the extent to which they are reabsorbed following filtration.

Glomerular-tubular balance refers to the fact that under steady-state conditions a constant fraction of the filtered Na^+ is reabsorbed in the proximal tubule despite variations in GFR and the Na^+ load. This occurs in order to protect Na^+ balance from fluctuations in GFR because even a small change in GFR can result in large changes in the amount of filtered Na^+. Similar adjustments are made in more distal segments as well. The mechanism for this is unknown.

In contrast to Na^+ and Cl^-, the excretion of K^+ is dependent primarily on the amount secreted in the distal nephron in response to aldosterone. Most of the K^+ filtered in the glomerulus is reabsorbed somewhere in the renal tubule. Furthermore, reabsorption of K^+ may occur anywhere in the tubule and an active K^+ reabsorptive pump has been demonstrated in the luminal membrane of both the proximal convoluted tubule and the collecting ducts in addition to the Na^+-K^+ exchange pump in the basolateral membrane. Under normal conditions, potassium is reabsorbed in the proximal tubules and in the loop of Henle, and it is either secreted or reabsorbed in the distal portion of the nephron. Ninety percent or more of the filtered K^+ is reabsorbed in the proximal convoluted tubule and in the loop of Henle before reaching the distal tubule. In states of K^+ deficiency, the remaining 10 percent is reabsorbed in the distal tubule and collecting system. However, K^+ excretion is the norm and it is the net secretion or lack of it in the distal portion of the distal tubule and collecting system that determines the rate of K^+ excretion in the urine. Normally, an amount equal to 10 to 20 percent of the filtered load of K^+ is excreted. Net tubular excretion of potassium is also possible, however, because the amount of potassium excreted can range from 3 to 150 percent of the amount filtered. The absorption of K^+ is active while the secretion is thought to be passive.

An H^+-K^+ exchange pump in the basolateral membrane enables H^+ to be exchanged for K^+. Increasing plasma pH, for instance, causes H^+ ions to leave cells along its concentration gradient in exchange for extracellular K^+ (thus maintaining electroneutrality). Decreasing plasma pH will cause K^+ to leave cells and H^+ to enter. Alkalosis, therefore, causes increased K^+ excretion by increasing the intracellular concentration of K^+ in the proximal tubular cells, thereby increasing the gradient for K^+ secretion into the tubules. Unfortunately, acidosis also increases the excretion rate of K^+, presumably by some other mechanism.

Pharmacologic doses of epinephrine and insulin (K^+ enters cells with glucose) drive K^+ into cells. High plasma pH, bicarbonate, and aldosterone also drive K^+ into cells. All of these, therefore, cause increased K^+ excretion by the kidneys by creating a greater gradient for K^+ secretion.

Renal Concentrating and Diluting Mechanisms

Antidiuretic Hormone

Antidiuretic hormone (ADH) or arginine vasopressin is an octapeptide synthesized in nerve cells in the hypothalamus. Release of ADH can occur in response to changes in both fluid volume and concentration, and is controlled by

1. Osmoreceptors responding to the effective osmolality of the ECF (Na^+ and its attendant anions Cl^- and HCO_3^-) but not to ineffective osmoles such as urea
2. Receptors sensitive to changes in effective ECF and circulating blood volumes
3. The action of circulating angiotensin II acting directly on the thirst center in the anterior hypothalamus
4. A number of stimuli in a manner reminiscent of renin release; nausea, vomiting, exercise, anxiety, fright, pain, nicotine, and anesthetic agents can all increase the secretion of ADH

Antidiuretic hormone increases the permeability of the entire collecting system to water. It also increases urea permeability of the medullary collecting ducts (Fig. 3-24D). These responses are very rapid, occurring within minutes of the stimulation of ADH secretion.

When an individual is deprived of water, the osmotically obligated excretion of water continues until the plasma is rendered slightly hyperosmotic. This stimulates osmoreceptors in the hypothalamus and possibly others in the distribution of the internal carotids to initiate signals that cause the secretion of ADH. The ADH increases the permeability of the entire collecting duct system to water. Water then flows out of the collecting systems into the hypertonic medullary interstitium, concentrating the urine up to a maximum (in humans) of 1200 mOsm per liter.

In contrast, after drinking a large amount of water, the plasma is rendered hypoosmotic, osmoreceptors call for decreased secretion of ADH, and the collecting system becomes impermeable to water once again. The dilute fluid entering the collecting system from the thick ascending limb and the distal convoluted tubule passes through the collecting system to be excreted as dilute urine (to a minimum of 50 mOsm/liter).

The Countercurrent Multiplier: Kuhn's Hypothesis

The countercurrent multiplier is really a countercurrent augmentor. Each successively deeper layer of the medullary interstitium augments the osmolality of the previous layer. This results in a continuously increasing osmolality or solute concentration parallel to the renal tubules that create it. By forming this osmotic gradient, the countercurrent multiplier enables the kidneys to excrete a concentrated urine. Furthermore, the same active process that creates the osmotic gradient simultaneously dilutes the TF and enables the kidneys to excrete a dilute urine (in the absence of ADH).

The term **countercurrent** refers to the flow in adjacent limbs of the U-shaped loop of Henle being in opposite directions (Fig. 3-24). The term **multiplier** is used because such countercurrent flow can establish large concentration gradients along the axis of the adjacent limbs. Concentration gradients in the transverse direction can remain relatively small and, therefore, within the capability of the renal tubules.

According to **Kuhn's hypothesis** there are three basic requirements for such a system to function:

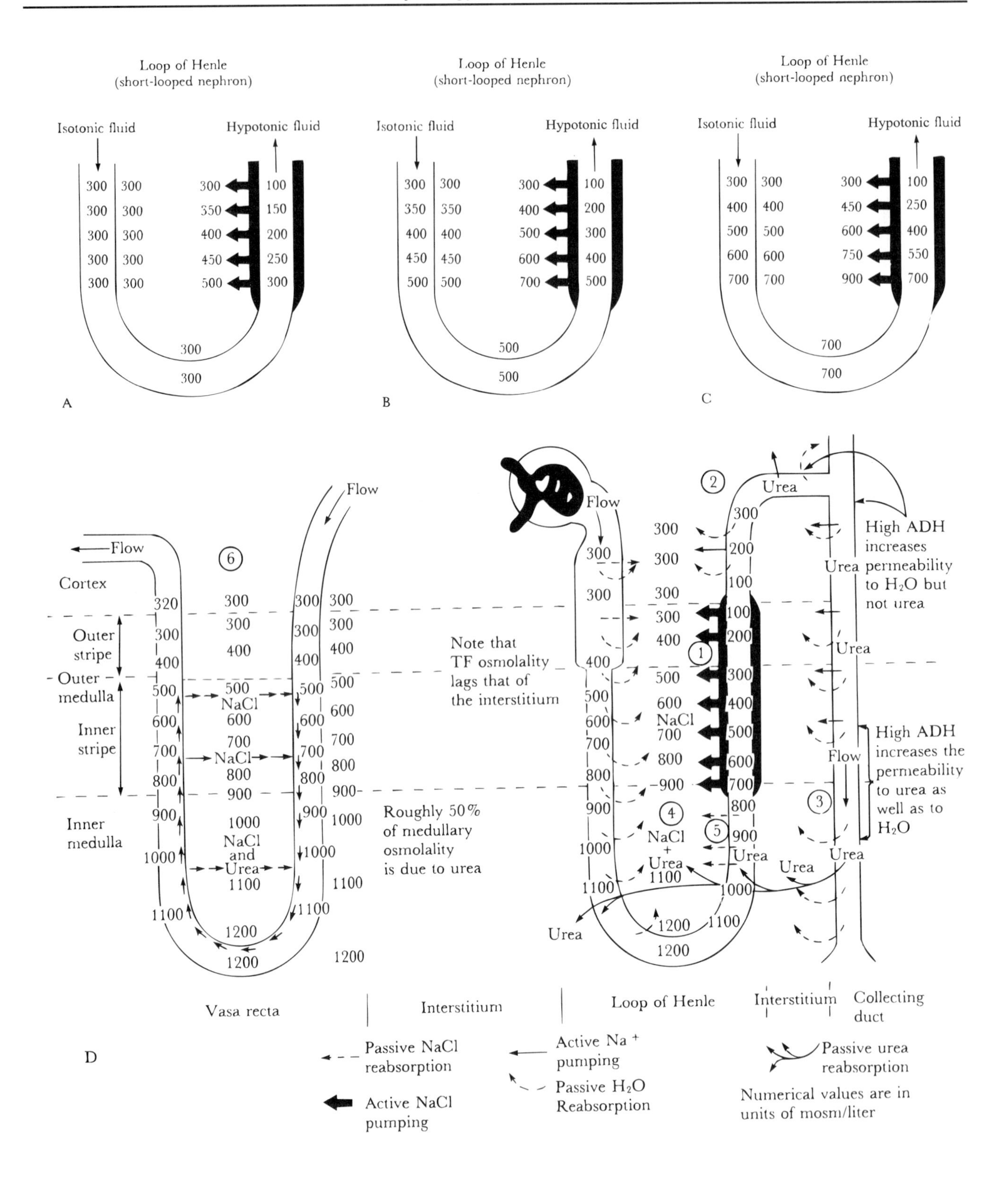

Loop of Henle (short-looped nephron)
Isotonic fluid
Hypotonic fluid
A
B
C
D
Flow
Cortex
Outer stripe
Outer medulla
Inner stripe
Inner medulla
NaCl
NaCl and Urea
Note that TF osmolality lags that of the interstitium
Roughly 50% of medullary osmolality is due to urea
Urea
High ADH increases permeability to H_2O but not urea
High ADH increases the permeability to urea as well as to H_2O
Vasa recta
Interstitium
Loop of Henle
Interstitium
Collecting duct
Passive NaCl reabsorption
Active NaCl pumping
Active Na^+ pumping
Passive H_2O Reabsorption
Passive urea reabsorption
Numerical values are in units of mosm/liter

1. Countercurrent flow
2. Differences in permeability between tubules carrying fluid in opposite directions
3. A source of energy

All three requirements are met by the kidneys: The loop of Henle provides the countercurrent flow, the descending limb is much more permeable to water than is the ascending limb, and active transport of sodium chloride from the ascending limb to the surrounding interstitium provides the energy.

The result of this arrangement is that active NaCl transport lowers the NaCl concentration (and osmolality) of fluid within the thick ascending limb and simultaneously raises the NaCl concentration in the interstitial fluid (Fig. 3-24A). Fluid that enters the descending limb still isoosmotic with plasma and glomerular filtrate now becomes more concentrated as water is extracted across the limb's permeable wall by the higher osmolality of the interstitium (Fig. 3-24B). As this more concentrated fluid rounds the bend of the loop and begins its ascent, it delivers a higher NaCl concentration to the site of the active NaCl pump, and the concentration of NaCl and hence the osmolality of the interstitium is raised still higher (Fig. 3-24B, C). Thus, the ability of the NaCl pump to establish a modest concentration difference across the wall of the ascending limb (perhaps 100 mM NaCl, equivalent to approximately 200

Fig. 3-24. Countercurrent multiplier — Kuhn's hypothesis. A. The initial formation of the hypertonic medulla occurs as NaCl is actively pumped out of the thick ascending limb. The pump is gradient limited to 200 mOsm/per liter (actual stoichiometry of the pump is $1Na^+: 1K^+: 2Cl^-$). B. Medullary tonicity is further augmented as the tubular fluid reaching the thin ascending limb becomes hyperosmotic. C. A medullary tonicity of approximately 900 mOsm/per liter is achieved. To increase medullary osmolality further using this method requires longer-looped nephrons with longer thick ascendng limbs; however, the long-looped juxtamedullary nephrons increase their length by incorporating thin descending and ascending limbs into their loops of Henle. No active pumping has ever been demonstrated in thin ascending limbs; however, medullary tonicity is increased by these nephrons to 1200 mOsm/per liter. This increased tonicity is attributed to the urea cycle. D. Recent modifications of the countercurrent hypothesis—the urea cycle. The thin ascending limb in the medulla, the thick ascending limb in the outer medulla, and the early distal tubule are impermeable to water. (1) Active pumping of NaCl into the interstitium by the thick ascending limb renders the tubular fluid dilute and the outer medullary interstitium hyperosmotic. This process provides the fuel for the urea cycle. Because the thick ascending limb is impermeable to both water and urea, urea remains behind in the diluted tubular fluid. (2) In the presence of ADH, the distal tubule and the cortical and other medullary collecting ducts are permeable to water but not to urea. Because the tubular fluid leaving the thick ascending limb is hypotonic, water now leaves the late distal tubule and cortical collecting ducts along its concentration gradient. This increases the concentration of urea that is left behind in the tubule. In the outer medulla, additional water is reabsorbed from the tubule due to the concentration gradient created by the thick ascending limb, further increasing the tubular fluid urea concentration. (3) In the presence of ADH, the medullary collecting duct is permeable to both water and urea. The concentrated urea reaching the medulla now diffuses out of the tubule into the interstitium along its concentration gradient, adding substantially to the tonicity of the medulla. Some urea reenters the thin ascending limb of the loop of Henle, which is somewhat permeable to urea. This medullary recycling of urea, in addition to the trapping of urea by countercurrent exchange in the vasa recta, causes urea to accumulate in large quantities in the medullary interstitium (large type) in the presence of ADH. Urea accounts for roughly 50 percent of the osmolality of the inner medulla in antidiuresis. (4) The thin descending limb is permeable to water but impermeable to NaCl and solute in general. The high concentration of urea in the interstitium osmotically extracts water from the thin descending limb, thus concentrating NaCl in the descending limb fluid (i.e., the high concentration of NaCl in the tubular fluid is balanced osmotically by the high concentration of urea in the interstitium). (5) When the fluid rich in NaCl enters the NaCl-permeable but water-impermeable thin ascending limb, NaCl moves passively out of the tubule along its concentration gradient, rendering the tubule fluid relatively hypoosomotic relative to the surrounding interstitium. The urea cycle can, therefore, account for both the reabsorption of water by the thin descending limb and the passive reabsorption of NaCl by the thin ascending limb. Note, however, that this process is dependent on the active pumping of solute by the thick ascending limb, and the configuration and differential permeabilities of the different parts of the nephron. (6) The countercurrent exchange system in the vasa recta (see also Fig. 3-25).

mOsm/kg water) is augmented by countercurrent flow to achieve a large difference between the isoosmotic fluid entering the descending limb and the hyperosmotic fluid at the tip of the papilla.

Fluid from the thick ascending limb enters the distal tubule in the cortex hypoosmotic to the cortical interstitium. In the presence of ADH, which increases the water permeability of the entire collecting tubule, water is reabsorbed until osmotic equilibrium is achieved between the contents of the collecting tubule and the isoosmotic cortical interstitium. As fluid enters the medullary portion of the collecting duct to be redirected for a second and final time through the medulla, more water is reabsorbed until equilibrium is established between luminal fluid and the hyperosmotic medullary interstitium. This results in the excretion of a concentrated urine. Additional active Na^+ reabsorption in the distal tubule and collecting system also contributes to the reabsorption of water.

Moreover, the same process that drives the multiplier also accounts for urinary dilution. In the absence of antidiuretic hormone the collecting tubule has a very low osmotic water permeability, so that fluid coming from the ascending limb remains dilute throughout the remainder of the distal and collecting tubule. Additional active Na^+ reabsorption in the distal tubule and collecting system now acts to dilute the tubular fluid.

Under normal circumstances, the regulation of urinary concentration is thus made elegantly simple: The degree to which urine is diluted or concentrated is controlled directly by the level of antidiuretic hormone in circulating plasma.

Only one major problem remains with this hypothesis. The thin ascending limb, while impermeable to water, does not actively pump out NaCl or any other substrate. Because the thick ascending limb is located in the outer medulla and the osmotic gradient continues to increase to reach a maximum in the inner medulla, some other process must be contributing to the renal concentrating mechanism, at least for the long-looped juxtaglomerular nephrons.

The Role of Urea

Urea is concentrated in the medullary interstitium. The active pumping of NaCl in the thick ascending limb provides the energy for the concentration of urea. The differential permeabilities of the tubules to water, solute, and urea provide the proper conditions. The resulting high concentrations of urea in the medullary interstitium can then account for the high osmolality of the inner medulla despite the absence of active solute transport there. The recycling of urea by the renal tubules accounts for the reabsorption of water by the descending thin limb. It also accounts for the "passive" reabsorption of NaCl by the ascending thin limb.

The concentration of urea in the renal medulla occurs in the following fashion (Fig. 3-24D). The thick ascending limb is impermeable to water and urea. Hence, both urea and water remain behind in the lumen of the thick ascending limb as NaCl is actively pumped out. When, in the presence of ADH, the TF reaches the late distal tubule and the cortical and outer medullary collecting ducts, water is reabsorbed into the cortex and outer medulla. Because urea cannot penetrate, however, it remains behind in the TF. The concentration of urea in the TF, therefore, increases and, consequently, a high concentration of urea reaches the inner medulla. The urea is concentrated as a result of the osmotic gradient created by the thick ascending limb. The inner medullary collecting tubule is permeable to urea when ADH is present. Urea diffuses into the medullary interstitium along its concentration gradient and is trapped there, adding substantially to the tonicity of the medulla. Roughly 50 percent of the tonicity of the inner medulla is attributed to urea.

Because the descending thin limb is relatively impermeable to urea and NaCl, urea in the medullary interstitium acts to osmotically extract water from the thin descending limb, increasing the concentration of the NaCl in the luminal fluid above that in the interstitium (i.e., the high concentration of NaCl in the tubule is balanced osmotically by the high concentration of NaCl and urea in the interstitium). When the fluid now enters the water-impermeable but NaCl-permeable thin ascending limb, NaCl moves passively down its concentration gradient into the interstitium, rendering the TF in the thin ascending limb hypoosmotic relative to the surrounding hypertonic interstitium. Because the thin ascending limb is somewhat permeable to urea, some of the urea now diffuses along its concentration gradient into the thin ascending limb to begin the cycle again.

Countercurrent Exchange and the Vasa Recta

The vasa recta perform two functions. They remove fluid from the medullary interstitium that has been reabsorbed from the thin descending limb and collecting tubules and they minimize solute uptake from the medulla to preserve medullary hypertonicity. The vascular architecture is designed to facilitate transcapillary exchange of water and solute between the ascending and descending vasa recta, that is, to maximize countercurrent exchange (Figs. 3-24D and 3-25). This exchange occurs in such a fashion as to minimize both the amount of water retained in the medulla and the loss of solute out of the medulla.

In the descending vasa recta, the balance of Starling's force is such that the net hydrostatic pressure favoring movement out of the capillaries and the opposing high oncotic pressure (resulting from filtration at the glomerulus) are nearly equal. Fluid

Fig. 3-25. The countercurrent exchange system in the vasa recta (dotted arrows), including passive transport of solutes and passive movement of water (dashed arrows). The numbers refer to osmolalities (mOsm/kg water) in the blood or interstitial fluid. (Adapted from R. W. Berliner, N. G. Levinsky, D. G. Davidson, and M. Eden. *Am. J. Med.* 2 4:730, 1958.)

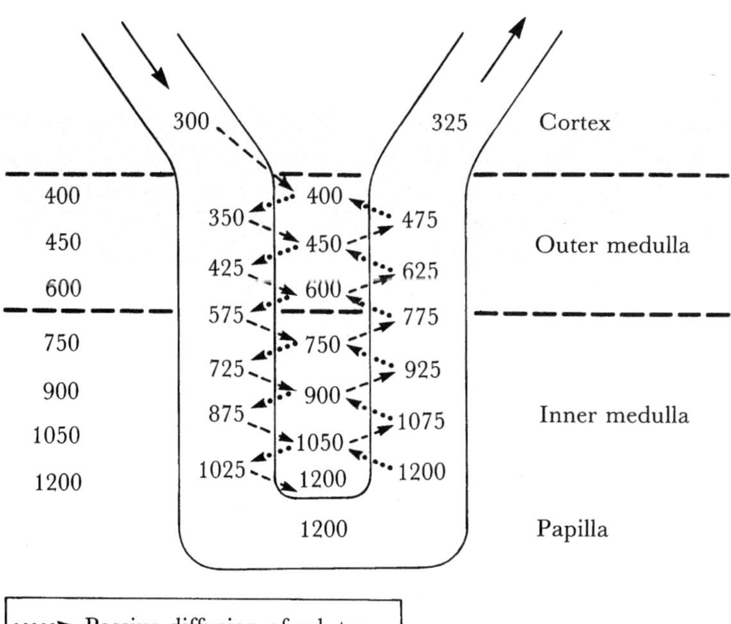

Renal

nonetheless leaves the descending vasa recta because of the high interstitial osmolality due to NaCl and urea. Normally NaCl and urea do not exert an osmotic force across capillary membranes. However, due to the rapid flow in the capillaries and the high, but nonetheless finite, membrane permeability to these solutes, water is able to leave the capillaries faster than solute is able to enter (i.e., water is more permeable than solute, so the solute concentration in the capillary lags behind that in the interstitium). Thus, there is a net egress of water from the descending vasa recta.

In the ascending vasa recta, the solute concentration of the tubular fluid is now higher than that of the interstitium and, rather than favoring egress of fluid, now combines with oncotic pressure to favor influx of fluid. Meanwhile, the hydrostatic pressure has also decreased somewhat. The balance is such that, overall, more fluid is taken up by the ascending vasa recta than was lost by the descending vasa recta. This net uptake is ultimately attributable to the increased oncotic pressure resulting from loss of ultrafiltrate in the glomerulus. The net volume reabsorbed by the ascending vasa recta is equal to the volume of fluid taken up from the ascending thin limb and collecting system.

Renal Tubular Transport: Reabsorption and Secretion of Organic Compounds

Glucose, amino acids, and other solutes are reabsorbed in the early proximal tubule by cotransport with sodium (symport). The energy for this reabsorption is derived indirectly from the active Na^+-K^+ pump and it is therefore called secondary active transport. They then diffuse out of the proximal tubular cells into the lateral intercellular space by either simple or facilitated (e.g., glucose) diffusion (see Figs. 3-14 and 3-16). Five or more separate carriers have been described for amino acid transport. These include carriers for neutral amino acids, dibasic amino acids, dicarboxylic amino acids, imino acids and glycine, and beta-amino acids. Like glucose, all of these are symports.

Sodium transport interacts with that of glucose, amino acids, bicarbonate, citrate, and lactate. Furthermore, the carriers mediating the reabsorption of Na^+ and each of these solutes are much less efficient at transport of either sodium or the solute alone. Reabsorption of either solute or sodium is decreased if its cotransportee is absent.

The transport maximum (T_{max}) for glucose is approximately 375 mg per minute in a male adult and somewhat less in a female adult. The renal threshold for spilling glucose into the urine should then be equal to the T_{max} divided by the GFR, or 375 mg per minute divided by 125 ml per minute (or 3 mg/ml). The threshold would be surpassed, therefore, when the blood glucose exceeds 300 mg/dl. Measurements of the actual renal threshold show it to be only 200 mg/dl arterial blood or 180 mg/dl venous blood. This discrepancy from the theoretical value is called splay and is shown graphically in Figs. 3-26 and 3-27. It can be explained by two observations. First, the transport maximum is not identical for all nephrons since it represents the average value for approximately two million nephrons. Consequently some nephrons may begin to spill glucose at values considerably above or below this value. Glucose is spilled into the urine as soon as the concentration of glucose in the plasma and therefore in the filtrate exceeds the ability of those nephrons least able to reabsorb glucose. The second explanation involves the fact that carriers, like enzymes, require supersaturating levels of substrate to operate maximally.

As the plasma glucose level exceeds the T_{max} for glucose, the clearance of glucose will increase from zero to asymptotically approach the clearance of inulin (Fig. 3-28). This is demonstrated by the following equations:

$$\text{Amount glucose excreted} = U_{PAH}\dot{V} = (C_{in})(P_{Glu}) - T_{max}$$

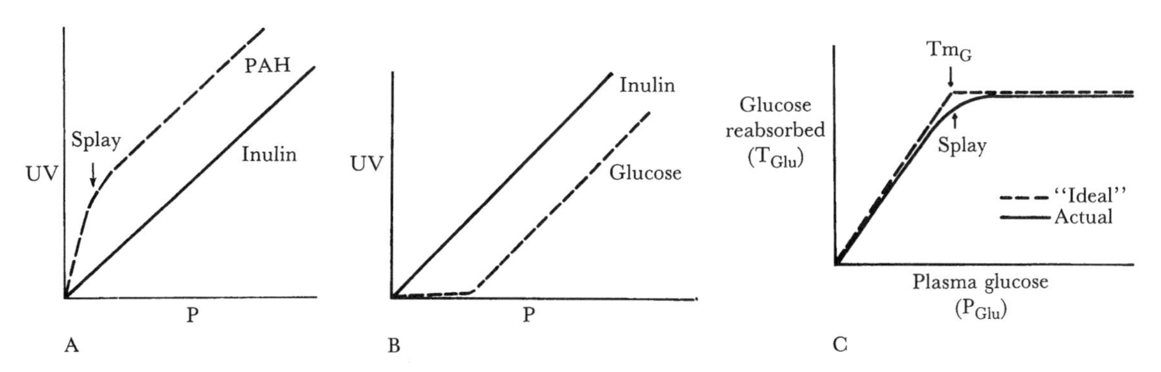

Fig. 3-26. A. Relation between plasma levels (P) and excretion ($U\dot{V}$) of PAH and inulin. B. Relation between plasma levels (P) and excretion ($U\dot{V}$) of glucose and inulin. C. Relation between plasma glucose level and amount of glucose reabsorbed. (From W. F. Ganong. *Review of Medical Physiology* [12th ed.]. Pp. 582–583. © 1985 by Lange Medical Publications, Los Altos, CA.)

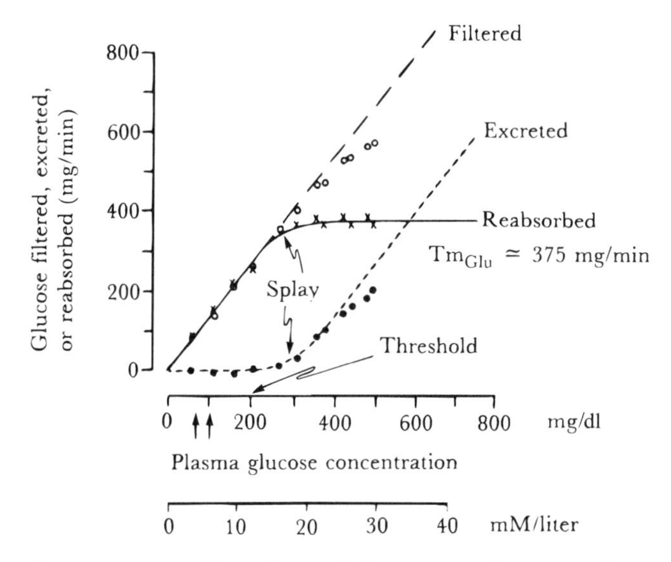

Fig. 3-27. Renal handling of glucose as a function of increasing plasma glucose concentrations. The curve for reabsorption is known as the glucose titration curve because it determines the plasma concentration at which the carrier for glucose becomes saturated. Tm_{Glu} refers to the maximal amount of glucose that can be transported per unit time. The range of normal plasma glucose concentration (70–100 mg/dl) is spanned by the arrows on the abscissa; note that normally virtually all the filtered glucose is reabsorbed. The significance of the splay is explained in the text. In the clinical laboratory, plasma glucose concentrations are ordinarily expressed as milligrams per deciliter of plasma; corresponding concentrations, in millimoles per liter, are given on the second abscissa. (From H. Valtin. *Renal Function: Mechanisms Preserving Fluid and Solute Balance in Health* [2nd ed.]. Boston: Little, Brown, 1983. P. 72.)

Dividing both sides by P_{Glu} yields

$$C_{Glu} = C_{in} - T_{max}/P_{Glu}$$

The ratio of the T_{max} of glucose to the plasma concentration of glucose approaches zero as the plasma glucose concentration increases and the amount of glucose filtered becomes many times greater than the amount reabsorbed.

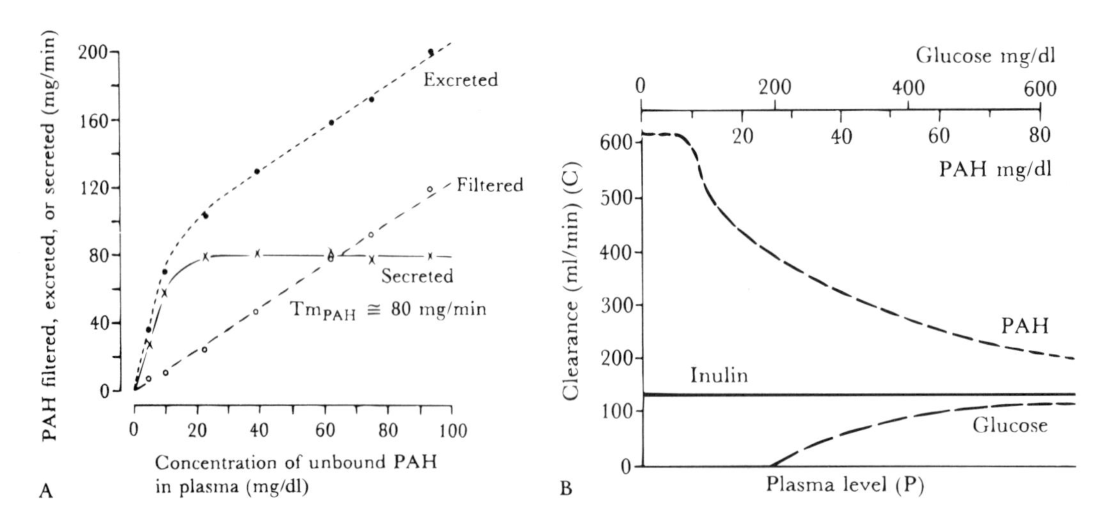

Fig. 3-28. A. Rates of filtration, excretion, and secretion at increasing plasma concentrations of PAH. (Modified from R. F. Pitts. *Physiology of the Kidney and Body Fluids* [3rd ed.]. Chicago: Year Book, 1974.) B. Clearance of inulin, glucose, and PAH at various plasma levels of each substance in humans. (From W. F. Ganong. *Review of Medical Physiology* [12th ed.]. P. 574. © 1985 by Lange Medical Publications, Los Altos, CA.)

There are at least two separate active mechanisms involved in secretion: an organic anion secretory pump and an organic cation secretory pump. The weak anions predictably compete with one another for secretion. Acetate and lactate facilitate this secretion while a number of substances inhibit it (including Krebs cycle intermediates, probenecid, dinitrophenol, and mercurial diuretics).

Para-aminohippurate is excreted by a combination of filtration in the glomerulus and secretion in the proximal tubule. The filtered load of PAH, like that of all filterable solutes, is a linear function of the plasma level while PAH secretion increases only until its transport maximum for secretion ($T_{max_{PAH}}$) is reached (Fig. 3-28B). Thus, at low concentrations of PAH, the clearance of PAH is high (Fig. 3-28A). As the plasma concentration of PAH increases, however, C_{PAH} asymptotically approaches the clearance of inulin (i.e., the GFR) as the amount of PAH secreted becomes less significant:

$$\text{Amount PAH excreted} = U_{PAH}\dot{V} = (C_{in})(P_{PAH}) + T_{max}$$

Dividing by P_{PAH}

$$C_{PAH} = C_{in} + T_{max}/P_{PAH}$$

Renal Metabolism and Oxygen Consumption

Renal blood flow per gram tissue is very large, and, therefore, the arteriovenous oxygen difference (A-V O_2) is quite small: only 14 ml oxygen per liter of blood flow versus 60 ml per liter for skeletal muscle, 62 ml per liter for brain, and 140 ml per liter for heart. Renal oxygen consumption varies directly with the rate of active transport of sodium. The cortex therefore has the highest oxygen consumption because it is there that Na^+ transportation is most active. The medulla has an oxygen consumption one-twentieth that of the cortex.

Questions

Directions: Each group of items in this section consists of lettered options followed by a set of numbered items. For each item, select the one lettered option that is most closely associated with it. Each lettered option may be selected once, more than once, or not at all.

A. Filtered and (normally) *entirely* reabsorbed by cotransport with Na^+
B. Filtration and active reabsorption
C. Filtration and renal tubule catabolism
D. Primarily filtered with little secretion or reabsorption

1. Creatinine
2. Glucose, amino acids
3. Inulin
4. Low-molecular-weight proteins, such as insulin, glucagon, PTH, and ADH
5. Na^+

A. Net excretion primarily due to secretion
B. Filtration and passive reabsorption
C. Normal excretory products
D. Diffuses out of renal tubule cells along its concentration gradient and becomes "trapped" in the tubular fluid

6. Ammonia
7. H^+, organic acids (e.g., penicillin, salicylates, PAH), organic bases (e.g., procainamide, choline)
8. H^+, phosphates, sulfates, uric acid, creatinine
9. K^+
10. Urea, water

Questions 11–15. Choose the most appropriate lettered answer: A, B, C, or D.

A. Water-conserving state (antidiuresis)
B. Water excess state (diuresis)
C. Isoosmotic excretion
D. Anuric state

11. $CH_2O > 0$
12. $CH_2O < 0$
13. $CH_2O = 0$
14. $TH_2O > 0$
15. $TH_2O < 0$

Questions 16–18. Each of the numbered items or incomplete statements is followed by answers or by completions of the statement. Select the one lettered answer or completion that is best in each case.

16. Precipitous decreases in hydrostatic pressure occur in the renal circulation
    A. At the afferent and efferent arterioles
    B. At the arcuate and interlobular arteries
    C. At the ascending and the descending vasa recta
    D. At the glomerular tuft of capillaries and the peritubular plexus
    E. At the renal and accessory renal arteries
17. Renal failure (uremia) is associated with all of the following **except**
    A. Anemia
    B. Ca^{2+} deficiency and bone disease

C. Hyperkalemia
D. Metabolic acidosis
E. Hyperaldosteronism

18. In a denervated kidney (i.e., the renal nerves are sectioned), what would be the anticipated changes in renal function and blood flow?
    A. Inability of the kidney to regulate glomerular filtration rate (GFR), renal vasoconstriction, and decreased renal blood flow (RBF)
    B. Normal physiologic function of the kidney, renal vasodilation, and increased RBF
    C. Inability of the kidney to regulate GFR, renal vasodilation, and increased RBF
    D. Normal physiologic function of the kidney, renal vasoconstriction, and decreased RBF
    E. Abnormal physiologic function with unopposed renal vasoconstriction, leading to a precipitous drop in GFR and eventually renal shutdown

Questions 19–25. Each group of items consists of lettered options followed by a set of numbered items. For each item, select the one lettered option that is most closely associated with it. Each lettered option may be selected once, more than once, or not at all.

A. Renal vasoconstriction and decrease in renal blood flow
B. Renal vasodilatation and increase in renal blood flow
C. No change in renal vasomotor tone or renal blood flow
D. Either renal vasoconstriction or vasodilatation can be initiated

19. Exercise, rising from a supine to a sitting position or administration of catecholamines
20. Hypoxia with a PaO_2 of less than 50 percent normal
21. A decreased discharge from baroreceptor fibers resulting from a decrease in systemic blood pressure
22. Stimulating the vasomotor center in the medulla, the cortex, or the anterior tip of the temporal lobe
23. Bed rest, changing from a standing to a reclining position, abolition of renal sympathetic input
24. An increased discharge from baroreceptor fibers resulting from an increase in systemic blood pressure
25. Placement of a Swan-Ganz catheter in an anesthetized patient

Questions 26–36. Each of the numbered items or incomplete statements is followed by answers or by completions of the statement. Select the one lettered answer or completion that is best in each case.

26. A 70-kg man is given 1 millicurie (mCi) of tritiated water (HTO) intravenously. During the period of equilibrium, 1/2 percent of the dose administered is lost in the urine. After equilibration a sample of the man's plasma contains 0.024 mCi per liter plasma. The volume of distribution of the marker is approximately
    A. 0.4 liters
    B. 4 liters
    C. 24 liters
    D. 34 liters
    E. 40 liters
27. Tritiated water is a marker for what body water compartment?
    A. Extracellular water (ECF)
    B. Interstitial fluid
    C. Intracellular water (ICF)

D. Plasma water
E. Total body water (TBW)

28. A 70-kg man is administered 12.5 g mannitol intravenously. Assuming sufficient time has passed for equilibration to occur and that 1.5 g mannitol has been lost in the urine, what is the volume of distribution of the marker if the serum mannitol concentration is 58.8 mg/dl?
A. 3.9 liters
B. 18.7 liters
C. 26.5 liters
D. 34.6 liters
E. 41.0 liters

29. Chloride is normally passively reabsorbed except in the thick ascending limb. In cases of severe chloride deficiency, an active chloride pump can also be demonstrated in which of the following?
A. Pars recta and distal convoluted tubules
B. Proximal convoluted and distal convoluted tubules
C. Distal convoluted and collecting tubules
D. Connecting piece and collecting tubules

30. On intravenous injection, mannitol would be expected to distribute evenly throughout the
A. Total body water (TBW)
B. Extracellular fluid (ECF)
C. Intracellular fluid (ICF)
D. Plasma water
E. Transcellular water

31. Thiocyanate is used to determine the ECF volume of a male patient who weighs 70 kg. If the patient is of normal build and his ECF volume as determined by injection of thiocyanate is 17 liters, what is his ICF volume?
A. 45 liters
B. 35 liters
C. 25 liters
D. 15 liters
E. 5 liters

32. The same man in question 31 is now injected with 0.01 g T-1824 dye (Evans blue) intravenously. A few minutes later, a blood sample has a concentration of 3.2 mg T-1824 per liter plasma. Which of the following indicates the resultant marker distribution and corresponding fluid compartment?
A. 31 liters, ECF
B. 31 liters, ICF
C. 3.1 liters, interstitial
D. 3.1 liters, plasma
E. 31 liters, TBW

33. If the volume of distribution of a 60-kg woman as determined by inulin is 18 liters, and the volume of distribution as determined by chromicized Cr-51 erythrocytes is 3.1 liters, what is her interstitial fluid volume?
A. 14.9 liters
B. 21.1 liters
C. 15.1 liters
D. 8.9 liters
E. Cannot be determined from available data

34. A 24-year-old previously healthy businessman is brought into the emergency room in a coma. His serum sodium is 141 mEq per liter, his BUN is 28 mg/dl, and his blood glucose is 180 mg/dl. The molecular weight of glucose is 180 and that of BUN is 28. His serum osmolality would be

A. 285
B. 302
C. 320
D. 404
E. 606

35. The comatose businessman in question 34 has a serum osmolality of 360 on his chemistry panel. Which of the following is the most likely explanation?
A. Hyperkalemia
B. Hyperlipidemia
C. Hypernatremia
D. Multiple myeloma with markedly increased serum protein
E. Alcohol intoxication

36. $[K^+]ICF = 156$, $[K^+]ECF = 4.0$ Calculate the size and direction of the electric potential.
A. −90 mV
B. −98 mV
C. −110 mV
D. 90 mV
E. 98 mV

Questions 37–41. Each group of items consists of lettered options followed by a set of numbered items. For each item, select the one lettered option that is most closely associated with it. Each lettered option may be selected once, more than once, or not at all.

A. Nernst equation
B. Starling's equation
C. Clearance
D. Countercurrent multiplier (Kuhn's hypothesis)
E. Countercurrent exchange

37. Vasa recta
38. Loop of Henle
39. GFR
40. UV/P
41. Resting membrane potential, single univalent ion

Questions 42–44. Each of the numbered items or incomplete statements is followed by answers or by completions of the statement. Select the one lettered answer or completion that is best in each case.

42. Which of the following is **not** a determinant of GFR?
A. Q_A
B. P_c
C. K_f
D. π_c
E. E

43. Volume contraction can be accompanied by each of the following **except**
A. Hyperosmolality
B. Isoosmolality
C. Hypoosmolality
D. Alkalosis
E. Acute weight gain

44. Volume expansion can be accomplished by each of the following **except**
A. Hyperosmolality
B. Isoosmolality

C. Hypoosmolality
D. Alkalosis
E. Acute weight gain

Questions 45–62. Each group of items consists of lettered options followed by a set of numbered items. For each item, select the one lettered option that is most closely associated with it. Each lettered option may be selected once, more than once, or not at all.

Questions 45–53 refer to the figure below. For each numbered section of the nephron (1–9), indicate the status of tubular fluid removed from that section as compared to plasma in the presence of ADH.

A. Hypertonic
B. Hypotonic
C. Isotonic
D. Maximally hypertonic (greater than 900–1200 mOsm/liter)
E. Maximally hypotonic (less than 50–100 mOsm/liter)

45. Section 1
46. Section 2
47. Section 3
48. Section 4
49. Section 5
50. Section 6
51. Section 7
52. Section 8
53. Section 9

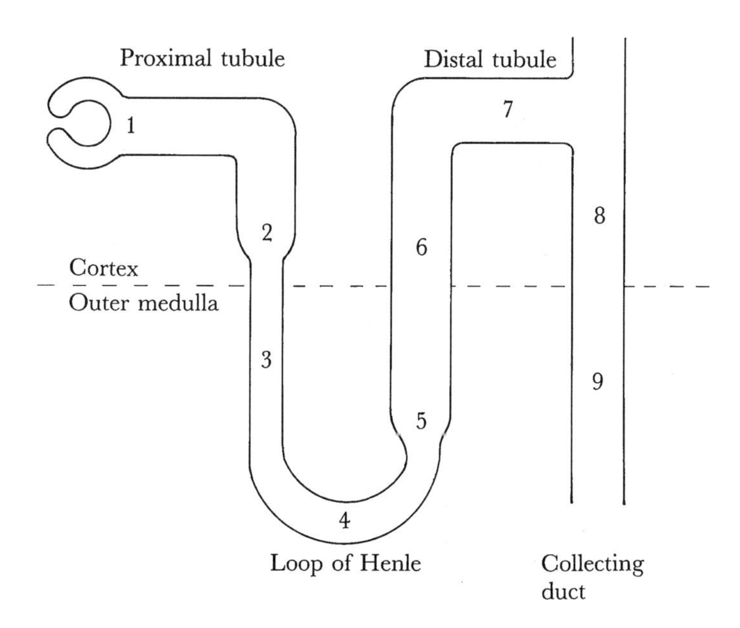

Questions 54–62 refer to the figure above. For each numbered section of the nephron (1–9), indicate the status of tubular fluid removed from that section as compared to plasma in the absence of ADH.

A. Hypertonic
B. Hypotonic
C. Isotonic

D. Maximally hypertonic (greater than 900–1200 mOsm/liter)
E. Maximally hypotonic (less than 50–100 mOsm/liter)

54. Section 1
55. Section 2
56. Section 3
57. Section 4
58. Section 5
59. Section 6
60. Section 7
61. Section 8
62. Section 9

Questions 63–66. Each of the numbered items or incomplete statements is followed by answers or by completions of the statement. Select the one lettered answer or completion that is best in each case.

63. Urea is a good choice of substrate for use by the renal concentrating mechanism for all of the following reasons **except**
    A. Urea is the major excretory by-product of metabolism
    B. Excretion of urea would otherwise osmotically obligate the excretion of large amounts of water were it not for its paradoxical use to concentrate the urine
    C. Urea is actively pumped by Na^+-K^+ ATPase in the thin ascending limb
    D. Urea is concentrated in the medulla as a result of active pumping of NaCl in the thick ascending limb
    E. Medullary recycling of urea maximizes the contribution of urea to the maintenance of hyperosmolality in the medullary interstitium
64. Which of the following regions of the kidney has the highest oxygen consumption?
    A. Cortex
    B. Medulla
    C. Columns of Bertin
    D. Perinephric fat
65. Fluid entering the distal tubule is
    A. Isoosmotic at all times
    B. Hyperosmotic at all times
    C. Hypoosmotic at all times
    D. Can be isoosmotic, hyperosmotic, or hypoosmotic
    E. Hyperosmotic or isoosmotic but never hypoosmotic
66. In the absence of ADH all of the following occur **except**
    A. The medullary osmolality and urea concentration decrease (the medullary interstitium is "washed out")
    B. Diuresis occurs
    C. Dilute fluid leaving the thick ascending limb is diluted even further by active NaCl pumping in the distal tubule and collecting system
    D. The medullary collecting duct becomes impermeable to urea
    E. Urea excretion decreases

Questions 67–76. Each group of items consists of lettered options followed by a set of numbered items. For each item, select the one lettered option that is most closely associated with it. Each lettered option may be selected once, more than once, or not at all.

For Questions 67–72, select the probable effect on urinary potassium excretion of each of the following perturbations.

A. Increase
B. Decrease
C. Not predictable
D. No change

67. Acute metabolic acidosis
68. Low-salt diet
69. Acute metabolic alkalosis
70. Diuretic that inhibits NaCl reabsorption
71. Oliguria
72. Administration of an aldosterone antagonist (e.g., spironolactone)

For Questions 73–76, select the section of the nephron most closely associated with each description.

A. The thick ascending limb, distal convoluted tubule, and collecting tubule
B. Thin descending limb and the thin ascending limb
C. Distal and collecting tubules
D. Thick ascending limb

73. Na^+-K^+-$2Cl^-$ pump (active)
74. No active sodium pumping has ever been demonstrated
75. Active chloride pumping has been demonstrated
76. Aldosterone acts here

Answers

In reference to Questions 1–10, see Fig. 3-8.

1. **D** Creatinine is primarily filtered with only a small amount, approximately 20 percent, secreted. Creatinine is frequently used in determinations of renal clearance.
2. **A** Glucose and amino acids are normally entirely reabsorbed.
3. **D** Inulin is freely filtered and neither reabsorbed nor secreted. Because of this, inulin clearance is the gold standard for determining renal clearance.
4. **C** Small proteins easily pass the size- and charge-selective barriers and are catabolized by the renal tubular cells.
5. **B** Sodium is actively reabsorbed everywhere in the tubule except in the descending and ascending thin limbs.
6. **D** The ammonia (NH_3) in the tubular fluid is derived from the breakdown of glutamine and other amino acids by renal tubular cells. Being lipid soluble, the ammonia diffuses out of the cells and into the tubular fluid along its concentration gradient. In the tubular fluid, H^+ combines with NH_3 to form ammonium (NH_4^+), which is not lipid soluble and becomes trapped in the tubular fluid (ion trapping).
7. **A** Organic acids (e.g., diuretics, salicylates, antibiotics, and PAH) are excreted by an anionic pump located in the basolateral membrane. Organic bases (e.g., procainamide and choline) are secreted using a cationic pump located in the basolateral membrane. Hydrogen ions are secreted and ultimately titrate urinary buffers or combine with filtered bicarbonate to form carbon dioxide and water.
8. **C** H^+, phosphates, sulfates, uric acid, and creatinine are by-products of metabolism that are excreted by the kidneys.
9. **A** Virtually all the potassium filtered is reabsorbed somewhere in the tubule. Net excretion is by passive secretion in the distal tubule.

10. B Urea and water are filtered and then passively reabsorbed. Approximately half the filtered load of urea may be passively reabsorbed into the medullary interstitium.

11. B Free water is water free of solute. Free water clearance (C_{H_2O}) is a calculation of the amount of distilled water that must be added to or removed from the urine to render the urine isoosmotic with plasma. It is an attempt to quantify the states of water diuresis and conservation and is not a true clearance (see text). It is calculated by subtracting the clearance of osmoles (C_{Osm}) from the minute urine flow. If C_{H_2O} is greater than zero, then water is being excreted in excess of solute (diuresis).

12. A If C_{H_2O} is less than zero, then solute is being excreted in excess of water (antidiuresis).

13. C If C_{H_2O} is equal to zero, then the urine is being excreted isoosmotically.

14. A Negative free water clearance (T_{H_2O}) is free water clearance (C_{H_2O}) with the sign reversed for convenience when discussing water-conserving states. If C_{H_2O} is less than zero, then solute is being excreted in excess of water (antidiuresis).

15. B Free water is water free of solute. Free water clearance (C_{H_2O}) is a calculation of the amount of distilled water that must be added to or removed from the urine to render the urine isoosmotic with plasma. It is an attempt to quantify the states of water diuresis and conservation and is not a true clearance (see text). It is calculated by subtracting the clearance of osmoles (C_{Osm}) from the minute urine flow. If C_{H_2O} is greater than zero, then water is being excreted in excess of solute (diuresis). Negative free water clearance (T_{H_2O}) is free water clearance (C_{H_2O}) with the sign reversed for convenience when discussing water-conserving states.

16. A There are large hydraulic pressure drops at the afferent arteriole and the efferent arteriole. The precapillary sphincter muscles, present in both the afferent and efferent arterioles, are important in the autoregulation of GFR. Increasing or decreasing the size of the afferent arteriole allows more or less of the systemic pressure to be felt by the glomerulus (altering the pressure head driving filtration), while changes in the diameter of the efferent arteriole can alter the filtration rate as well as help govern flow.

17. E Renal failure does not cause hyperaldosteronism. It does cause anemia (because of decreased production of erythropoietin), calcium deficiency (from vitamin D deficiency), and hyperkalemia and metabolic acidosis (as the kidneys' excretory functions fail).

18. B The kidney is able to function normally after transection of the renal nerves. Section of the nerves would result in renal vasodilation and an increase in renal blood flow because renal sympathetic nerve activity normally predominates at rest and sectioning the nerves therefore results in a net decrease in renal sympathetic tone.

19. A In a manner reminiscent of ADH release, all of these actions may result in a decrease in renal blood flow and increase in renal vasomotor tone. The decrease in renal blood flow and increase in renal vasomotor tone is presumed to be mediated by reflex increases in sympathetic nerve activity in response to baroreceptor input and by direct action of circulating catecholamines.

20. A Hypoxia acts via an increase in sympathetic nerve discharge to cause renal vasoconstriction and a decrease in renal blood flow.

21. A The decrease in renal blood flow and increase in renal vasomotor tone is presumed to be mediated by reflex increases in sympathetic nerve activity in response to baroreceptor input and by direct action of circulating catecholamines.

22. D Vasomotor centers in the medulla, cortex, and anterior tip of the temporal lobe can cause renal vasoconstriction or vasodilation.

23. B Bed rest, changing from a standing to a reclining position, and abolition of renal sympathetic input would all result in decreased renal sympathetic tone. This decrease in renal sympathetic input would result in renal vasodilation, increased renal blood flow, and an increase in GFR.

24. B An increase in baroreceptor firing would result in decreased sympathetic tone, increased renal blood flow, and decreased renal vasomotor tone.

25. C Placement of a Swan-Ganz catheter in a previously anesthetized patient should have no effect on renal vasomotor tone or blood flow (the anesthesia could have profound effects, however).

26. E Volume of distribution of marker using indicator-dilution method = (1.0 – 0.005)/0.024 = 41.5 liters (see Table 3-5).

27. E Tritiated water distributes throughout TBW (see Table 3-5).

28. B volume of distribution of marker using indicator-dilution method = (12.5 – 1.5)/0.588 = 18.7 liters. Mannitol distributes throughout total body water (see Table 3-5).

29. C The thick ascending limb contains the active Na^+-K^+-$2Cl^-$ active pump while the distal convoluted and collecting tubules have an active chloride pump that can be unmasked in severe chloride depletion. The connecting piece is a portion of the tubule connecting the distal convoluted tubule to the collecting system.

30. B Mannitol is a marker for the extracellular fluid. It is a sugar that distributes throughout the extracellular fluid but is not able to enter cells and therefore does not equilibrate in the ICF. Similarly, inulin, sucrose, and raffinose are sugars that cannot enter cells and are therefore useful in calculating ECF volume. Radiochloride, radiobromide, radiosodium, and thiocyanate distribute through the ECF (see Table 3-5).

31. C A male human of normal build is approximately 60 percent water by weight (depending primarily on percentage body fat, since fat contains very little water). A 70-kg man would therefore consist of 42 liters of water. If the ECF volume of this man is 17 liters (one liter water is equal to 1 kg in mass), then his ICF volume must be 25 liters, that is, TBW = ECF + ICF or ICF = TBW – ECF = 42 – 17 = 25.

32. D Volume of distribution = (10 – 0)/0.0032 = 3100 ml = 3.1 liters; Evans blue is a marker for the plasma since after intravenous injection it remains in the intravascular space (see Table 3-5). Note that only three markers are commonly used in determining plasma volume and that two are labeled normal constituents of the intravascular space, that is, chromicized Cr-51 erythrocytes and iodinated I-131 albumin.

33. A Interstitial fluid volume = ECF volume – plasma volume. Inulin is a marker for the ECF and Cr-51 red cells are a marker for the plasma volume; therefore, interstitial volume = 18 – 3.1 = 14.9 liters.

34. B Serum osmolality = $2[Na^+]$ + glucose mg%/18 + BUN mg%/2.8 = 302 mOsm per liter.

35. E A serum osmolality of 360 suggests the presence of an abnormally high concentration of some osmole other than glucose, sodium, or urea. Answer C, hypernatremia, is therefore incorrect since sodium is used to calculate the predicted osmolality and therefore cannot, by definition, be the culprit. Since the patient was apparently healthy before his lapsing into a coma, answers B and D are unlikely because both lipids and proteins (in the case of multiple myeloma) can cause hyperosmotic states and could account for the hyperosmolality but probably not the coma. Hyperkalemia, answer A, cannot be correct because a serum potassium of 78 would be totally incompatible with life. Answer E, alcohol intoxication, is therefore the most likely explanation since alcohol is an abnormal osmole, acute ingestion of which can lead to dehydration, hyperosmotic state, and coma. The legal limit in many states for alcohol intoxication is 0.1 g%. Three to four times this concentration is close to the lethal limit. Given that the molecular weight of alcohol is 46 and assuming a blood alcohol content of three times the legal limit,

0.3 g% or 300 mg%, the osmolality due to alcohol is 300 mg%/4.6 or 65 mOsm per liter. Alcohol intoxication could therefore explain the coma and hyperosmotic state.

36. B Using the Nernst equation, potential (millivolts) = −61.5 log 156/4 = −97.9 mV cell interior relative to cell exterior (by convention, potential is always given interior relative to exterior).

37. E The vasa recta engage in countercurrent exchange to maintain the hyperosmotic medullary interstitium.

38. D The loop of Henle is the site of the countercurrent multiplier or augmentor, that is, the site where the osmotic gradient is established.

39. B Starling's forces (K_f, π, and P), described in Starling's equation, are major determinants of the glomerular filtration rate (GFR). The fourth determinant is initial or afferent glomerular capillary flow (Q_A).

40. C Clearance is equal to urine concentration times minute urine flow divided by the plasma concentration (UV/P).

41. A The Nernst equation is used to calculate electric potential differences across cell membranes for ions at equilibrium.

42. E All are determinants of GFR except for E, which is the symbol for electric potential gradient.

43. E Acute loss of volume has to be accompanied by weight loss, not gain. Depending on the form that the volume loss takes and what kind of volume replacement is attempted, however, the end result of the volume loss can be hyperosmolality (e.g., loss of hypotonic sweat), isoosmolality (e.g., acute loss of isotonic blood), or hypoosmolality (e.g., loss of gastric contents or blood with oral water replacement). Contraction alkalosis occurs because volume contraction favors secretion of H^+ by the kidneys.

44. D Volume expansion can result in hyperosmolality, isoosmolality, or hypoosmolality depending on the mechanism causing the volume expansion. Volume expansion causes weight gain acutely. Volume expansion, however, does not favor alkalosis and answer D is therefore correct.

Questions 45–62 (see figure below)

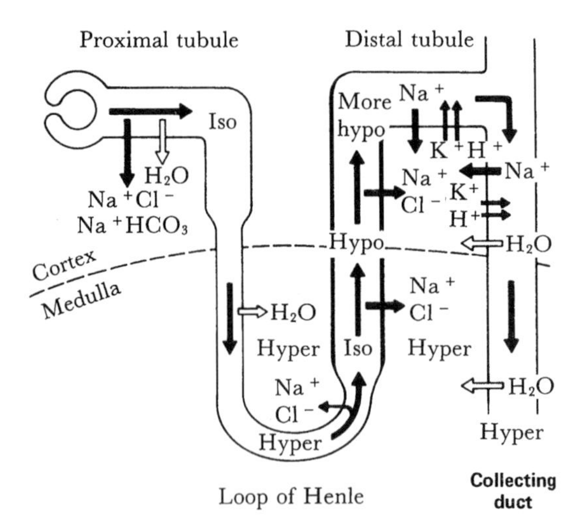

ADH present

45. C Isotonic
46. C Isotonic
47. A Hypertonic

48. A Hypertonic
49. C Isotonic
50. B Hypotonic
51. A Hypertonic
52. A Hypertonic
53. D Maximal hypertonicity

ADH absent

54. C Isotonic
55. C Isotonic
56. A Hypertonic
57. A Hypertonic (note that the osmolality of sections 1–4 is identical whether or not ADH is present)
58. C Isotonic
59. B Hypotonic
60. B Hypotonic
61. B Hypotonic
62. E Maximal hypertonicity
63. C Urea is not actively transported.
64. A The cortex has the highest oxygen consumption because it is there that Na^+ transport is most active. The medulla has an oxygen consumption one-twentieth that of the cortex.
65. C Both renal concentrating and diluting mechanisms depend on active pumping of solute from the thick ascending limb and, therefore, tubular fluid reaching the distal tubule is always hypoosmotic.
66. E Urea excretion increases because urea can no longer be reabsorbed in the inner medulla.
67. A Both acidosis and alkalosis increase potassium excretion.
68. A A low-salt (Na^+) diet will cause the kidney to conserve sodium and excrete more potassium.
69. A Both acidosis and alkalosis increase potassium excretion.
70. A Excretion of potassium increases as urinary flow increases because there is less time and opportunity for potassium to be reabsorbed.
71. B Excretion of potassium decreases as urinary flow decreases because there is ample time for potassium to be reabsorbed.
72. B Spironolactone is a potassium-sparing diuretic and decreases potassium excretion.
73. D The active Na^+-K^+-$2Cl^-$ pump is located only in the thick ascending limb.
74. B Active sodium pumping has never been demonstrated in the thin descending and the thin ascending limbs.
75. A Active chloride pumping has never been demonstrated in the thick ascending limb and distal and convoluted tubules.
76. C Aldosterone acts on the distal tubule and collecting ducts.

Bibliography

Berne, R. M., and Levy, M. N. *Physiology* (3rd ed.). St. Louis: Mosby–Year Book, 1993.

Ganong, W. F. *Review of Medical Physiology* (15th ed.). Los Altos, CA: Lange, 1991.

Guyton, A. C. *Textbook of Medical Physiology* (8th ed.). Philadelphia: Saunders, 1991.

Koeppen, B. M., and Stanton, B. A. *Renal Physiology* (1st ed.). St. Louis: Mosby, 1992.

Valtin, H. *Renal Function: Mechanisms Preserving Fluid and Solute Balance in Health* (2nd ed.). Boston/Toronto: Little, Brown, 1983.

Vander, A. J. *Renal Physiology* (4th ed.). New York: McGraw-Hill, 1991.

West, J. B. *Best and Taylor's Physiological Basis of Medical Practice* (12th ed.). Baltimore: Williams & Wilkins, 1990.

4

Respiratory Physiology

Charles H. Tadlock

The primary function of the lungs is gas exchange. Oxygen utilized by the tissues is replaced and carbon dioxide generated during metabolism is excreted. Oxygen is absorbed and carbon dioxide is excreted in the gas-exchanging units of the lungs. The gas-exchange unit is the **alveolus.** Normal human lungs are composed of approximately 300 million alveoli, which average 110 microns (μ) in diameter at rest. Gas transport is largely accomplished by two bulk (convective) transport processes: **ventilation (V)** and blood flow or **perfusion (Q).** At two key interfaces, however, bulk transport ceases and simple diffusion becomes the predominant transport mechanism: between the alveolus and the pulmonary capillary, and between the systemic capillary and the mitochondria.

Eighty percent of the alveolar surface area is covered by pulmonary capillaries. Gas exchange occurs across a thin, aqueous fluid layer that contains surfactant, plasma, and connective tissue separating the pulmonary capillaries from the alveoli. The average distance between alveolar gas and the hemoglobin in the red cell is only 1.5 μ. The pulmonary capillary bed contains approximately 150 ml blood spread out over a surface of 70 m^2 (1 m^2/kg), creating a film 10 μ (or approximately one red cell diameter) thick that surrounds the alveoli.

At basal heart rates, the red cells spend approximately three-fourths of a second in the pulmonary capillaries, three times the amount of time needed to equilibrate with the alveolar gases. This results in a gas-exchange system that is so efficient that over half the lung volume may be lost before any abnormality of gas exchange appears. In fact, increases in pulmonary vascular resistance from loss of vascular tree rather than decreases in gas exchange limit survival. During peak exercise the red cell remains in the capillaries for approximately 0.3 seconds, more than the period required for equilibration to occur. In this manner 250 cc oxygen is absorbed and 200 cc carbon dioxide is excreted per minute at rest. The process of gas exchange at the level of the lungs is referred to as respiration. Respiration also refers to oxidative metabolism at the cellular level.

Structure and Function of the Lung

The Airways

Air enters the lungs through the nose and oropharynx, where it is warmed and humidified. The inspired air is conducted to the alveoli through a series of branching tubes. The **trachea, mainstem bronchi,** and their major branches contain cartilage, which acts to maintain a patent lumen during changes in intrathoracic pressure, and are lined by a pseudostratified columnar epithelium that rests on spiral bands of smooth muscle. The **bronchi** decrease in diameter and length with each successive branching but the sum of their cross-sectional areas actually increases. Thus, the bronchi are able to constrict or dilate independently of lung volume. More distally the cartilage support gradually disappears until the cartilage completely disappears in airways of approximately 1 mm diameter. By convention all subsequent branches are referred to as membranous **bronchioles.** The bronchioles are small, have no cartilage support or muscle, and, most importantly, are embedded in the connective tissue framework of the lung so that their diameter (and in some circumstances their patency) depends on lung volume. The airways divide 23 times; the first 16 divisions of the tracheobronchial tree comprise the conduction system. The final division of the conduction system is called the **terminal bronchiole.** These areas of the lung are supplied by the bronchial circulation and do not participate in gas exchange. The volume of gas contained within these tubes is therefore referred to as the **anatomic dead space** and normally accounts for 150 of the 500 cc of tidal ventilation. The division between the conducting zone and the gas exchange areas of the lung is referred to as the transitional zone.

The gas-exchange portion of the lungs consists of the final seven divisions of the respiratory tree. It is supplied by branches of the pulmonary arteries. The portion of lung distal to the terminal bronchiole forms an anatomic unit (the **terminal respiratory unit,** also called the primary lobule or acinus; Fig. 4-1). This so-called respiratory zone is composed of the respiratory bronchioles, alveolar ducts, and alveoli, and is where gas exchange actually occurs. Each unit contains approximately 500 anatomic alveoli and 250 alveolar ducts. These multiple divisions increase the cross-sectional area of the airways from 2½ cm^2 at the trachea to 12,000 cm^2 at the alveolus.

Smooth muscle in the walls of the larger airways can alter regional airflow by modifying the resistance properties of the airways. Resistance to flow occurs primarily in the larger airways, with only minor resistance encountered in the small airways (Fig. 4-2). Resistance to laminar flow is given by the equation

Fig. 4-1. Normal lung structure-function relationships emphasizing the anatomic relationship between the pulmonary artery and the airway. (Redrawn from R. M. Berne and M. N. Levy. *Physiology* [3rd ed.]. P. 553. © 1993 Mosby–Year Book, St. Louis.)

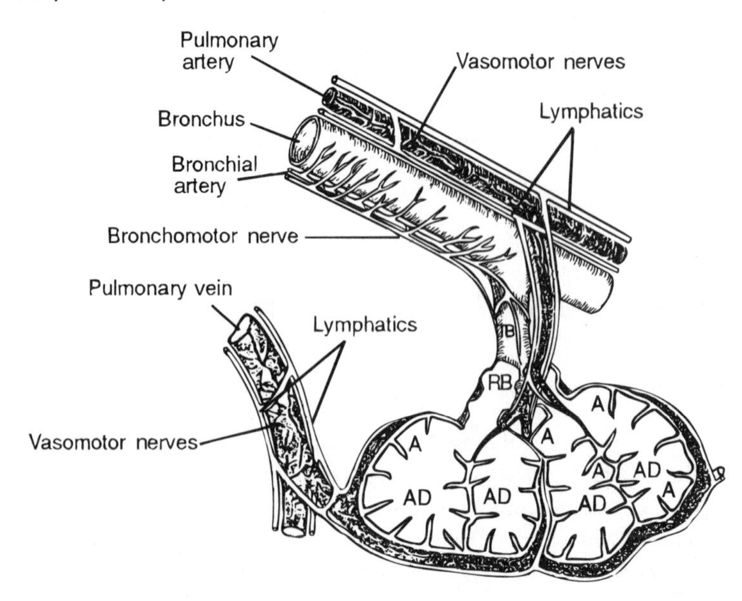

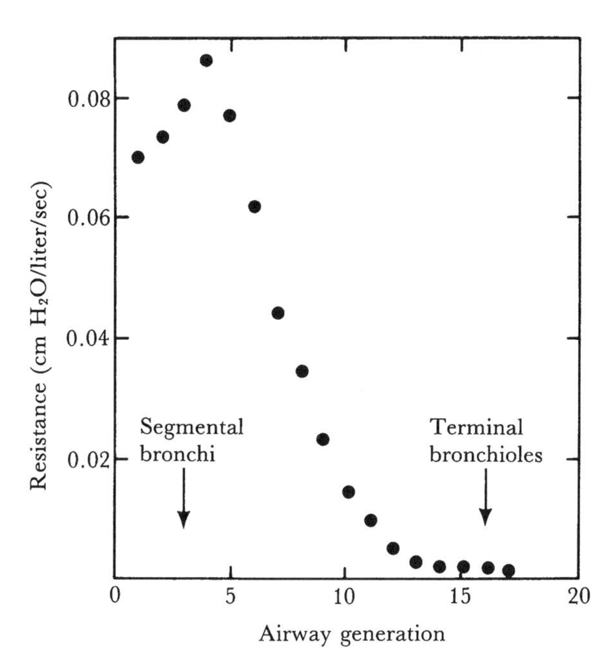

Fig. 4-2. Location of the chief site of airway resistance. Note that the intermediate-sized bronchi contribute most of the resistance and that relatively little is located in the very small airways. (Redrawn from T. J. Pedley, R. C. Schroter, and M. F. Ludlow. The prediction of pressure drop and variation of resistance within the human bronchial airways. *Respir. Physiol.* 9: 387, 1970.)

$$R = \frac{8\eta l}{\pi r^4}$$

where η is fluid viscosity, l is length, and r is the radius of the airway.

This equation would suggest that resistance to airflow would be greatest in the smaller-diameter airways; however, this is not the case. The total cross-sectional area of the smaller airways is so great that resistance actually decreases in the smaller airways of the lungs (compare Figs. 4-2 and 4-3).

Bulk Flow and Diffusion

During respiration air moves by bulk flow from the atmosphere into the tracheobronchial tree to the level of the terminal bronchioles (the last nonexchanging branch of the tracheobronchial tree). At the level of the terminal bronchioles the total cross-sectional area of the airways becomes so great that bulk flow essentially ends. Diffusion of gases is responsible for gas transport from terminal bronchioles through the respiratory ducts, the alveolar ducts, the alveoli, and into the blood.

Blood-Gas Barrier

As illustrated in Fig. 4-4, the blood-gas barrier consists of

1. Aqueous fluid layer containing pulmonary surfactant
2. Alveolar epithelium
3. Epithelial basement membrane
4. Interstitium

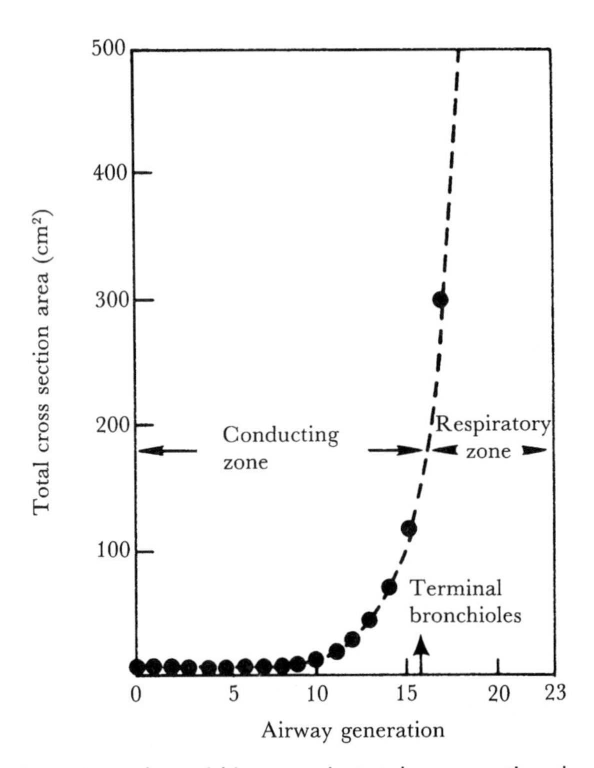

Fig. 4-3. Diagram to show the extremely rapid increase in total cross-sectional area of the airways in the respiratory zone. As a result, the forward velocity of the gas during inspiration becomes very small in the region of the respiratory bronchioles, and gaseous diffusion becomes the chief mechanism of ventilation. (From J. B. West. *Respiratory Physiology: The Essentials* [4th ed.]. P. 7. © 1990 Williams & Wilkins, Baltimore.)

5. Endothelial basement membrane
6. Capillary endothelium
7. Plasma
8. Red cell membrane
9. Intracellular fluid

Oxygen must traverse these structures in the order shown as it diffuses from the alveolar air sacs into the red cells to ultimately combine with hemoglobin. Carbon dioxide diffuses along its concentration gradient in the opposite direction.

Cell Types

There are four major cell types represented in the lungs.

1. The capillaries are lined by fenestrated endothelial cells.
2. The alveoli contain two types of epithelial cells. Type I epithelial cells are flat and line the alveoli.
3. Type II epithelial cells or granular pneumocytes are thicker, contain lamellar inclusion bodies, and secrete surfactant.
4. The fourth major cell type in the lungs is the pulmonary alveolar macrophage (PAM), but also present are mast cells, lymphocytes, plasma cells, and others.

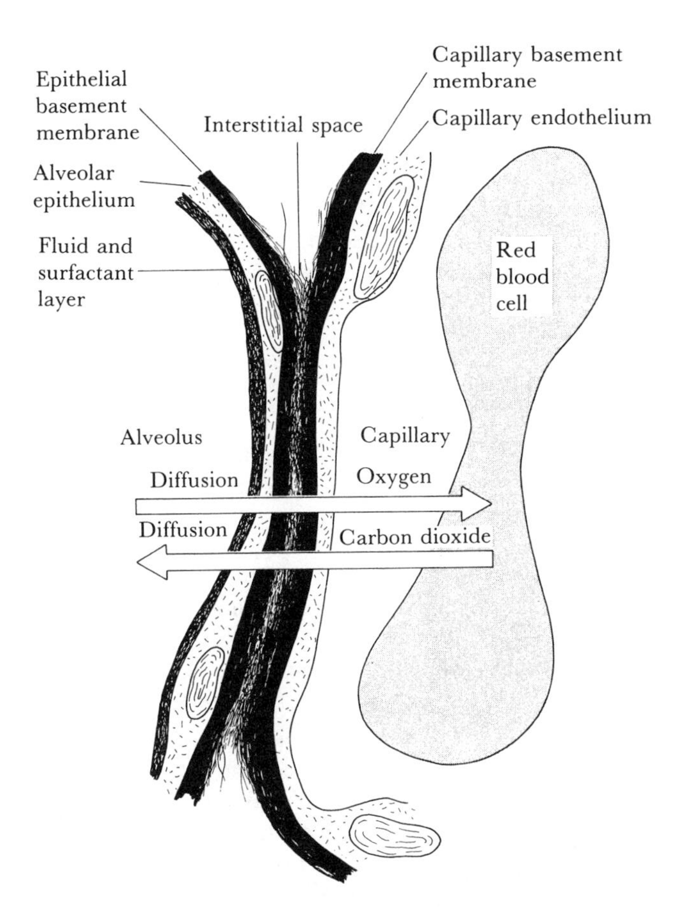

Fig. 4-4. Ultrastructure of the respiratory membrane as shown in cross section. (From A. C. Guyton. *Textbook of Medical Physiology* [8th ed.]. Philadelphia: Saunders, 1991. P. 429.)

Pulmonary Circulation

The lungs are supplied by two independent arterial circulations: the pulmonary circulation and the bronchial circulation. The pulmonary arteries carry deoxygenated venous blood from the right ventricle to supply the gas-exchanging units of the lungs. The pulmonary arteries are larger and thinner than their systemic counterparts. Thus, the pulmonary circulation is a high-compliance and low-resistance system. The pulmonary artery branches accompany branches of the bronchi. Small pulmonary arteries surrounded by alveoli respond to changes in alveolar gas composition. Alveolar hypoxia (i.e., a decrease in P_{AO_2}) causes pulmonary artery hypoxia and pulmonary vasoconstriction. Pulmonary arteries and veins larger than about 50 μ in diameter contain smooth muscle and are capable of active regulation of their diameters, which alters resistance to blood flow. This results in the shunting of blood from poorly ventilated areas of lung to better ventilated areas. This is in contrast to systemic circulations in which hypoxia causes vasodilatation.

Bronchial Circulation

In contrast, the bronchial circulation accounts for only 1 to 2 percent of pulmonary blood flow and originates from branches of the aorta. The bronchial arteries are high-pressure, high-resistance systemic vessels with thick muscular walls and a vasodilatatory response to hypoxia. The bronchial circulation supplies the supporting tissues of the lung, including the connective tissue, septa, and the large and small bronchi from the trachea to the level of the terminal bronchioles, with oxygenated arterial blood. Besides providing nutrition to the airways and the larger pulmonary blood vessels, the bronchial arteries warm and humidify the inspired air. After supplying the supporting tissue, the bronchial venous blood drains directly into the pulmonary veins, diluting the well-oxygenated blood that has just passed through the pulmonary capillaries with poorly oxygenated venous blood. This results in a small physiologic right-to-left shunt.

Lymphatic Circulation

The pulmonary lymphatic system extends from the perivascular and peribronchial spaces in the supporting tissues of the lung to the hilum of the lung. From the hilum the pulmonary lymphatics drain primarily into the right lymphatic duct.

Response to Exercise

Roughly 10 percent of the body's total blood volume is present in the lungs at any given time. At resting cardiac outputs an erythrocyte would spend an average of 0.75 seconds in the pulmonary capillaries. During heavy exercise, the cardiac output may increase to several times basal levels. The erythrocyte transit time decreases to approximately 0.3 seconds, still leaving adequate time for equilibration with alveolar gases to occur. Further decreases in transit time are avoided as additional pulmonary capillaries are recruited to handle the increased flow and as those vessels already open are distended. Indeed, increases in cardiac output cause no increase in pulmonary artery pressure since recruitment of additional capillary beds and distention of the capillaries result in a decrease in pulmonary vascular resistance.

Zones 1, 2, and 3

Gravity causes uneven distribution of blood in the lung as illustrated in Fig. 4-5. The lung can be divided into three zones. In zone 1, the apex of the upright lung may have a pulmonary alveolar pressure that is greater than the pulmonary capillary pressure, causing collapse of the pulmonary capillaries and cessation of blood flow (Q). This may occur only intermittently, for example, during expiration or not at all. The result is ventilation-perfusion mismatch (V/Q mismatch). In zone 2, pulmonary capillary pressure is greater than pulmonary alveolar pressure, which, in turn, is greater than pulmonary venous pressure. Blood flow therefore proceeds normally and a normal V/Q ratio results. In zone 3, the capillaries are engorged due to the increased pressure in the pulmonary capillaries resulting from the action of gravity on the column of blood in the lungs. Perfusion now exceeds ventilation and V/Q mismatch again results. Pulmonary vessels can be divided into alveolar and extraalveolar based on their responses to alveolar pressure. High alveolar pressures during positive pressure ventilation, for example, can squeeze blood out of most of the pulmonary capillaries (the alveolar vessels) but not the pulmonary arteries and veins (the extraalveolar vessels), which are tented open by the surrounding lung tissue.

Pulmonary Innervation and the Mucociliary Blanket

The bronchi and trachea, especially the carina, are sensitive to irritation. This stimulates the cough reflex, which is mediated by the vagus nerve and a regulatory center in the medulla. Initially, 2 to 3 liters of air is inspired. The epiglottis and vocal cords

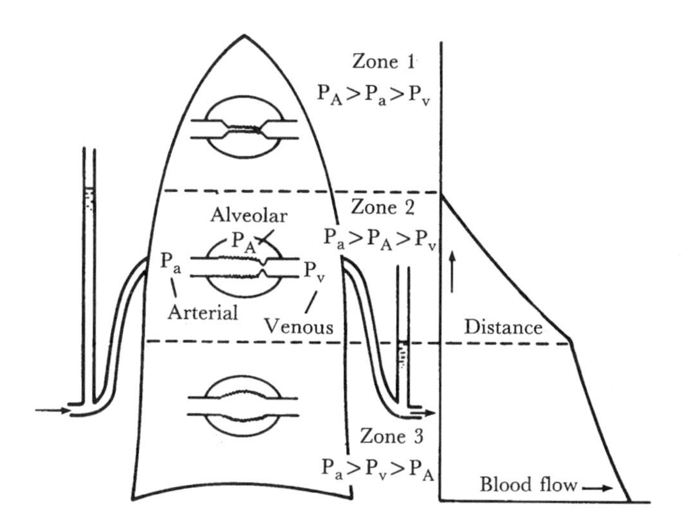

Fig. 4-5. Model to explain the uneven distribution of blood flow in the lung based on the pressures that affect the capillaries. A graph of blood flow versus height is appended on the right. (From J. B. West, C. T. Dollery, and A. Naimark. Distribution of blood flow in isolated lung; relation to vascular and alveolar pressures. *J. Appl. Physiol.* 19:713, 1964.)

are then closed and contraction of the expiratory muscles causes rapid development of up to 100 mm Hg of positive pressure in the lungs. The vocal cords and epiglottis open suddenly and air rushes out at 75 to 100 miles per hour. During this explosive expiration the noncartilaginous portions of the bronchi and trachea collapse forming narrow slits, all of which serves to propel foreign objects from the respiratory passageways. The sneeze reflex is similar to the cough reflex except that the uvula is depressed, shunting air flow through the nose. The sneeze reflex is mediated by the trigeminal nerve and the medullary regulatory center. The pulmonary artery walls are richly innervated with branches of the sympathetic nervous system. No clear function has yet been elucidated for these.

Air is warmed, humidified, and filtered during its journey through the nasal passageways and the trachea. The numerous obstructions to airflow in the nasal passageways, the septa, turbinates, and hair serve to increase the surface area available for equilibration of temperature and humidity, and to remove large particles by turbulent precipitation. Ciliated epithelium propels the precipitated particles and the mucus secreted by the mucous membrane covering of the nose and oropharynx toward the pharynx, where it is either swallowed or expectorated. Similarly, the **mucociliary blanket** of the tracheobronchial tree propels particles out of the lungs and into the oropharynx to meet the same fate.

Particles greater than 4 to 6 μ are removed by turbulent precipitation in the nose and tracheobronchial tree as described previously. Smaller particles, 1 to 5 μ in size, settle out by gravitational precipitation in the smaller bronchioles. Particles less than 1 μ in size diffuse against the walls of the alveoli and adhere to the alveolar fluid, where they are ultimately removed by alveolar macrophages. Particles less than 0.5 μ in diameter may remain suspended in the alveolar air to be expired later.

Mechanics of Respiration

Visceral and Parietal Pleura

The lungs are encased within a double sac. The inner sac is the **visceral pleura,** which is tightly adherent to the lung surface. The outer sac is tightly adherent to the chest wall and is called the **parietal pleura.** The **pleural space** is a potential space located between these two sacs. Normally, it contains only a small amount of fluid, which serves both to lubricate the pleurae as they move relative to one another during breathing and to cement the two together, much as a film of water will cause two panes of glass to adhere.

The Thoracic Pump

A negative intrapleural pressure causes air to enter the lungs. The negative intrapleural pressure results from (1) downward movement of the diaphragm as it contracts, lengthening the chest cavity, and (2) elevation of the ribs by the inspiratory muscles (primarily the external intercostals), causing the anterior-posterior diameter of the chest to increase (pail-handle motion).

These two processes create a vacuum or bellows effect, sucking air into the lungs. Because of the extremely high compliance of the lung and the chest wall, only a few millimeters of mercury of negative intrapleural pressure is sufficient to produce a normal inspiration.

During quiet breathing, relaxation of the diaphragm and chest wall combined with the elastic recoil properties of the lungs, chest wall, and abdominal structures causes the lung to be compressed resulting in expiration.

Heavy breathing requires active expiratory effort with the abdominal muscles contracting to force the abdominal contents against the bottom of the diaphragm and the internal intercostals contracting to decrease the volume of the thoracic cage by rotating the ribs downward. The **accessory muscles of respiration,** the sternocleidomastoids and the scalenes, pull up on the rib cage and thereby assist inspiration.

Lung Volumes and Capacities

Lung Volumes

Figure 4-6 illustrates the use of a spirometer to determine lung volumes and capacities. The volumes and capacities quoted here and elsewhere in this chapter are those for a healthy 70-kg man. Normal values vary with age, sex, weight, and body surface area. Note that a lung capacity is equal to two or more lung volumes. The following are the common lung volumes used in physiology:

1. **Tidal volume (TV)** is the volume expired or inspired with each breath at rest and is normally equal to about 500 cc.
2. **Inspiratory reserve volume (IRV)** is the additional volume of air that can be inspired on maximal forced inspiration at the end of a normal tidal inspiration (i.e., in addition to the tidal volume). Normally, this is approximately 3000 cc.

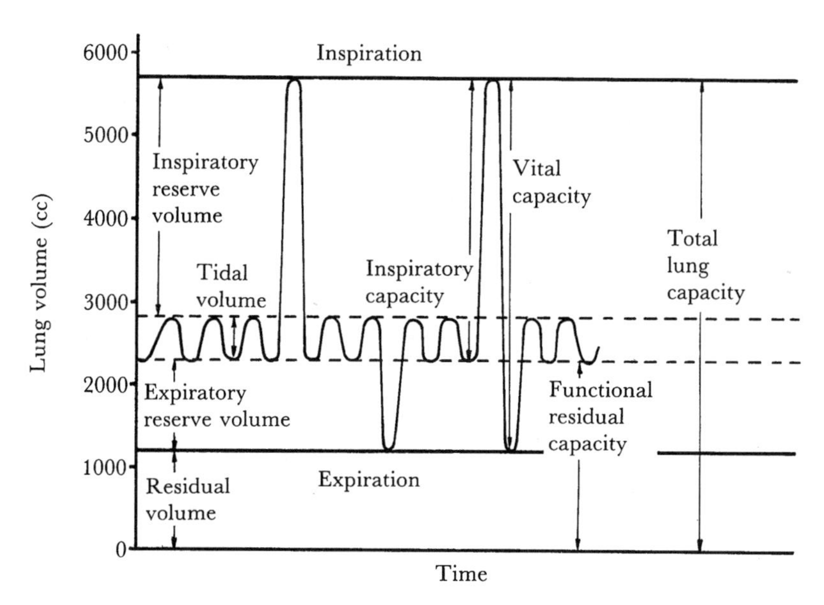

Fig. 4-6. Lung volumes and capacities. Note that the functional residual capacity and residual volume cannot be measured by spirometry alone. (From A. C. Guyton. *Textbook of Medical Physiology* [8th ed.]. Philadelphia: Saunders, 1991. P. 407.)

3. **Expiratory reserve volume (ERV)** is the volume of air that can still be expired by forceful expiration after the end of a normal tidal expiration. Normally, ERV is approximately 1100 cc.
4. **Residual volume (RV)** is the volume of air that remains in the lungs after a maximal expiration. Note that the residual volume cannot be measured by spirometry but must be determined by helium dilution or plethysmography.

Alveolar Ventilation Versus Dead Space Ventilation

Breathing is measured in terms of ventilation (frequency × depth of breathing). Approximately 500 cc air is inspired with each resting breath, TV. At a respiratory rate of 12 breaths per minute, the minute ventilation (volume respired in one minute, or $\dot{V}$) would be 6000 cc per minute. That portion of each TV that remains in the nonexchanging portions of the airways (the anatomic dead space) is referred to as the dead space ventilation, or V_D. The volume of the anatomic dead space is roughly 150 cc. The alveolar ventilation (V_A) is therefore 350 cc. Alveolar minute ventilation ($\dot{V}_A$) would therefore be equal to 350 × 12, or 4200 cc per minute. The ratio of dead space ventilation to total ventilation (V_D/V_T) is normally 0.2 to 0.35. Figure 4-7 illustrates typical lung volumes and flows.

Alveolar air is a mixture of freshly inspired air and the air remaining in the lungs between breaths (roughly 3000 cc). The alveolar air remaining in the lungs between breaths, **the functional residual capacity,** or **FRC,** serves to prevent the wide swings in alveolar and arterial gas composition that would occur if no air were present between breaths. If the FRC is small, for instance, P_AO_2 would approach that of mixed venous blood, $P\bar{v}O_2$ = 40 mm Hg, and rise to nearly that of moist tracheal air during breaths, PO_2 = 149. Instead, breath-to-breath fluctuations in PO_2 and PCO_2 are only 3 to 4 mm Hg and are generally ignored.

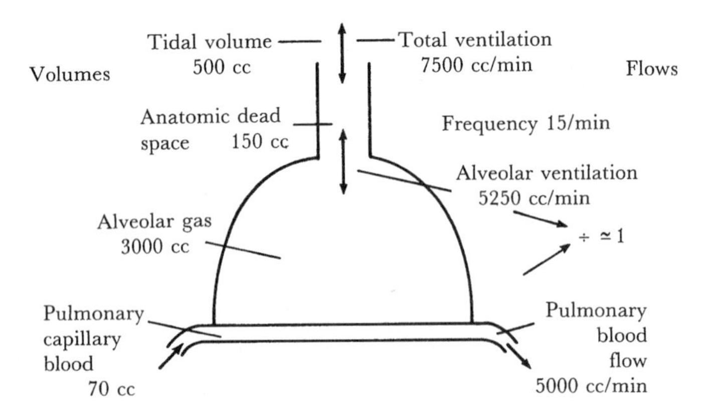

Fig. 4-7. Diagram of a lung showing typical volumes and flows. There is considerable variation around these values. (Modified from J. B. West. *Ventilation/Blood Flow and Gas Exchange.* Oxford: Blackwell, 1977. P. 3.)

Lung Capacities

The following are the common lung capacities used in physiology.

1. **Inspiratory capacity (IC)** = IRV + TV; the inspiratory capacity is therefore the maximal volume of air that can be inspired at the end of a normal tidal expiration.
2. **Functional residual capacity (FRC)** = ERV + RV; the functional residual capacity is the total volume of air remaining in the lungs at the end of a normal expiration. It may be calculated using either helium dilution or plethysmography.
3. **Vital capacity (VC)** = IRV + TV + ERV; vital capacity is the maximum volume of air that can be expelled from the lung following a maximal inspiration. It is equal to the inspiratory capacity plus the expiratory reserve volume.
4. **Total lung capacity (TLC)** = IRV + TV + ERV + RV = VC + RV; the total lung capacity is equal to the total volume of the lung following a maximal inspiration.

Measurement of Lung Volumes

Spirometry cannot be used to calculate the total lung volume because the lung never entirely empties of air. A certain amount of air, the RV, necessarily remains. Thus, RV, and therefore FRC and TLC, must be determined by utilizing some other technique.

The **helium dilution technique** utilizes a spirometer that contains a known concentration of helium (Fig. 4-8). Helium is virtually insoluble in blood and after a few breaths the helium concentration in the spirometer and the lungs equilibrate. Ignoring any loss of helium from the system, the amount of helium present before equilibrium (concentration × volume) is $C_1 \times V_1$ and is equal to the amount after equilibration, $C_2 \times (V_1 + V_2)$. The following equation can be utilized to determine FRC if the subject begins breathing from the spirometer at the end of a normal tidal expiration and if measurement of the helium concentration at equilibrium is done at the end of a normal tidal expiration:

$$C_1V_1 = C_2(V_1 + V_2)$$

$$V_2 = (C_1\, V_1/C_2) - V_1 = \text{FRC}$$

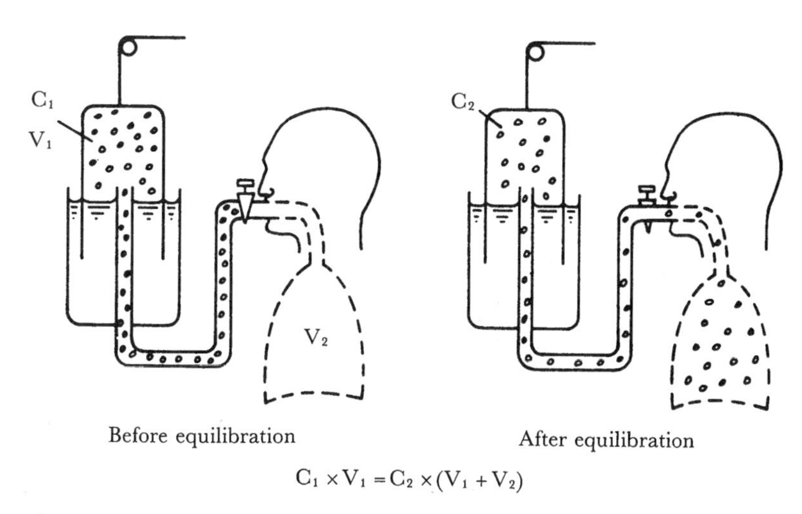

Fig. 4-8. Measurement of the functional residual capacity by the helium dilution method. (From J. B. West. *Respiratory Physiology: The Essentials* [4th ed.]. 1990. P. 13. © 1990 Williams & Wilkins, Baltimore.)

in which C_1 is the initial concentration of helium in the spirometer, C_2 is the concentration of helium after equilibration has occurred, V_1 is the volume of the spirometer, and V_2 is the volume of the lungs at end-tidal expiration (i.e., the FRC).

The technique of **total body plethysmography** uses a large airtight box like a telephone booth in which a person can sit (Fig. 4-9). The volume of the booth is known and the pressure within the booth can be monitored. A subject enters the booth and is instructed to make a respiratory effort at the end of a normal tidal expiration against a mouthpiece with an automatic shutter. As the subject tries to inhale, the automatic shutter closes. The subject expands the volume of the lungs as the intrathoracic pressure becomes increasingly negative in a fruitless attempt to move air against the closed mouthpiece. The pressure in the box, in contrast, rises as the thorax expands. **Boyle's**

Fig. 4-9. Measurement of the functional residual capacity by whole body plethysmography. When the subject makes an inspiratory effort against a closed airway, he slightly increases the volume of his lung. Airway pressure decreases and box pressure increases. (From J. B. West. *Respiratory Physiology: The Essentials* [4th ed.]. P. 14. © 1990 Williams & Wilkins, Baltimore.)

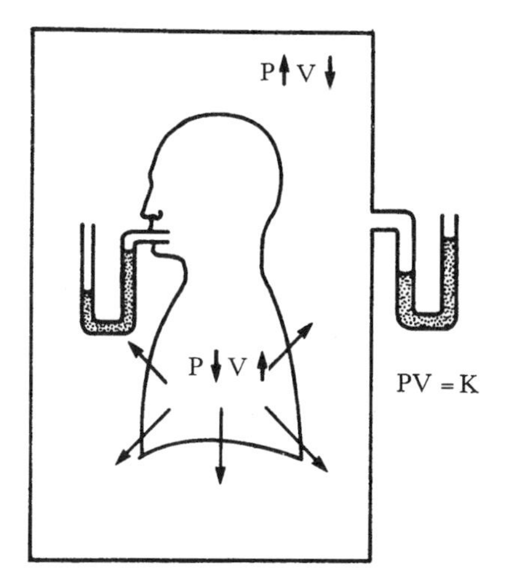

law states that pressure times volume is equal to a constant ($P \times V = C$). Thus, because the change in pressure can be measured, the change of volume of the lungs can be calculated as follows:

Change in volume of the lung = change in volume of the box = constant/change in box pressure during the inspiratory effort

or

$$\Delta V = C/\Delta P$$

Now, by simultaneously measuring pressures at the mouth as the subject attempts a tidal inspiration against a closed mouthpiece, and by again applying Boyle's law, the FRC of the lung can be calculated.

$$P_1V = P_2 (V + \Delta V)$$

or

$$V = P_2\Delta V/(P_1 - P_2)$$

in which P_1 is the mouth pressure at the beginning of inhalation, P_2 is the maximal pressure developed, V is the FRC, and ΔV is the change in volume of the box as calculated above.

Compliance, Elastance, and the Elastic Recoil Properties of the Lung

Compliance is defined as the change of volume per unit change of pressure ($\Delta V/\Delta P$) and is equal to the slope of the pressure-volume curve of the lung (Fig. 4-10). For normal lungs, this value is 0.200 liters per centimeter water. Elastance is the reciprocal

Fig. 4-10. The deflation pressure-volume curve of a normal lung demonstrating how compliance is obtained. Zero applied translung pressure is indicated by the thin vertical line. Pressures to the right are positive and act to distend the lung; pressures to the left are negative and act to compress it. (Redrawn from R. M. Berne and M. N. Levy. *Physiology* [3rd ed.]. P. 564. © 1993 Mosby–Year Book, St. Louis.)

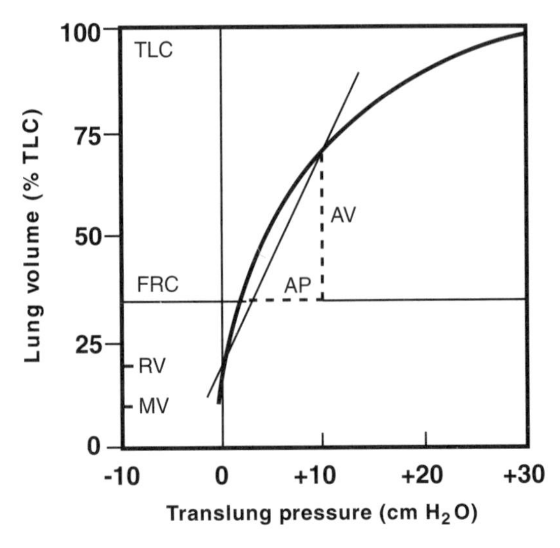

of compliance (1/C) and describes the resistance of an object to deformation by an external force. The pressure-volume curve of the lung is unique in several respects. It is curvilinear; its end points show sharp discontinuities, and inflation and deflation follow different paths (Fig. 4-11); the pressure-volume curve of the lung must therefore be read in the direction of the arrows, that is, it exhibits **hysteresis.** Hysteresis is primarily attributable to the interaction of **surface tension** and **surfactant** and only slightly due to the nonuniform structure of the lung (see below).

The lung's compliance characteristics are determined by its anatomy and the fact that it is not a perfectly elastic body but may instead be described as an elastic component in parallel with an inelastic component. As long as the elastic component has some stretch left, the lungs are very compliant. However, once the inelastic component is required to stretch, that is, at high lung volumes, the lungs become much less compliant. The other major determinant of lung elasticity is surface tension at the alveolar liquid-air interface.

The elastic recoil properties of the lung tend to collapse it while the action of the chest wall acts to keep the lungs expanded. The resting midposition of the lungs (FRC) is determined by the tendency of the chest wall to expand and the lung to collapse. If the chest wall is opened, the lung will collapse to approximately 10 percent of total gas capacity, at which point intralung pressure will equal atmospheric pressure (minimal volume; see Fig. 4-10) and further collapse will be prevented by air trapping in the collapsed airways. The elastic recoil properties result from the interaction of the stretched elastic elements of the lung and from the surface tension of the fluid lining the alveoli. The elastic fibers, like a rubber band, are stretched and tend to regain their resting length. The fluid lining the alveoli tends to minimize its surface area (i.e., the fluid attempts to become a sphere) as a result of the strong molecular attraction be-

Respiratory

Fig. 4-11. A. The inflation-deflation curve of the normal air-filled lung contrasted with the inflation-deflation curve of a saline-filled lung. Note the disappearance of hysteresis and the relative ease with which the saline-filled lung is inflated. Also note the high compliance (steep slope) over the lower portions of the normal inflation-deflation curve. This is beneficial because it means the lung can be easily inflated over the lower two-thirds of the curve and even in exercise one normally breathes over an area circumscribed by the lower 70 percent of the curve. (From R. M. Berne and M. N. Levy. *Physiology* [3rd ed.]. P. 565. © 1993 Mosby–Year Book, St. Louis.) B. Pressure-volume curve of the lung showing the inspiratory work done overcoming elastic forces (area OAECDO) and viscous forces (hatched area ABCEA). Line AEC is the compliance line. (Modified from J. B. West. *Respiratory Physiology: The Essentials* [4th ed.]. P. 112. © 1990 Williams & Wilkins, Baltimore.)

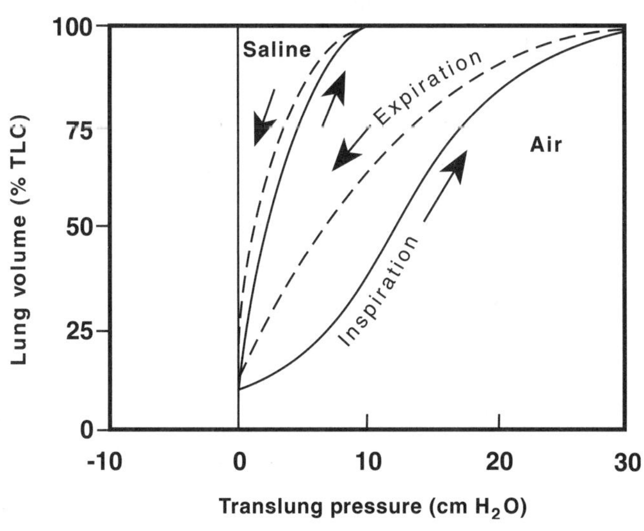

A

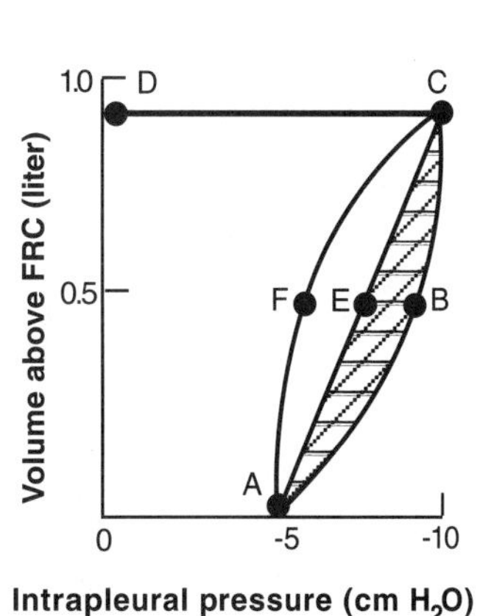

B

tween the water molecules. Both of these forces tend to collapse the lungs. Indeed, surface tension is such a potent force tending to collapse the lungs that surfactant (dipalmitoyl lecithin) is required. At rest, roughly two thirds of the tendency to collapse is attributable to surface tension and one third to elastic forces. At high lung volumes the inelastic supporting structures (e.g., collagen) must be stretched and elastic forces predominate.

Role of Surfactant in Decreasing Surface Tension

Normally, only a few millimeters of negative intrapleural pressure is required to keep the lungs expanded. In the absence of surfactant, however, lung expansion becomes much more difficult and much greater negative intrapleural pressures may be necessary to overcome the tendency of the alveoli to collapse.

The pressure required to keep the alveoli open can be determined using the **law of Laplace,** which states that the transalveolar pressure required to keep alveoli expanded is proportional to the tension in the alveolar wall divided by the diameter of the alveolus. Assuming that the alveoli are perfect spheres

$$\text{Transalveolar pressure} = 2\ (\text{wall tension})/\text{diameter}$$

This equation implies that the transalveolar pressure required to keep the alveoli open should increase linearly as diameter decreases. If wall tension remained constant, the pressure required to maintain alveolar patency would rapidly increase as the lung deflated, inevitably leading to collapse. This does not occur because surface tension is the major cause of wall tension and surface tension decreases as diameter decreases in the presence of surfactant. Therefore, the need for greater distending pressures at lower alveolar diameters is negated by the decrease in wall tension associated with increasing concentrations of surfactant. Surface tension in the alveoli of the lung has the special property of varying as a function of surface area, resulting in higher pressure requirements for inflation than for deflation (hysteresis). In the presence of surfactant, surface tension and consequently wall tension decrease as diameter decreases (instead of increasing as it does in a water-air interface) because the surfactant molecules become concentrated. The surface tension of a water-air interface is many times greater than the surface tension of a surfactant-air interface, which in turn is many times greater than an interface in which no surface tension interface exists. In a saline-filled lung the pressure-volume curve of the lung is shifted to the left (lower distending pressure) at any given volume, especially during inflation, because the alveolar air-liquid interface has disappeared and surface tension is nonexistent (see Fig. 4-11). In disease states (e.g., respiratory distress syndrome of the newborn) in which surfactant is absent, the pressure-volume curve of the lung is shifted to the right. Hysteresis is almost completely eliminated in the saline-filled lung, implicating the interaction between surface tension and surfactant as the major cause of hysteresis. Transorgan pressures are always measured from inside out; for example, translung pressure is equal to alveolar pressure minus pleural pressure.

Mechanism of Action of Surfactant

Surfactant works by forming a monomolecular layer between the fluid lining the alveoli and the air. The major surface tension reducing component of surfactant is dipalmitoyl phosphatidylcholine, which contains two 16-carbon fatty acid chains. The hydrophobic 16-carbon fatty acid chains are arranged vertically facing the alveolar gas and resist mechanical deformation; in particular they resist any attempt to be compressed during lung deflation. The hydrophilic phosphatidylcholine moiety faces the aqueous layer. This mechanically stabilizes the interface and lowers surface tension.

As the fatty acid chains become more crowded at lower lung volumes, they resist even more. During lung inflation surface tension increases as surfactant concentration decreases, opposing further inflation but also having the beneficial effect of pulling more surfactant to the air-liquid interface. As deflation begins, the surfactant immediately starts to be compressed at the air-liquid interface and this results in a rapid decrease in surface tension and therefore in the transpulmonary pressure required to keep the alveoli open at any given lung volume during deflation. This explains the difference in transpulmonary pressures seen between lung inflation and deflation. Since surface tension also tends to pull fluid into the alveoli from the alveolar wall, the decrease in surface tension seen with surfactant also serves to prevent pulmonary edema.

Other Factors That Affect Alveolar Stability

The decreasing surface tension seen during deflation in the presence of surfactant also stabilizes the smaller alveoli so that they do not shrink faster than the larger alveoli. In addition, alveoli share walls and are thus interdependent, which helps maintain patency of the alveoli because each helps stabilize the other.

Work of Breathing

The work of breathing refers to the energy expended to perform the following work (parenthetical remarks refer to Fig. 4-11B):

1. Expand the elastic tissues of the chest wall and lungs (compliance work = $P\Delta V$, area OAECDO)
2. Overcome the viscosity of the inelastic structures of the chest wall and lungs (i.e., move everything out of the way and then move it back; viscous work, see hatched area)
3. Move air against the resistance of the airways (resistance work)

The work of breathing normally accounts for about 2 to 3 percent of the body's total energy expenditures. At rest compliance work predominates while on exercise overcoming airway resistance is more important.

Pulmonary Gas Exchange

Alveolar Gas Exchange

Inspired air moves from the environment to the level of the terminal bronchioles by bulk flow. From there, the air diffuses into the alveoli to mix with air already present there. Oxygen then diffuses along its concentration gradient from the alveoli into the pulmonary capillaries. Carbon dioxide diffuses along its concentration gradient in the opposite direction.

At sea level the barometric pressure is equal to 760 mm Hg. Dry air is 79 percent N_2, 21 percent oxygen, 1 percent inert gases, and 0.04 percent carbon dioxide. **Dalton's law** states that each gas contributes to the total pressure in direct proportion to its relative concentration. The partial pressure of any gas can therefore be calculated by multiplying its percentage composition by the ambient barometric pressure. The partial pressure of N_2 at sea level is 600 mm Hg and that of oxygen is 160 mm Hg.

Important changes occur to the air as it traverses the conduction system to enter the gas-exchanging regions of the lungs. The temperature and humidity of the gas approach body temperature and 100 percent humidity. The contribution of the partial pressure or vapor pressure of water must then be taken into consideration when calculating the proportion of gases in the alveoli. The vapor pressure of water at 37°C is 47 mm Hg. The water vapor expands the volume of air and thereby dilutes the gases present in the inspired air. The sum of the partial pressures of the other gases present is now 760 – 47 = 713 mm Hg. Furthermore, carbon dioxide is being excreted into the lungs, causing further dilution of the inspired gases. The **alveolar gas equation** enables one to estimate the partial pressure of oxygen in the alveolus:

$$P_{A}O_2 = (BP - 47) \times F_{I}O_2 - (P_{A}CO_2/RER)$$

where $P_{A}O_2$ is the partial pressure of oxygen in the alveolus, BP is the barometric pressure, $F_{I}O_2$ is the fractional inspired oxygen concentration, and $P_{A}CO_2$ is the partial pressure of carbon dioxide in the alveolus. RER is the **respiratory exchange ratio** or **respiratory quotient** and is equal to the ratio of carbon dioxide given off by the body to the oxygen consumed by the body. This is usually calculated as the minute carbon dioxide production divided by the minute oxygen consumption ($\dot{V}CO_2/\dot{V}O_2$). The RER may change with alteration in diet and metabolism but is normally considered to average 0.8. The respiratory exchange ratio for glucose and other carbohydrates is 1.0, for fat 0.7, and for protein 0.8.

The **alveolar-arteriolar oxygen difference,** or **$P(A - a)O_2$,** is the difference between the partial pressure of oxygen in the alveoli as calculated by the alveolar gas equation and the arterial partial pressure of oxygen. Normally, this difference should not exceed 15 to 20 mm Hg. The $P(A{-}a)O_2$ results from the small amount of venous blood from the bronchial venules and thebesian (heart) veins that contaminates pulmonary venous outflow so that the PaO_2 is decreased, that is, physiologic shunt. The **arterial partial pressure of oxygen,** or PaO_2, is age dependent and can be estimated by the equation

$$PaO_2 = 103 - 0.4 \text{ (age in years)}$$

As oxygen diffuses from the alveoli into the capillaries the partial pressure of oxygen in the alveolar gas and that in the liquid equilibrate. As **Henry's law** states, the actual concentration of dissolved gas in a liquid is equal to the partial pressure of the gas in contact with the liquid multiplied by the solubility coefficient of the gas in that particular liquid. The solubility coefficient of oxygen in blood is 0.003 cc O_2/mm Hg/liter blood while that of carbon dioxide is 0.03 millimoles CO_2/mm Hg/liter blood. Note that the two solubilities use different units!

Diffusion Capacity

The gradient for diffusion is the partial pressure of the gas in one area minus the partial pressure in a second area divided by the distance (T) over which diffusion must occur:

$$\frac{P_1 - P_2}{T} = \Delta P / T$$

Factors that influence diffusion include

1. Temperature (assumed to remain constant at 37°C and, therefore, ignored hereafter)
2. Molecular weight of the gas (MW)
3. Distance over which diffusion must occur (T)

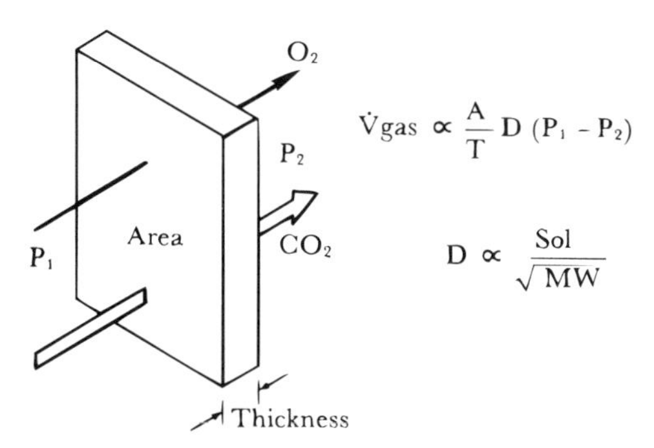

Fig. 4-12. Diffusion through a tissue sheet. The amount of gas ($\dot{V}_{gas}$) transferred is proportional to the area (A), a diffusion constant (D), and the difference in partial pressure ($P_1 - P_2$) and is inversely proportional to the thickness (T). The constant is proportional to the gas solubility (Sol) but inversely proportional to the square root of its molecular weight (MW). (From J. B. West. *Respiratory Physiology: The Essentials* [4th ed.]. P. 22. © 1990 Williams & Wilkins, Baltimore.)

4. Cross-sectional area available for diffusion to occur (A)
5. Solubility of the gas in the fluid (S)
6. The pressure gradient driving diffusion (ΔP)

The overall rate of diffusion at any given temperature is therefore proportional to

$$\frac{\Delta P \times A \times S}{T \times MW}$$

The **diffusing capacity** (or volume/mm Hg/min/pair of lungs) is the volume of a gas that diffuses through the respiratory membrane each minute with a pressure difference of 1 mm Hg (Fig. 4-12).

For normal lungs, the diffusing capacity for oxygen is 21 cc/mm Hg/min at rest. It may be increased by recruitment and dilatation of pulmonary capillaries. The diffusing capacity for carbon dioxide is so high that it cannot be measured due to its lipid solubility. Carbon monoxide is less lipid soluble and, therefore, more likely to be diffusion limited. Because of its high affinity for hemoglobin, carbon monoxide is often used to calculate the diffusion capacity of the lungs in an attempt to determine whether a limitation to diffusion exists. Only the inspiratory carbon monoxide pressure need be measured since carbon monoxide binds to hemoglobin with such great affinity that the partial pressure of CO in the blood is almost zero.

Ventilation-Perfusion (Mis)Match

Ventilation is better in the more dependent portions of the lung. Perfusion is better in the more dependent portions of the lung. The ventilation to perfusion ratio, V/Q, is best at the apex of the lung. Due to the effect of gravity, perfusion is greatest at the bases. Ventilation is also greatest at the lung bases, in this case because of the action of the diaphragm pulling the alveoli open at the bases. The ratio of ventilation to perfusion, however, is greatest at the apex of the lungs because perfusion is more affected by gravity than ventilation is by the action of the diaphragm. Ideally, a ventilation to perfusion ratio of 1 should exist with ventilation being perfectly matched to perfusion. However, as Fig. 4-13 illustrates, the normal relationship between minute alve-

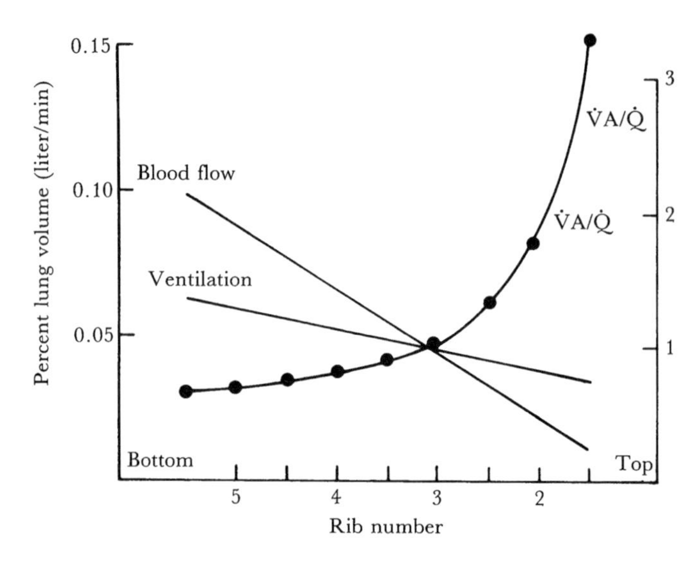

Fig. 4-13. Distribution of ventilation and blood flow down the upright lung. Note that the ventilation-perfusion ratio decreases down the lung. (From J. B. West. *Ventilation/Blood Flow and Gas Exchange* [4th ed.]. Oxford: Blackwell, 1985.) See Fig. 4–14.

olar ventilation ($\dot{V}_A$) and minute perfusion ($\dot{Q}$) in the upright lung is such that ventilation is roughly three times greater than necessary at the apex and only 60 percent of that required at the bases. During exercise, blood flow to the upper lung increases minimizing ventilation-perfusion mismatch.

Oxygen exchange is exquisitely sensitive to V/Q changes. Carbon dioxide excretion, while less sensitive to minor alterations in V/Q relationships (since increasing the rate and depth of ventilation can compensate to a large extent), is affected in extreme examples.

If the lung is perfused but not ventilated, then the V/Q ratio is zero and no gas exchange takes place. Similarly, if the lung is ventilated but not perfused then the ratio approaches infinity and no gas exchange takes place (Fig. 4-14). At one extreme a sampling of alveolar air will resemble venous blood; and at the other inspired air.

If ventilation and perfusion are ideally matched, then a ratio of one exists and the efficiency of gas transport is maximized. The normal response to either hypoxia or

Fig. 4-14. The normal PO_2–PCO_2, $\dot{V}/\dot{Q}$ diagram. (From A. C. Guyton. *Textbook of Medical Physiology* [8th ed.]. Philadelphia: Saunders, 1991. P. 431.)

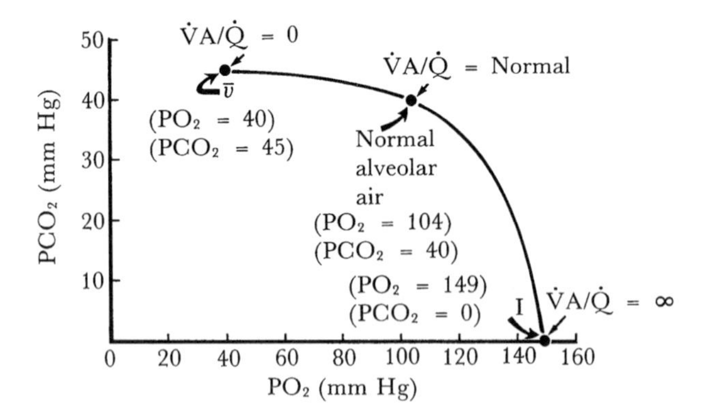

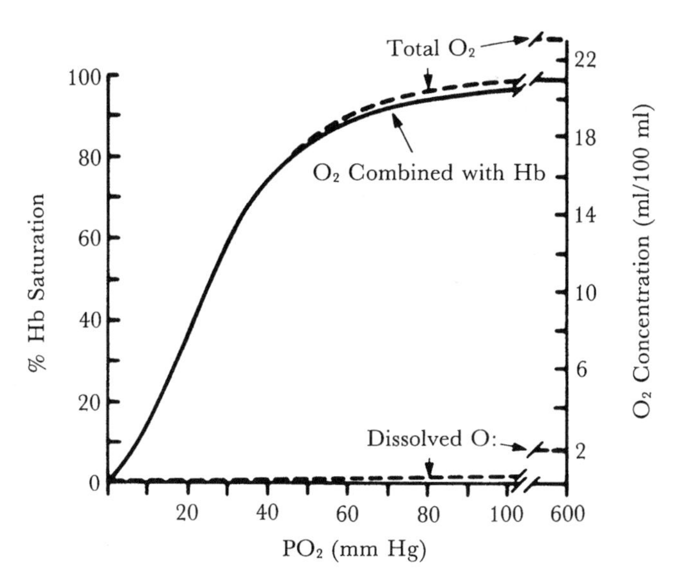

Fig. 4-15. Oxygen dissociation curve (solid line) for pH 7.4, PCO_2 40 mm Hg, and 37°C. The total blood oxygen concentration is also shown for a hemoglobin concentration of 15 g/dl blood. (From J. B. West. *Respiratory Physiology: The Essentials* [4th ed.]. P. 70. © 1990 Williams & Wilkins, Baltimore.)

Respiratory

hypercarbia is to increase the rate and depth of ventilation, that is, to hyperventilate. This works quite well for carbon dioxide since underventilation of one region of lung, perhaps secondary to obstruction of a bronchus, is readily compensated for by hyperventilation of other regions of the lung. The arterial carbon dioxide would then be the arithmetic average of the blood coming from the under- and overventilated regions of the lungs.

The **oxygen content** of blood, however, is only minimally affected by dissolved oxygen and is instead dependent on the amount of oxygen bound to hemoglobin (Fig. 4-15 and Table 4-1). Blood that has been exposed to a partial pressure of oxygen sufficient to saturate hemoglobin will not carry substantially more if either ventilation or FIO_2 is increased. Similarly, if blood with poorly oxygenated hemoglobin is mixed with well-oxygenated blood the partial pressure of oxygen that results will be remarkably affected, as oxygen-starved hemoglobin soaks up the small amount of oxygen dissolved in the well-oxygenated blood and then begins to pull oxygen off of the saturated hemoglobin until a new equilibrium is reached (Fig. 4-16).

Normally, the lungs are able to keep the V/Q ratio within reasonable limits. Underventilation of a region of lung causes blood to be shunted preferentially to other better ventilated regions of the lungs as the decrease in PAO_2 and increase in PCO_2 cause reflex constriction of the pulmonary vessels. Similarly, if perfusion is compromised to a certain region of the lungs, a reflex bronchoconstriction will result. Nonetheless, these changes will only minimize and not eliminate V/Q mismatch.

Physiologic Shunts and Physiologic Dead Space

A physiologic shunt occurs when the ratio of ventilation to perfusion is abnormally low. It refers to that fraction of the blood passing through the pulmonary vasculature that does not become fully oxygenated and assumes that all the hypoxemia results from blood passing through unventilated alveoli. The bronchial vessels constitute a normal component of physiologic shunt, accounting for roughly 2 percent of the cardiac output. The physiologic shunt can be calculated using the following equation:

Table 4-1. Oxygen content of whole blood

PO_2 (mm Hg)	% saturation of Hb[a]	O_2 content of Hb (ml/dl)[b]	Dissolved O_2 (ml/dl)
10	13.5	2.71	0.03
20	35	7.04	0.06
27	50	10.05	0.08
40	75	15.08	0.12
60	89	17.89	0.18
90	96.5	19.40	0.27
100	97.5	19.60	0.30
600	99+	19.90	1.80

[a]pH = 7.40, temperature = 38° C.
[b]Estimated using 1.34 ml O_2/g hemoglobin (Hb) and 15g Hb/dl blood.

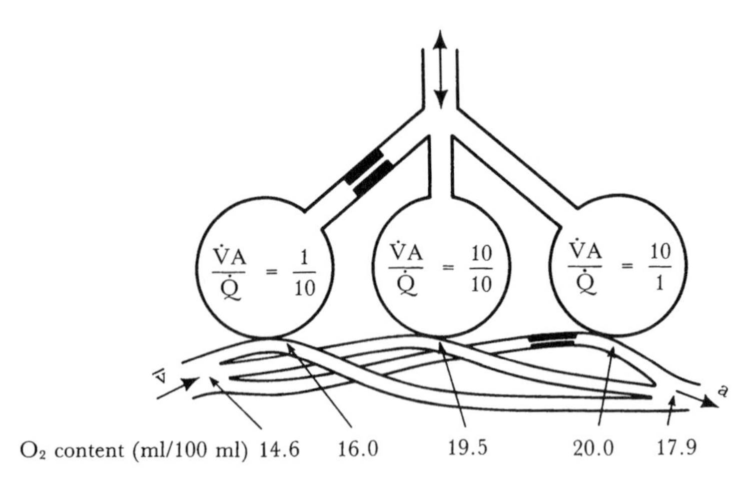

Fig. 4-16. Depression of arterial PO_2 by mismatching of ventilation and blood flow. The lung units with a high ventilation-perfusion ratio add relatively little oxygen to the blood compared with the decrement caused by alveoli with a low ventilation-perfusion ratio. $\bar{V}$ = mixed venous blood; a = arterial blood. (Modified from J. B. West. *Ventilation/Blood Flow and Gas Exchange* [4th ed.]. Oxford: Blackwell, 1985.)

$$\frac{Q_{ps}}{Q_T} = \frac{C_iO_2 - C_aO_2}{C_iO_2 - C_VO_2}$$

where

Q_{ps} is the physiologic shunt blood flow.

Q_T is the total blood flow through the lungs.

C_iO_2 is the total oxygen content of blood draining from an ideal alveolus (calculated using the alveolar gas equation to determine P_AO_2 and then using the calculated P_AO_2 to determine the total oxygen content of the blood, i.e., the total of the amount bound to hemoglobin and the amount free in solution at that PaO_2).

C_aO_2 is the arterial oxygen content.

C_VO_2 is the venous oxygen content.

The physiologic shunt equation does not differentiate between the three types of shunts: intrapulmonary shunts (e.g., pulmonary artery to pulmonary vein), anatomic shunts (e.g., right-to-left shunting of blood in the heart), and physiologic shunting due to inadequate ventilation relative to perfusion.

Physiologic dead space occurs when the V/Q ratio is greater than normal. It therefore refers to wasted ventilation much as the physiologic shunt refers to wasted perfusion. It is equal to the dead space ventilation (including unperfused alveoli) plus the amount of alveolar ventilation that exceeds that required to adequately supply the blood flowing through the lungs. The **Bohr equation** can be used to estimate the physiologic (or total) dead space. In normal adults, the anatomic dead space comprises most if not all of the physiologic dead space. The total ventilation (VT) is equal to the alveolar ventilation (VA) plus dead space ventilation (VD). A tidal breath of 500 ml normally consists of 350 cc VA and 150 cc VD.

The Bohr equation calculates the ratio of dead space to total ventilation by attributing all of the lowering of expiratory partial pressure of carbon dioxide ($PECO_2$) relative to arterial partial pressure of carbon dioxide ($PaCO_2$) to unperfused alveoli and to the anatomic dead space.

$$V_D/V_T = \frac{P_ACO_2 - P_ECO_2}{P_ACO_2} = \frac{PaCO_2 - P_ECO_2}{PaCO_2}$$

The reasoning is as follows. If no gas exchange is occurring in a portion of lung because it is not being perfused adequately, that portion will not add carbon dioxide to the alveolar air. Because carbon dioxide diffuses so readily across the respiratory membrane, its partial pressure in the expiratory air should equal that in the pulmonary capillary blood and any discrepancy must be due to air that has not been exposed to pulmonary capillary blood. Arterial blood is used in the equation since pulmonary capillary blood is difficult to obtain.

Blood Gas Transport and Tissue Gas Exchange

Carbon Dioxide Transport and Exchange

Carbon dioxide transport and exchange are covered in detail in Chap. 5. Briefly, $PaCO_2$ is directly proportional to P_ACO_2 and to the effective ventilation, with the lungs excreting the equivalent of 10,000 mEq **carbonic acid (H_2CO_3)** per day as carbon dioxide. Carbon dioxide is in equilibrium with carbonic acid, hydrogen ion (H^+), and bicarbonate (HCO_3^-) in the blood, the initial reaction of which is catalyzed by the enzyme carbonic anhydrase (CA), located in the red blood cells but not in the plasma (Fig. 4-17):

$$CO_2 + H_2O \overset{CA}{\Longleftrightarrow} H_2CO_3 \Longleftrightarrow H^+ + HCO_3^-$$

As carbon dioxide from tissues diffuses into the plasma and subsequently into the RBCs along its concentration gradient, some of the carbon dioxide combines with hemoglobin facilitating the offloading of oxygen (the **Bohr effect**), some combines with plasma proteins, and some remains dissolved in solution, but the overwhelming majority undergoes the reaction catalyzed by CA in the RBC to form H^+ and HCO_3^-. (Note: Very little H_2CO_3 is formed in the plasma because in the absence of CA the hydration of carbon dioxide proceeds very slowly. Furthermore, carbonic acid is

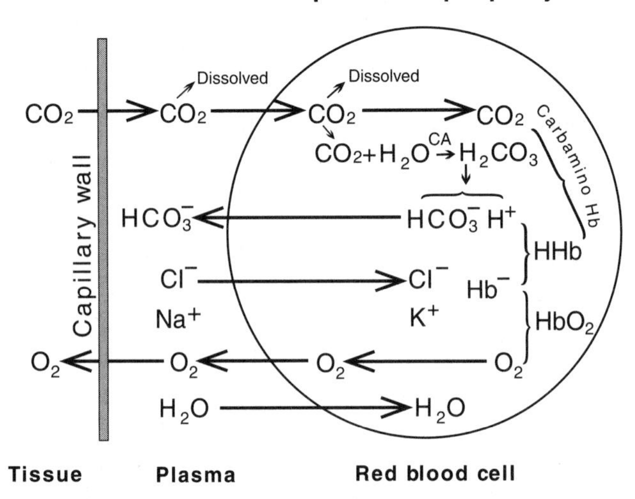

Fig. 4-17. Uptake of carbon dioxide and liberation of oxygen in systemic capillaries. Exactly opposite events occur in the pulmonary capillaries. (From J. B. West. *Respiratory Physiology: The Essentials* [4th ed.]. P. 75. © 1990 Williams & Wilkins, Baltimore.)

rapidly ionized in the second reaction, which does not require a catalyst.) Bicarbonate then diffuses out of the RBC along its concentration gradient and chloride ion diffuses into the RBC (**the chloride shift**) in a process facilitated by a special **bicarbonate-chloride carrier protein** in the RBC membrane. The chloride shift acts to maintain electroneutrality as predicted by the **Gibbs-Donnan equilibrium** (see Chap. 3).

When the RBC enters the pulmonary capillaries, oxygen diffuses into the cell and combines with hemoglobin to form oxyhemoglobin, which releases carbon dioxide (the **Haldane effect**) and H^+ (since oxyhemoglobin is a weaker base than deoxyhemoglobin). As carbon dioxide diffuses out of the RBC and into the alveolus along its concentration gradient, HCO_3^- and H^+ combine to form additional carbon dioxide, which is also free to follow the concentration gradient out of the cell. The bicarbonate concentration in the RBC falls and chloride therefore leaves the cell to maintain electroneutrality.

The carbon dioxide content of the arterial blood is carried in three forms: 5 percent as dissolved gas, 90 percent as bicarbonate, and 5 percent combined with proteins, particularly hemoglobin. Dissolved carbon dioxide, like oxygen, obeys Henry's law but is 20 times as soluble as oxygen. It is extremely lipid soluble and it therefore equilibrates rapidly across cell membranes. In venous blood, substantially more carbon dioxide is carried in dissolved form and in combination with hemoglobin, but the majority is still carried as HCO_3^-. Of the carbon dioxide that is ultimately excreted in the lung, that is, the arterial-venous carbon dioxide difference, 7 percent is transported in dissolved form, 23 percent as carbaminohemoglobin, and 70 percent as bicarbonate.

Oxygen Transport and Exchange

Oxygen is transported in the blood in two forms—as dissolved gas and bound to hemoglobin. Oxygen is less soluble in blood than carbon dioxide. The solubility of oxygen in blood is equal to 0.003 ml oxygen per mm Hg/dl blood. Assuming an arterial

PaO_2 of 100 mm Hg, only 0.30 ml oxygen as dissolved gas is carried per deciliter of blood (for comparison, venous blood contains 2.7 ml CO_2/dl and arterial blood 2.4 ml CO_2/dl as dissolved gas).

In a person with a normal hematocrit, the hemoglobin concentration in blood is 15 g/dl blood. Because hemoglobin can carry 1.34 ml oxygen per gram hemoglobin, 20.1 ml oxygen can theoretically be carried bound to hemoglobin. At a PaO_2 of 100 mm Hg, hemoglobin is over 97 percent saturated (see Table 4-1). Thus, increasing the partial pressure of oxygen above this level can add only a small fraction of additional oxygen. Similarly, the solubility of oxygen in blood is so low that little benefit is gained by increasing the amount dissolved in blood. The actual amount of oxygen carried by hemoglobin is calculated using the following equation:

$$1.34 \text{ ml } O_2/\text{g Hb} \times \text{g Hb/dl blood} \times \%O_2 \text{ saturation}$$

The total oxygen content of blood is therefore equal to the amount bound to hemoglobin plus that dissolved as gas. At a PaO_2 of 100 mm Hg and a hemoglobin concentration of 15 g/dl, 19.7 ml oxygen is bound to hemoglobin and 0.30 ml is dissolved in the blood.

At rest, one quarter of the oxygen content of blood is extracted by the tissues in one circulation or 5 ml oxygen per deciliter of a total of approximately 20 ml/dl present in arterial blood. This results in a drop of oxygen partial pressure from 100 mm Hg (97% saturation) in the arteries to 40 mm Hg (75% saturation) in the venous blood. On maximal exercise in a normal nonathlete, the extraction ratio increases to approximately 50 percent, enabling the tissues to utilize 10 ml oxygen per deciliter (resulting in a PvO_2 of 27 and percent saturation of 50 percent). Because cardiac output also increases approximately threefold, total oxygen delivery to the tissues can increase sixfold.

Hemoglobin

Hemoglobin is composed of two alpha polypeptide chains and two beta polypeptide chains, each of which contains a separate heme group. A total of four oxygen molecules can, therefore, be carried by a single hemoglobin molecule.

Binding of the initial oxygen molecule to hemoglobin results in a change in configuration of hemoglobin. This change increases the affinity of hemoglobin for oxygen and is an example of positive cooperativity. Such enhanced binding yields the sigmoidal curve characteristic of such an interaction (Fig. 4-15). The $P_{50}O_2$ is defined as the PO_2 at which hemoglobin is 50 percent saturated. At pH 7.4 the $P_{50}O_2$ is 27 mm Hg.

The oxyhemoglobin dissociation curve is shifted to the right by acidosis, hypercapnia, elevated temperature, or increased levels of 2,3-diphosphoglycerate (2,3-DPG). The rightward shift caused by increased levels of carbon dioxide is referred to as the **Bohr effect** (Fig. 4-18).

The hemoglobin dissociation curve is shifted to the left by alkalosis, hypocapnia, hypothermia, and decreased levels of 2,3,-DPG.

At PO_2s greater than 90 mm Hg hemoglobin is saturated irrespective of any shift in the hemoglobin dissociation curve. Similarly, at extremely low PO_2s shifts in the curve have little effect. The rightward shift that occurs in the tissues as the pH decreases and carbon dioxide levels increase improves tissue oxygen unloading. The leftward shift that occurs in the pulmonary capillaries does not adversely affect oxygen uptake because oxygen levels are still sufficiently high to saturate hemoglobin.

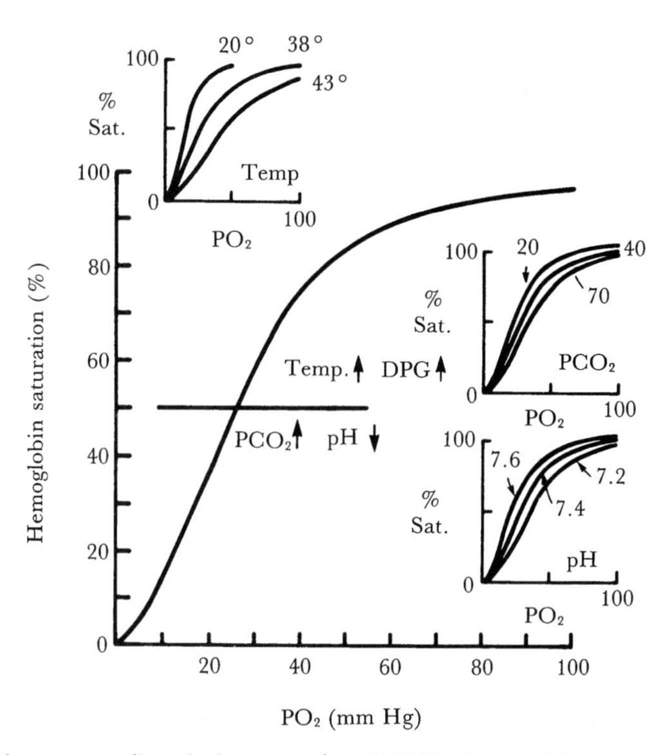

Fig. 4-18. Shift of the oxygen dissociation curve by pH, PCO_2, temperature, and 2,3-diphosphoglycerate (DPG). A shift to right indicates increased hydrogen ion, increased carbon dioxide, increased temperature, and/or increased 2,3-DPG. (From J. B. West. *Respiratory Physiology: The Essentials* [4th ed.]. P. 73. © 1990 Williams & Wilkins, Baltimore.)

Deoxyhemoglobin takes up oxygen and releases H^+ and carbon dioxide as it traverses the pulmonary capillaries. As oxygen binds to hemoglobin, hemoglobin loses its affinity for carbon dioxide (the **Haldane effect**). The H^+ released inside the erythrocyte combines with bicarbonate to form additional carbonic acid, which is then dehydrated to form additional carbon dioxide and water. Carbon dioxide diffuses out of the red cells into the alveoli and is excreted by the lungs. This results in the conversion of venous blood with a pH of 7.2, PCO_2 of 45, and PO_2 of 40 to arterial blood with a pH of 7.40, PCO_2 of 40, and PO_2 of 97.

In the tissue capillaries the arterial blood enters an area of low PO_2, high PCO_2, and low pH, all of which act to facilitate oxygen unloading. Deoxyhemoglobin then combines with H^+ (by protonation of the imidazole group of histidine) and with carbon dioxide (by formation of carbaminohemoglobin). The low pH and high PCO_2 facilitate oxygen unloading by causing a rightward shift in the hemoglobin dissociation curve. Thus, hemoglobin acts both to supply the tissues with oxygen and to buffer the excretory acid load of the tissues.

Regulation of Respiration

General Principles

Regulation of respiration includes coordinated activity on the part of the central regulatory centers, the central and peripheral chemoreceptors, lung receptors, and the respiratory muscles. Both the rate and depth of breathing are regulated so that $PaCO_2$

(used here interchangeably with P_ACO_2) is maintained close to 40 mm Hg. This automatically sets PaO_2 at an appropriate level as predicted by the alveolar gas equation given an adequate inspiratory PO_2.

Central Regulatory Centers

Respiration is a reflex activity in humans. The rate, depth, and rhythm of inspiration and expiration are controlled by diffuse nerve collections in the medulla and pons. The neurons responsible for inspiration and expiration appear to be organized in diffuse networks. Central control of ventilation can be divided into conscious or voluntary control and unconscious or reflex control.

Voluntary Control

Little is known about the voluntary control system except that higher centers in the thalamus and cerebral cortex are involved and that these are necessary to coordinate breathing in relation to such volitional actions as talking, singing, suckling, swallowing, coughing, sneezing, defecation, parturition, anxiety, and fear. Section of the brainstem above the medulla will leave rhythmic breathing essentially intact. The cortex can override these lower centers for voluntary control on a temporary basis, as in voluntary breath holding or hyperventilation. Projections from the cerebral motor cortex that subserve volitional control descend to the respiratory neurons in the brainstem via the corticobulbar tracts and to the spinal motor neurons via the corticospinal tracts, which are located in the dorsolateral columns of the spinal cord. Discrete lesions in the spinal cord may abolish reflex respiratory activity but leave volitional control intact **(Ondine's curse).** Sectioning the brain of an experimental animal just above the level of the lower pons results in inspiratory gasps or apneuses interrupted by transient expiratory efforts. This finding resulted in the initial description of an **apneustic center,** but few investigators still believe that such a discrete center exists.

Reflex Control of Ventilation in the Medulla and Pons

The medulla contains respiratory neuronal groups that predominantly affect the respiratory rhythm. These neuronal groups are intimately associated with the reticular activating system (which affects the state of alertness of the brain). The respiratory neurons are located mainly in two ill-defined areas, called the nucleus of the tractus solitarius, located dorsally near the exit of the ninth cranial nerve, and the nucleus retroambiguus, located ventrally and extending from the level of the first cervical spinal segment to the caudal border of the pons. The dorsal group discharges mainly in inspiration. The ventral group contains both inspiratory and expiratory neurons. Both act to control muscles of the chest wall and abdomen. Efferent fibers from both groups cross the midline and synapse on contralateral spinal motor neurons. Afferent nerve fibers from the peripheral chemoreceptors, baroreceptors, and pulmonary mechanoreceptors synapse in the dorsal motor group via the vagus and glossopharyngeal nerves. The mechanism by which these networks mediate rhythmic breathing is unknown. However, the dorsal respiratory group continues to generate the inspiratory action potentials even when all sources of nerve input are severed and the medulla is sectioned above and below.

A network of neurons located in the upper pons, the **pneumotaxic center,** acts to inhibit inspiration and regulate the respiratory rate. When the pneumotaxic center is inactivated, inspiration becomes prolonged (apneustic breathing with inspirations lasting tens of seconds).

In order to function properly, the central regulatory systems depend on central and peripheral sensory receptors for information regarding the effectiveness of the respiratory efforts and on the neuromuscular effector organs to actually implement changes in respiratory effort.

Central Chemoreceptors

The central chemoreceptors respond to changes in hydrogen ion concentration $[H^+]$, and are involved in the minute-by-minute control of ventilation. An increase in the concentration of H^+ in the brain extracellular fluid (ECF) stimulates ventilation while a decrease inhibits it. Carbon dioxide levels affect the central chemoreceptors only to the extent that H^+ ion concentration is affected. Central chemoreceptors are located in the medulla.

Cerebrospinal fluid (CSF) composition is the most important determinant governing the hydrogen ion and carbon dioxide concentrations seen by the central receptors. Local metabolism and adequacy of perfusion can also affect the composition of the extracellular fluid bathing the central receptors. Respiratory acclimatization appears to be mediated by changes in the composition of the CSF.

The composition of the CSF, in turn, is governed by the blood-brain barrier and by active pumping of ions by the choroid plexi. The blood-brain barrier separates the blood and cerebrospinal fluid compartments. This barrier is relatively impermeable to H^+ and HCO_3^- although carbon dioxide readily diffuses across the barrier. When blood PCO_2 rises, carbon dioxide diffuses into the CSF from the cerebral blood vessels and liberates H^+, which stimulates the central chemoreceptors and ultimately results in hyperventilation.

Normally, the pH of CSF is 7.32 and is much more labile than blood pH for a given change in PCO_2. The CSF has less protein and therefore less buffering capacity. Because the choroid plexus contains active HCO_3^- secretory pumps, chronic elevations in PCO_2 may be compensated for by increased pumping of HCO_3^- into the CSF. Furthermore, this pumping occurs more rapidly and more completely than does renal compensation.

Peripheral Chemoreceptors

The **carotid body** chemoreceptors located at the bifurcation of the common carotids and the aortic bodies located above and below the aortic arch constitute the peripheral chemoreceptors.

The carotid body chemoreceptors respond to decreases in arterial PO_2 and pH, and to increases in PCO_2, by increasing the rate and depth of respiration. The carotid bodies have the greatest blood flow per unit tissue of any tissue in the body and are therefore able to maintain a very low arteriovenous (A-V) oxygen difference despite a high metabolic rate. Because of the very small A-V oxygen difference, the carotid receptors respond to arterial PO_2 rather than venous PO_2.

The carotid body chemoreceptors are responsible for all increases in ventilation in response to arterial hypoxia. Responses to decreases in PaO_2 are nonlinear and little response of any kind is seen until the arterial PO_2 is less than 100 mm Hg (Figs. 4-19 and 4-20). Complete loss of hypoxic drive has been shown with bilateral carotid body resection.

The **aortic body** chemoreceptors respond to increased levels of carbon dioxide by increasing the rate and depth of respiration. In humans the carotid but not the aortic bodies respond to a fall in arterial pH. The response occurs regardless of the etiology of the acid-base disturbance, whether respiratory or metabolic.

Only 20 percent or less of the ventilatory response to increased carbon dioxide is attributed to peripheral chemoreceptors. However, they may be important in match-

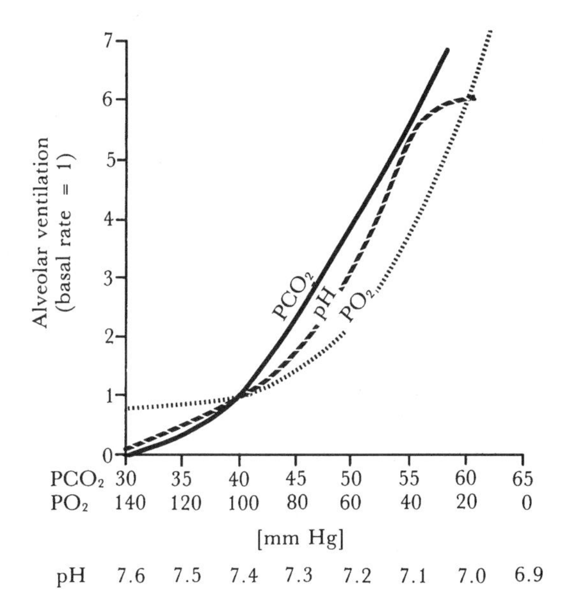

Fig. 4-19. Approximate effects of arterial PCO_2 and arterial pH on alveolar ventilation. (From A. C. Guyton. *Textbook of Medical Physiology* [8th ed.]. Philadelphia: Saunders, 1991. P. 447.)

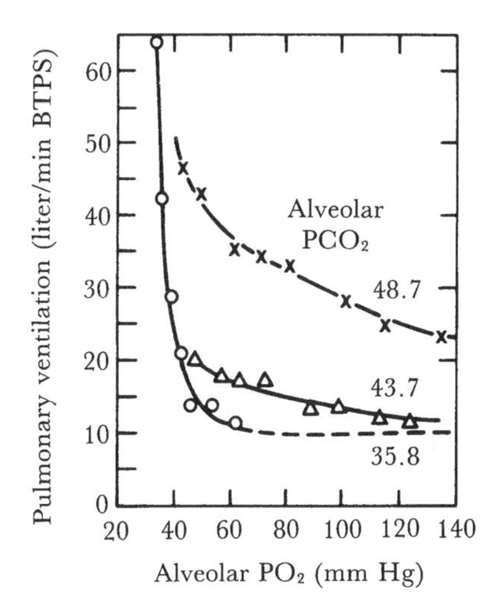

Fig. 4-20. Hypoxic response curves. Note that when the PCO_2 is 36 mm Hg, almost no increase in ventilation occurs until the PO_2 is reduced to about 50 mm Hg. (Modified from H. H. Loeschke and K. H. Gertz. *Arch. Ges. Physiol.* 267:460, 1958.)

ing ventilatory responses to abrupt changes in PCO_2 and they do account for the entire ventilatory response to hypoxia.

Ventilatory responses to changes in PO_2, PCO_2, and pH may be potentiated or abrogated by changes in the other two components of ventilatory drive.

Pulmonary Receptors

There are three types of pulmonary receptors described. Afferent impulses from all three receptor types travel to the CNS in the vagus nerve.

1. **Pulmonary stretch receptors** are located in the smooth muscle and respond to distention of the lung with sustained firing. Stimulation of these receptors results in slowing of the respiratory rate due to an increase in expiration time. This is known as the **Hering-Breuer reflex.** The normal inspiration to expiration ratio (I : E) is 1 : 2.
2. **Irritant receptors** are believed to lie interposed between airway epithelial cells and are stimulated by noxious gases, cigarette smoke, dusts, and cold air. Reflex effects include bronchoconstriction and hyperpnea.
3. **Pulmonary C fiber** stimulation results in rapid shallow breathing with more intense stimulation resulting in apnea. Engorgement of pulmonary capillaries and increases in the interstitial fluid volume of the alveolar walls activate these receptors. Bronchial C fibers perform an analogous function in the bronchi.

Ventilatory Responses to Alterations in PCO_2, PO_2, and pH

Under normal conditions, PCO_2 is the most important variable in the regulation of ventilation. $PaCO_2$ is normally maintained to within a few mm Hg of normal value.

Figure 4-21 illustrates the ventilatory response to rising concentrations of inspiratory carbon dioxide (with alveolar PO_2 held constant). Note that alveolar end-tidal PO_2 and PCO_2 are generally accepted to reflect arterial levels since the end-tidal air is presumably alveolar air that has equilibrated with arterial blood.

Altering alveolar PO_2 while maintaining alveolar PCO_2 at 36 mm Hg results in the ventilatory response curve shown in Fig. 4-20. This illustrates that alveolar PO_2 can be reduced to approximately 50 mm Hg before any appreciable increase in ventilation. Increasing the PCO_2 increases the ventilation at any given PO_2; however, when the $PaCO_2$ is increased, any reduction in PO_2 below 100 mm Hg will cause some stimulation of ventilation, in contrast to the situation when PCO_2 is normal.

The role of hypoxia in the day-to-day control of ventilation is therefore fairly small. However, in some patients with pulmonary disease the hypoxic drive becomes very important (as indeed it does in normal people when ascending to high altitudes) and in some patients it is absent.

The cause of increased ventilation with exercise is unknown. PO_2 actually increases slightly with exercise, PCO_2 actually decreases, and pH is generally unchanged until lactate accumulation occurs.

Fig. 4-21. Ventilatory response to different concentrations of inspired carbon dioxide. Note that airway obstruction reduces the response. (From J. B. West. *Respiratory Physiology: The Essentials* [4th ed.]. P. 70. © 1990 Williams & Wilkins, Baltimore.)

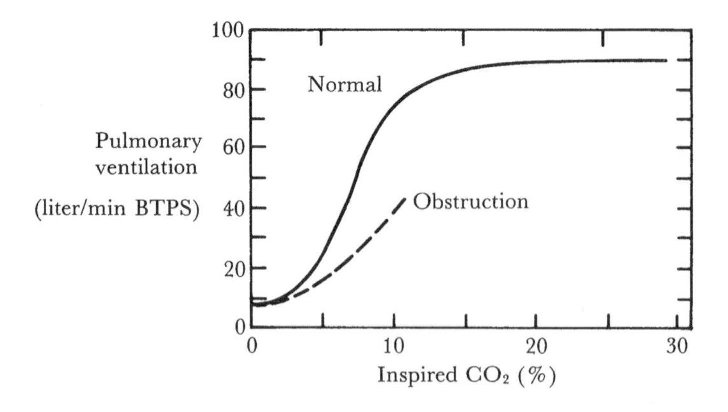

Acclimatization to High Altitude

The following responses occur on ascent to high altitude:

1. Acutely, alveolar ventilation increases by up to 65 percent (the response is limited by the resulting decrease in PCO_2 and increase in pH).
2. Chronically, alveolar ventilation may increase by three- to sevenfold as the body apparently develops tolerance for the resulting alkalosis and decreased PCO_2.
3. Hematocrit increases from 40 to 45 percent to 60 to 65 percent **(polycythemia).**
4. Hemoglobin concentration increases by 20 to 30 percent.
5. Circulatory blood volume increases by up to 50 to 90 percent.
6. Levels of 2,3-DPG increase, causing decreased affinity of hemoglobin for oxygen, thus enhancing tissue oxygen unloading.
7. Diffusing capacity for oxygen increases in a manner similar to that seen in exercise. This effect may be secondary to the increased capillary blood volume, to increased number and size of pulmonary capillaries, or to increased pulmonary artery pressure.
8. Increased tissue vascularity can be demonstrated.
9. Acclimatization at the cellular level, at least in native animals, may also occur with increased numbers of mitochondria and increased levels of certain cellular oxidative enzymes.

General and Cellular Nonrespiratory Lung Functions

The lungs have metabolic, immunologic, and filtration functions. They filter out small blood clots (small pulmonary emboli), preventing the serious complications of systemic embolization. Bronchial secretions contain **secretory immunoglobulins** (IgA) and other substances that serve to resist infection. **Pulmonary alveolar macrophages** are actively phagocytic and remove bacteria and small particles inhaled by the lungs as well as performing the other functions of macrophages (e.g., attraction of polymorphonuclear leukocytes, release of vasoactive and chemotactic substances). The **metabolic and endocrine functions** of the lung include production of factor VIII and surfactant, conversion of angiotensin I to angiotensin II by converting enzyme, and synthesis of prostaglandins, histamine, and kallikrein in addition to a number of other substances. The lungs contain a **fibrinolytic system** that lyses clots in the pulmonary veins. Metabolic breakdown of certain substances also occurs in the lungs, the significance of which is unknown.

Questions

Directions: Each of the numbered items or incomplete statements in this section is followed by answers or by completions of the statement. Select the one lettered answer or completion that is best in each case.

1. Decreasing the diameter of a bronchus by 50 percent will cause the resistance to change by a factor of
   A. 2
   B. 4
   C. 8

D. 16
E. 32

2. Half the length of a bronchus is surgically excised. Simultaneously, a constriction occurs at the site of the surgery and the diameter of the bronchus is decreased by 50 percent. What is the combined result on resistance of the airway?
   A. Increase by 3-fold
   B. Increase by 2-fold
   C. Increase by 4-fold
   D. Increase by 16-fold
   E. Increase by 8-fold
3. Alveolar ventilation is acutely decreased to half of normal. Which of the following statements most accurately describes the resulting change in $PaCO_2$?
   A. $PaCO_2$ doubles immediately
   B. $PaCO_2$ halves immediately
   C. $PaCO_2$ doubles over the next several minutes
   D. $PaCO_2$ halves over the next several minutes
   E. $PaCO_2$ doubles, then continues to rise until ventilation normalizes
4. A patient has a unilateral pulmonectomy. What is the predicted effect on the compliance and elastance of the remaining lung?
   A. Compliance and elastance are halved
   B. Compliance is doubled and elastance is halved
   C. Compliance is doubled and elastance is unchanged
   D. Compliance is halved and elastance is doubled
   E. Compliance is halved and elastance is unchanged
5. The partial pressure of oxygen in the alveolus of a normal lung at sea level (when $PaCO_2$ is 40 mm Hg and the respiratory quotient is one) is
   A. 149
   B. 110
   C. 100
   D. 95
   E. 75
6. Which of the following statements is **not** correct?
   A. Carbon monoxide diffuses more rapidly than oxygen and carbon dioxide
   B. Carbon monoxide has a higher affinity for hemoglobin than does oxygen
   C. Low concentrations of carbon monoxide are used to calculate the diffusing capacity of the lung by the single-breath method
   D. In carbon monoxide poisoning, the victim's PaO_2 remains normal and breathing continues
   E. Only the partial pressure of carbon monoxide in the blood must be measured to calculate the diffusion gradient between alveolar air and the blood
7. The largest possible tidal volume is defined as
   A. Functional residual capacity
   B. Inspiratory capacity
   C. Inspiratory reserve volume + expiratory reserve volume
   D. Total lung capacity
   E. Vital capacity
8. A patient undergoes pulmonary function testing. The results include the following: Total lung volume = 2.4 liters; rate = 16/min; change in intrapleural pressure = 3 cm H_2O; tidal volume = 75 cc. The compliance of the lung is
   A. 0.04 cm H_2O/cc
   B. 1.2 liters/min
   C. 25 cc/cm H_2O
   D. 800 ml/cm H_2O
   E. 0.05 liters/cm H_2O

9. Tuberculosis develops in areas of the lung with high oxygen tension. In what lung area would a bat most likely develop tuberculosis?
   A. Apex
   B. Base
   C. Carotid sinus
   D. Midlung
   E. Trachea and mainstem bronchi
10. Dyspnea is frequently associated with left heart failure and interstitial lung disease. It is associated with
    A. Bronchial C-fibers
    B. Irritant receptors
    C. Pulmonary C-fibers
    D. Pulmonary stretch receptors
    E. None of the above
11. A previously healthy patient on 100% oxygen has a PaO_2 of 100 mm Hg. Which of the following is probably true?
    A. Hemoglobin Kansas (an abnormal hemoglobin with increased affinity for oxygen) is present
    B. Mild to moderate pulmonary fibrosis is present
    C. Right-to-left shunting is present
    D. The patient has normal lungs
    E. The patient has carbon monoxide poisoning
12. Ascent to high altitude may result in
    A. Hyperventilation and respiratory acidosis
    B. Hyperventilation and respiratory alkalosis
    C. Hypoventilation and respiratory acidosis
    D. Hypoventilation and respiratory alkalosis
    E. Metabolic acidosis
13. The acute respiratory response to ascent to altitude is followed by normalization of blood pH as additional compensatory mechanisms activate. The mechanism primarily responsible for normalizing acid-base status is
    A. Enhanced erythropoiesis resulting in polycythemia and increased buffering by hemoglobin
    B. Increased levels of 2,3-DPG
    C. Retention of bicarbonate by the kidneys
    D. Bicarbonate pumping into the CSF
    E. Increased excretion of bicarbonate by the kidneys
14. An acclimatized person returns to sea level from a mountain top. Which of the following is most likely to occur?
    A. Hyperventilation and respiratory acidosis
    B. Hyperventilation and respiratory alkalosis
    C. Hypoventilation and respiratory acidosis
    D. Hypoventilation and respiratory alkalosis
    E. Metabolic acidosis
15. The Hering-Breuer reflex is associated with
    A. Bronchial C-fibers
    B. Irritant receptors
    C. Pulmonary C-fibers
    D. Pulmonary stretch receptors
    E. None of the above
16. A patient with long-standing carbon dioxide retention and chronic obstructive pulmonary disease (COPD) has an elevated $PaCO_2$, acidic arterial pH, and an abnormally low ventilatory rate for his blood gases. Which of the following is true of his CSF?

A. CSF pH is acidotic
B. CSF pH is alkalotic
C. CSF pH is normal
D. CSF pH is not predictable

17. If inspiratory CO is 0.001%, CO uptake by the blood is 28 ml/min, and barometric pressure is 700 mm Hg, what is the diffusing capacity of the lung for carbon monoxide?
A. 100 mm Hg
B. 40 ml/mm Hg/min
C. 0.7 mm Hg
D. 10 ml/mm Hg/min
E. 0.007 mm Hg

18. Each of the following is a component of the respiratory membrane **except**
A. Alveolar epithelium
B. Capillary epithelium
C. Epithelial basement membrane
D. Red cell membrane
E. Visceral pleura

Directions: Each group of items in this section consists of lettered options followed by a set of numbered items. For each item, select the one lettered option that is most closely associated with it. Each lettered option may be selected once, more than once, or not at all.

19–20. Match the following statements concerning the responses of the bronchi to changes in H^+, PO_2, and PCO_2.

A. Increase in H^+
B. Decrease in H^+
C. Decrease in PCO_2
D. Increase in PCO_2
E. Increase in PO_2

19. Bronchoconstriction
20. Bronchodilation

21–23. Match the following statements concerning the responses of the pulmonary, systemic, and bronchial arteries to changes in H^+, PO_2, and PCO_2.

A. Increase in H^+, increase in PCO_2, decrease in PO_2
B. Increase in H^+, increase in PCO_2, increase in PO_2
C. Increase in H^+, decrease in PCO_2, decrease in PO_2
D. Decrease in H^+, increase in PCO_2, increase in PO_2
E. Decrease in H^+, decrease in PCO_2, increase in PO_2

21. Pulmonary artery vasoconstriction, bronchial artery vasodilation
22. Pulmonary artery vasodilation, bronchial artery vasoconstriction
23. Systemic arterial vasoconstriction

24–27. Match the lettered cell types to the numbered item that most closely approximates their function.

A. Pulmonary alveolar macrophage
B. Type I pneumatocyte
C. Type II pneumatocyte
D. Endothelial cell

24. Line alveolar surface
25. Line capillaries
26. Phagocytosis
27. Produce surfactant

28–30. Match the pressure gradient determining pulmonary blood flow with the appropriate lung zone.

A. Zone 1 (apex of the lung)
B. Zone 2 (midlung)
C. Zone 3 (base of the lung)

28. $P_A > Pa > P_V$
29. $Pa > P_A > P_V$
30. $Pa > P_V > P_A$

Questions 31–35 refer to the illustration below.

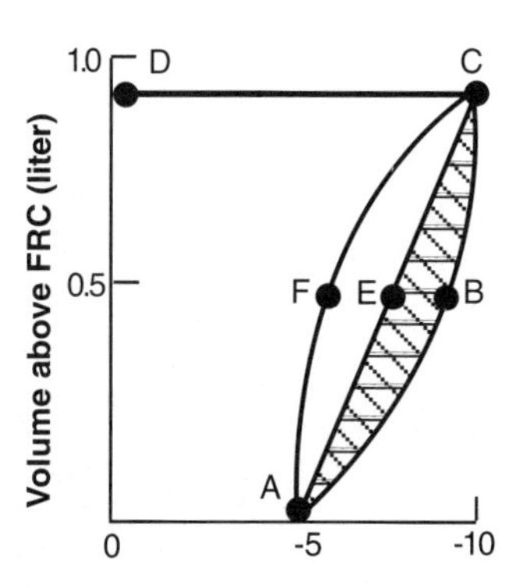

A. Line AEC
B. Hatched area ABCEA
C. Area OAECDO
D. AFCBA
E. AFC
F. ABC

31. Compliance of the lung
32. Inflation pressure-volume curve of the lung
33. Deflation pressure-volume curve of the lung
34. Viscous work
35. Inspiratory work required to overcome the elastic forces of the lung

36–39. Match each patient's respiratory quotient ($\dot{V}CO_2/\dot{V}O_2$) to the diet the patient is most likely consuming.

A. 0.98
B. 0.81
C. 0.69
D. 1.1
E. 1.92

36. The fat-burner diet (the all-cholesterol diet)
37. Vegetarian diet
38. Average American diet
39. IV dextrose

40–43. Match each value or range of values to the most appropriate variable.

A. 0
B. 0–20
C. 0.2–0.35
D. 27
E. 100

40. Normal A-a oxygen gradient
41. Normal A-a carbon dioxide gradient
42. VD/VT ratio
43. $P_{50}O_2$ (in mm Hg)

44–48. One deciliter of blood with a PaO_2 of 27 mm Hg (50% O_2 saturation) is mixed with one deciliter of blood with a PaO_2 of 100 mm Hg (97% saturation).

A. 75
B. 19.8
C. 10.13
D. 0.3
E. 0.081

44. What is the amount of oxygen in solution in the first deciliter?
45. What is the amount of oxygen in solution in the second deciliter?
46. What is the total amount of oxygen in the first deciliter if the deciliter has a hematocrit of 15 g Hb/dl?
47. What is the total amount of oxygen in the second deciliter if it has the same hematocrit as in the first deciliter: 15 g Hb/dl?
48. What is the approximate oxygen saturation of the mixture? Feel free to look at Table 4-1 and Fig. 4-17.

Directions: Each of the numbered items or incomplete statements in this section is followed by answers or by completions of the statement. Select the one lettered answer or completion that is best in each case.

49. One liter of blood with a $PaCO_2$ of 20 is mixed with a second liter with a $PaCO_2$ of 40. What is the $PaCO_2$ of the mixture?
A. 10
B. 20
C. 30
D. 40
E. 60

50. At standard temperature and pressure, 1 mole of a gas occupies 22.3 liters volume. If the solubility of carbon dioxide in plasma is 0.03 mmole/mm Hg/liter plasma, calculate its solubility in ml/mm Hg/liter plasma.
A. 0.10
B. 0.67
C. 1.37
D. 2.34
E. 11.89

51. A normal subject has a PaO_2 of 95 and a P_ACO_2 of 40. If the subject is metabolizing carbohydrate with a respiratory exchange ratio (RER) of 1.0, what would be the $P(A\text{-}a)O_2$ of this subject?
A. 50
B. 25
C. 15

D. 5
E. 0

52. If minute oxygen consumption ($\dot{V}O_2$) at rest is equal to 200 ml per minute, approximately what would the resting cardiac output have to be to provide that amount of oxygen to the periphery when hemoglobin is absent (assume that the tissues can extract 100% of the dissolved oxygen and that PaO_2 is 100 mm Hg)?
   A. 7000 liters per minute
   B. 700 liters per minute
   C. 70 liters per minute
   D. 7 liters per minute
   E. 0.7 liters per minute
53. If minute oxygen consumption ($\dot{V}O_2$) at rest is equal to 200 ml per minute, approximately what would be the resting cardiac output required to provide that amount of oxygen to the periphery when hemoglobin is present at 15 g Hb/dl (assume that the tissues can extract 100% of the dissolved oxygen, that PaO_2 is 90 mm Hg, and that PvO_2 is 40 mm Hg)?
   A. 16 liters per minute
   B. 8 liters per minute
   C. 4 liters per minute
   D. 2 liters per minute
   E. 1 liter per minute
54. If oxygen consumption doubles and blood flow remains constant, which of the following is true?
   A. $P(a\text{-}v)O_2$ is doubled
   B. $C(a\text{-}v)O_2$ is doubled
   C. $P(a\text{-}v)O_2$ is halved
   D. $C(a\text{-}v)O_2$ is halved
   E. Cardiac output doubles

Respiratory

Answers

1. D Resistance is inversely proportional to the radius to the fourth power. A decrease of 50 percent is equal to $1/(0.5)^4 = 16$.

2. E Halving the length results in halving the resistance since resistance is directly proportional to length. Decreasing the radius by 50 percent results in a 16-fold increase in resistance (see above), but overall the change is equal to an 8-fold increase in resistance as given by the equation $R = 8\eta l/\pi r^4$.

3. C Changes in PCO_2 do not occur instantaneously. A good rule of thumb is that if breathing entirely ceases, PCO_2 will increase by 4 mm Hg per minute for the first 3 minutes and 3 mm Hg per minute thereafter. It would therefore take approximately 12 minutes for PCO_2 to double if, for example, the patient was paralyzed and received oxygen by insufflation (i.e., no tidal exchange) but was not ventilated. Obviously, it takes longer to reach equilibrium if some ventilation is occurring, but equilibrium should be reached by 20 to 30 minutes after an acute change in ventilation. Since PCO_2 is directly proportional to alveolar ventilation, halving ventilation should, all else being equal, double PCO_2.

4. D Halving the lung volume, either through surgery or disease, or simply because one is younger and/or smaller, causes calculated compliance to decrease (which is why compliance and most pulmonary variables are adjusted for size). Since compliance is the change in volume/change in pressure, halving the volume will halve the compliance. Similarly, elastance is the inverse of compliance and elastance therefore doubles.

5. B (760 – 47)0.21 – (40/1) = 109.73.
6. E The partial pressure of carbon monoxide in the inspired air is the critical variable. The method works because the pressure in the blood is known and is zero. Therefore, the gradient is equal to the inspired carbon monoxide pressure.
7. E See Fig. 4-6. The vital capacity is the maximum volume of air that can be expelled from the lung following a maximal inspiration. That is the biggest tidal volume possible by definition.
8. C Compliance is equal to the change in volume divided by the change in pressure. In this case, 75 cc/3 cm H_20 = 25cc/cm H_2O.
9. B In humans, the ratio of ventilation to perfusion is greatest in the upper lobes. Bats hang around upside down and therefore oxygen tension is highest in their lung bases.
10. C Pulmonary C-fibers. Engorgement of pulmonary capillaries and increases in interstitial fluid volume of the alveolar walls are seen in left heart failure and result in an increased firing rate of pulmonary C-fibers.
11. C Right-to-left shunting (or ventilation-perfusion mismatch) is the most likely explanation. A patent ductus arteriosus in an infant or collapsed lung segment in an adult would be two possible explanations. In both cases well-oxygenated blood mixes with poorly oxygenated blood, leading to precipitous drops in the oxygen partial pressure because of the shape of the oxyhemoglobin dissociation curve. Answer A is incorrect because abnormal hemoglobins with altered affinities to oxygen change the oxygen saturation and content at any given partial pressure but would not be expected to change PaO_2. A patient with mild to moderate fibrosis would not exhibit problems with oxygenation; therefore, answer B is incorrect. If the patient had normal lungs his P_AO_2 and consequently his PaO_2 should be in the hundreds, and therefore answer D is wrong also. In carbon monoxide poisoning, answer E, the victim's PaO_2 remains normal as long as the patient continues to breathe. A similar condition occurs with anemic hypoxia.
12. B Ascent to high altitude is accompanied by a decrease in barometric pressure. The percentage composition of inspired gas remains 21% oxygen but the absolute partial pressure of oxygen decreases resulting in hypoxia. Recall that it is the partial pressure of oxygen, not the percentage, that determines oxygen content. The patient therefore hyperventilates, blowing off carbon dioxide, leading to acute respiratory alkalosis.
13. E The kidneys excrete bicarbonate thus normalizing acid-base status. Answers A and B do occur at high altitude, but their effects on acid-base status are minimal. Answer C is wrong since retention of bicarbonate would result in a mixed acid-base disorder with respiratory and metabolic alkalosis. Similarly, bicarbonate pumping into the CSF by the choroid plexus would worsen the alkalosis centrally. Instead, the pH of bulk CSF normalizes with time.
14. E Metabolic acidosis due to low total body bicarbonate is the most likely result. pH may or may not be acidic, but bicarbonate levels are depressed until the kidney can reclaim enough bicarbonate to replace that excreted on the mountain.
15. D Pulmonary stretch receptors mediate the Hering-Breuer reflex. Distention of the lung results in lowering of the respiratory rate.
16. C Bicarbonate pumping into the CSF by the choroid plexus has compensated for his chronic respiratory acidosis. Since central drive accounts for 80 percent of total respiratory drive, ventilation decreases as CSF pH normalizes.
17. B $P_ACO = (700 - 47) \times 0.001 \approx 0.7$ mm Hg (ignoring any dilution by carbon dioxide). PaCO = 0 mm Hg; therefore, the gradient is 0.7 mm Hg and the diffusing capacity is 28/0.7 = 40 ml/mm Hg/min.
18. E Visceral pleura is a component of the chest wall and not of the respiratory membrane.

19. C Alveolar hypocapnia occurring secondary to decreased perfusion causes ventilation to be shunted to better perfused areas of the lung. Alveolar hypoxia can also cause bronchoconstriction.

20. D Alveolar hypercapnia results in bronchodilatation and increased ventilation of that region of lung. Increasing PO_2 has little effect on bronchial tone unless hypoxia is present.

21. A Increases in H^+ and PCO_2 and decreases in PO_2 cause systemic arteries, including the bronchial arteries, to vasodilate and pulmonary vessels to vasoconstrict. This results in increased blood flow to hypoxic, hypercarbic, and acidotic tissues in the periphery and decreased blood flow to underventilated alveoli in the lung.

22. E Decreases in H^+ and PCO_2 and increases in PO_2 cause systemic arteries, including the bronchial arteries, to vasoconstrict and the pulmonary vessels to vasodilate.

23. E Decreases in H^+ and PCO_2 and increases in PO_2 cause systemic arteries, including the bronchial arteries, to vasoconstrict and the pulmonary vessels to vasodilate.

24. B Type I pneumatocytes or alveolar epithelial cells line the alveolar surface.

25. D Capillary endothelial cells line the capillaries.

26. A Pulmonary alveolar macrophages are phagocytic.

27. C Type II pneumatocytes produce surfactant.

28. A Gravity causes uneven distribution of blood flow in the lung (see Fig. 4-5). In zone 1, the apex of the upright lung may have a pulmonary alveolar pressure greater than the pulmonary capillary pressure, causing the capillaries to collapse.

29. B In zone 2 of the lung, pulmonary artery pressure is greater than pulmonary alveolar pressure, which is greater than pulmonary venous pressure, and blood flow proceeds normally.

30. C In zone 3, the capillaries are engorged due to the pressure of the column of blood overhead and pulmonary venous pressure exceeds pulmonary alveolar pressure.

31. A Line AEC is the compliance line.

32. F Hatched area ABCEA corresponds to the inspiratory work done to overcome the viscous forces of the lung.

33. E Area OAECDO corresponds to the inspiratory work done to overcome elastic forces of the lung.

34. B Curve AFC is the deflation curve of the lung.

35. C Curve ABC is the inflation curve of the lung.

36. C The respiratory quotient (RQ) or respiratory exchange ratio (RER) is the ratio of the amount of carbon dioxide given off relative to the amount of oxygen consumed. For fat that is 0.7.

37. A The RQ for carbohydrates is 1.0

38. B The RQ for an average person is 0.8.

39. D The RQ for dextrose is 1.1 and complex polysaccharides such as those found in total parenteral nutrition can be even higher. This can be important in weaning patients from a ventilator (or keeping them off of one) since they may already have difficulty excreting carbon dioxide (i.e., breathing). Thus, increasing the amount of carbon dioxide they are producing relative to the amount of nutrition they are getting certainly will not help.

40. B Texts disagree as to the exact number. However, most anesthesiologists will assure you that an A-a oxygen gradient of less than 20 won't get you intubated.

41. A Carbon dioxide is 20 times more soluble than oxygen and is essentially never limited by diffusion. Carbon monoxide diffuses even more rapidly.

42. C Normally, roughly one third of a tidal ventilation is wasted on dead space.

43. D Normal adult hemoglobin is 50% saturated at a PO_2 of roughly 27 mm Hg.

44. E Oxygen in solution = 27 mm Hg (0.003 ml O_2/mm Hg/dl blood) = 0.081 ml oxygen

45. D Oxygen in solution = 100(0.003) = 0.3 ml oxygen

46. C Oxygen content of first liter = (15 g Hb/dl blood) (1.34 ml O_2/g Hb) (0.5) + (27/mm Hg) (0.003 ml O_2/mm Hg/dl blood) = 10.13 ml O_2/dl blood

47. B Oxygen content of second liter = (15 g Hb/dl blood) (1.34 ml O_2/9 Hb) (0.97) + (27 mm Hg) (0.003 ml O_2/mm Hg/dl blood) = 19.8 ml O_2/dl blood

48. A Oxygen saturation of 75 percent (approximately that of venous blood); see Table 4-1 and Fig. 4-17.

49. C (20 + 40)/2 = 30. For carbon dioxide, taking the mean works!

50. B (22.3 liters/mole) (1000 ml/liter) (0.03 mmole CO_2/mm Hg/liter plasma) (1 mole/1000 mmole) = 0.67 ml CO_2/mm Hg/liter plasma.

51.

$$P_AO_2 = F_IO_2\,(BP - 47) - (P_ACO_2/RER)$$

$$= 0.21\,(760 - 47) - (P_ACO_2/RER)$$

$$= 150 - (P_ACO_2/1.0) = 150 - 40/1 = 110$$

$$P(A\text{-}a)O_2 = 110 - 95 = 15$$

For protein with an RER of 0.8, the answer would be 5 mm Hg; for fat with an RER of 0.7, the answer is 0.

52. C In the absence of hemoglobin, the oxygen content of blood (which in this case is the same as plasma) is equal to the amount dissolved in blood, that is, 0.003 ml O_2/mm Hg/dl blood. Assuming a PaO_2 of 100 mm Hg and using the Fick equation:

$$\text{Cardiac output} = \text{minute oxygen consumption}/C(a\text{-}v)O_2$$

$$= 200 \text{ ml/min} \div 0.3 \text{ ml } O_2\text{/mm Hg/dl blood}$$

$$= 666.67 \text{ dl blood/min}$$

$$= 66.67 \text{ liters/min}$$

53. C At a PaO_2 of 90 mm Hg, hemoglobin is roughly 97 percent saturated; hence,

Total arterial oxygen content = (15) (0.97)(1.34) + (0.003)(90) = 19.77 ml O_2/dl blood

Total venous oxygen content = (15)(0.75)(1.34) + (0.003)(40) = 15.2 ml O_2/dl blood

$C(a\text{-}v)O_2$ = 19.77 − 15.2 = 4.57 ml O_2/dl blood

Cardiac output = minute oxygen consumption/$C(a\text{-}v)O_2$ = 200 ÷ 4.57 = 43.76 dl/min = 4.376 liters/min

54. B Using the Fick equation: $C(a\text{-}v)O_2 = \dot{V}O_2$/cardiac output; if $\dot{V}O_2$ doubles and cardiac output is held constant, then $C(a\text{-}v)\ O_2$ must double.

Bibliography

Berne, R. M., and Levy, M. N. *Physiology* (3rd ed.). St. Louis: Mosby–Year Book, 1993.

Ganong, W. F. *Review of Medical Physiology* (15th ed.). Los Altos, CA: Lange, 1991.

Guyton, A. C. *Textbook of Medical Physiology* (8th ed.). Philadelphia: Saunders, 1991.

West, J. B. *Best and Taylor's Physiological Basis of Medical Practice* (12th ed.). Baltimore: Williams & Wilkins, 1990.

West, J. B. *Respiratory Physiology: The Essentials* (4th ed.). Baltimore: Williams & Wilkins, 1990.

5

Acid-Base Physiology

Charles H. Tadlock

General Concepts

Acids and Bases Defined

Brønsted defined acids as proton donors and bases as proton acceptors. For each acid or base, there is a **conjugate** base or acid from which it differs by the gain or loss of a proton.

Regulation of pH

Three basic mechanisms constitute the body's ability to regulate acid-base status.

1. Excretion of hydrogen ions and reabsorption of bicarbonate ions by the kidneys
2. Excretion of carbon dioxide by the lungs
3. Buffering of hydrogen ions by the weak acids and bases that comprise the body's buffering systems

These regulatory mechanisms are necessary because **free hydrogen ion concentration, $[H^+]$**, is important in the regulation of enzyme activity and in the maintenance of fluid and electrolyte balance.

Body hydrogen ion concentrations range from 10^{-1} moles per liter in the gastric juice to 10^{-9} moles per liter in the pancreatic fluid. In the plasma, normal hydrogen ion concentration is approximately 4×10^{-8} moles per liter. The negative logarithm to the base ten of the $[H^+]$ in moles per liter is called the **pH.**

$$pH = -\log [H^+]$$

Normal arterial pH ranges from 7.35 to 7.45, with venous blood being predictably slightly more acidic. Estimates place intercellular fluid (ICF) pH at approximately 7.0.

Henderson-Hasselbalch Equation

If HA represents an **acid** and A^- its **conjugate base,** then the **acid dissociation reaction** is

$$HA \Longleftrightarrow H^+ + A^-$$

A **strong acid** dissociates, completely liberating one mole of H^+ for each mole of acid present in solution. Conversely a **weak acid** dissociates only partially. Each acid has a **dissociation constant (K_a)**, which describes its tendency to dissociate.

$$K_a = \frac{[H^+][A^-]}{[HA]} = \frac{[products]}{[reactants]}$$

By solving for $[H^+]$

$$[H^+] = \frac{K_a [HA]}{[A^-]}$$

One can now solve this equation for pH by taking the negative logarithm of each side of the equation.

$$-\log [H^+] = -\log K_a + \log \frac{[A^-]}{[HA]}$$

$$pH = pK_a + \log \frac{[A^-]}{[HA]}$$

This form of the acid-base equilibrium is called the **Henderson-Hasselbalch equation.** It can also be written as

$$pH = pK_a + \log \frac{[proton\ acceptor]}{[proton\ donor]}$$

Daily Endogenous Acid Production

There are two major sources of daily acid production: (1) **metabolism of carbohydrates, fat, and proteins,** leading to the production of **fixed organic acids,** and (2) **oxidative metabolism,** leading to the production of the **volatile acid anhydride carbon dioxide.** Normal metabolism yields a variety of organic acids. Breakdown of carbohydrate, fat, and protein yields **lactic** and **pyruvic acid.** Lactic acid may accumulate in large quantities as a consequence of **anaerobic metabolism** during heavy exercise or due to some cause of tissue hypoxia. Oxidation of sulfur-containing amino acids yields H^+ and SO_4^{2-} and oxidation of phosphoproteins including nucleoproteins yields H^+ and PO_4^{3-}. These sources of fixed organic acids contribute approximately 1 mEq/kg/day of H^+ ions to the body (an average of 20–70 mEq/day). This daily acid production is normally excreted into the urine by the kidneys.

A far larger source of acid production is **oxidative metabolism.** Oxidative metabolism results in the formation of approximately 12,500 mEq H^+ per day, as illustrated by the breakdown of glucose in the **Krebs cycle:**

$$C_6H_{12}O_6 + 6\ O_2 \Longleftrightarrow 6\ H_2O + 6\ CO_2$$

$$CO_2 + H_2O \Longleftrightarrow H_2CO_3 \Longleftrightarrow H^+ + HCO_3^-$$

The carbon dioxide produced is predominantly excreted by the lungs (i.e., it is a **volatile or nonfixed acid**). The partial pressure of carbon dioxide in the arterial plasma ($PaCO_2$) is thereby maintained at 40 mm Hg, with venous blood having an average partial pressure ($PvCO_2$) of 45 mm Hg.

Protein metabolism results in production of excess fixed acid, which is eliminated by the kidneys via formation of an acidic urine. Consumption of fruits, in contrast, results in excess production of alkali, which must be eliminated via the formation of an alkaline urine.

Role of Buffers

Buffers Defined

A buffer is a solution of two or more chemicals that minimizes changes of pH in response to addition of acid or base. Many buffers consist of a weak acid and its conjugate base. Such a buffering system is able to bind or release H^+ in solution and thus minimize fluctuations in $[H^+]$ over its effective buffering range (Fig. 5-1).

A buffer is most effective (i.e., the change in pH per unit of acid or base added is least) when the pH of the solution equals the pK_a of the buffer. The **pK_a** of a buffer is that pH at which equal amounts of the buffer exist as the acidic and the basic forms. This can be shown using the Henderson-Hasselbalch equation:

$$pH = pK_a + \log [base]/[acid]$$

If the $pH = pK_a$, then,

$$pH - pK_a = 0 = \log [A^-]/[HA]$$

and

$$10^0 = 1 = [A^-]/[HA] \text{ or } [A^-] = [HA]$$

Intuitively, such a buffer couple should be most effective for buffering addition of either an acid or a base when half of the buffer exists as the acidic and half as the basic form, that is, when the $pH = pK_a$. It follows that the ideal extracellular buffer would

Fig. 5-1. Titration curves for the bicarbonate and inorganic phosphate buffers in a closed system. Under these conditions, when the concentration of carbon dioxide cannot be kept low through diffusion to the outside, the bicarbonate system is less efficient as a buffer than phosphate. It is the ability of the lungs to eliminate the carbon dioxide that makes the bicarbonate system such an efficient physiologic buffer. (From R. F. Pitts. *Physiology of the Kidney and Body Fluids* [3rd ed.]. Copyright © 1974 by Year Book Medical Publishers, Inc., Chicago.)

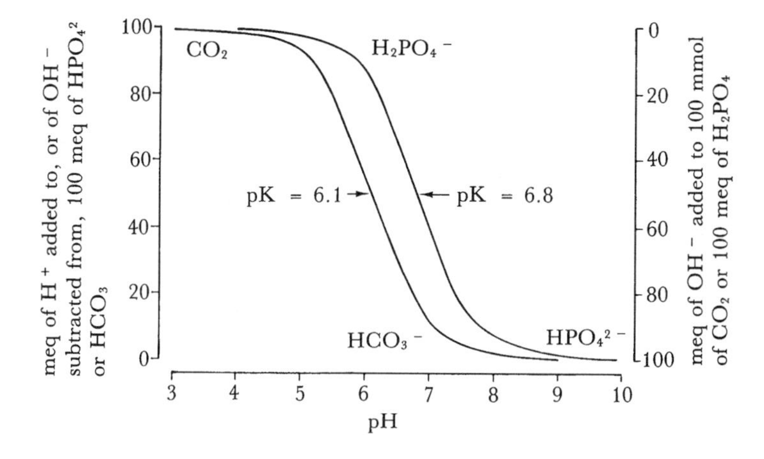

have a pK_a of 7.4. In fact, buffers are quite effective when the pH of the solution is within 1 pH unit of the pK_a Of the buffer and somewhat effective when the pK_a is within 2 pH units. Figure 5-1 illustrates the typical **sigmoidal titration curve** of a buffer. See Table 5-1 for the pK_as of important physiologic buffers.

The efficacy of a buffered system over that of a nonbuffered system in minimizing pH changes in the physiologic range is apparent when comparing the titration curves in Fig. 5-2. If a strong acid such as HCl is added to a nonbuffered system, then for each molecule of a strong acid added to the solution, H^+ also increases by one. Hence, relatively small additions of acid will lead to marked decreases in pH. In the presence of a buffer, the increase in $[H^+]$ resulting from the addition of the strong acid acts by mass action to protonate the conjugate base of the buffer, forming more of the buffer acid. The protons or hydrogen ions that bind to the buffer are no longer free in solution and therefore do not contribute to the $[H^+]$. The pH does drop slightly, of course, because the ratio $[A^-]$ to [HA] has decreased as A^- is converted to HA by the strong acid, and a small amount of the additional H^+ remains free. Conversely, if a strong base is added, the buffer releases H^+. Thus, fluctuations in $[H^+]$ are minimized over the effective range of the buffer.

The Isohydric Principle

The **isohydric principle** states that buffers in a common solution must be in equilibrium with one another because they are all in equilibrium with the same $[H^+]$.

Buffering Systems

Carbonic and Noncarbonic Acid Loads

When discussing the buffering capacities of various buffering systems, it is important to define what type of acid load is being discussed. By definition, a buffer system cannot buffer an acid-base disturbance for which it itself is responsible. Because the HCO_3^--CO_2 system is not only the major extracellular buffer but also the major

Table 5-1. pK_as of representative buffers

Name of buffer	Buffer pair	pK_a
Acetic acid	CH_3COO^-/CH_3COOH	4.7
Acetoacetic acid	$CH_3COCH_2COO^-/CH_3COCH_2COOH$	3.8
Ammonia	NH_3/NH_4^+	9.2
β–Hydroxybutyric acid	$CH_3CH_2CHOHCOO^-/CH_3CH_2CHOHCOOH$	4.8
Carbonic acid*	HCO_3^-/H_2CO_3	3.57
	HCO_3^-/CO_2	6.1
	CO_3^{2-}/HCO_3^-	9.8
Hemoglobin		
Deoxyhemoglobin	$Hb^{n-}/HHb^{(n-1)-}$	7.9
Oxyhemoglobin	$HbO_2^{n-}/HHbO_2^{(n-1)-}$	6.7
Lactic acid	$CH_3CHOHCOO^-/CH_3CHOHCOOH$	3.9
Phosphoric acid	$H_2PO_4^-/H_3PO_4$	2.0
	$HPO_4^{-2}/H_2PO_4^-$	6.8
	PO_4^{3-}/HPO_4^{-2}	12.4

*Carbon dioxide does not fit the Brønsted definition of an acid. It is an acid anhydride. Because carbon dioxide is found in such high concentrations in blood, it is potentially a much larger source of H^+ than its intermediary, H_2CO_3. The pK_a of H_2CO_3 above is 3.57. However, in the presence of CO_2 the apparent pK_a (pK_a') is 6.1.

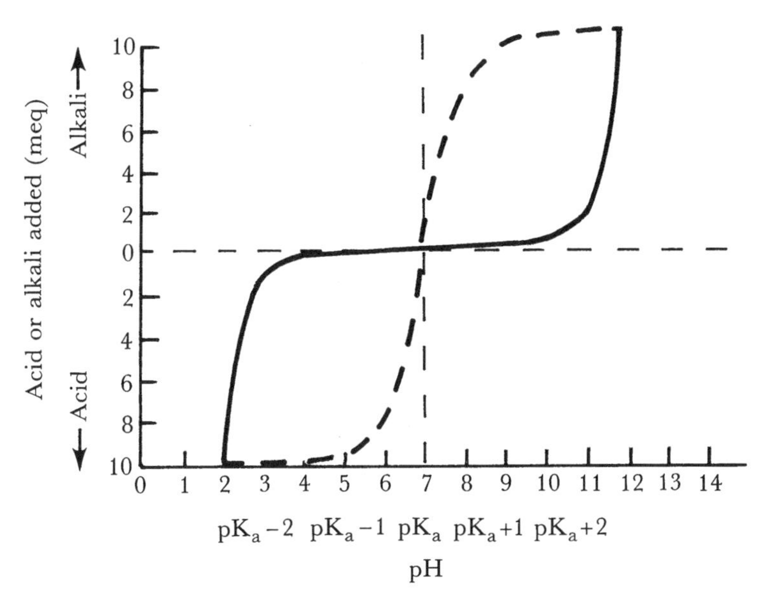

Fig. 5-2. Titration curves resulting from addition of strong acid or base to water (*solid line*) or a weak buffer solution with pK_a of 7.0 (*dashed line*). Note that over a wide pH range (including the range compatible with life) small changes in $[H^+]$ lead to large changes in pH in an unbuffered system, hence the almost flat line between pH 3.0 to 11. In contrast, the buffered system is quite effective in minimizing pH changes in response to addition of acid or alkali over its effective buffering range.

end product of oxidative metabolism, acid-base disturbances are generally divided into those in which a disturbance is due to carbon dioxide **(respiratory disturbance)** and those in which the disturbance is due to some other cause **(metabolic disturbance).** Because carbon dioxide is the culprit in respiratory disturbances, the HCO_3^--CO_2 system cannot contribute toward buffering a respiratory disorder and must be ignored when determining the contribution of various buffering systems to correcting a respiratory disorder.

The Plasma Buffers: Bicarbonate and Protein

The principle plasma buffers are **bicarbonate** and the **plasma proteins** with a very small contribution by ammonia and phosphate because these are present in quite small concentrations.

The **bicarbonate-carbonic acid system** is the most important plasma and extracellular buffer despite its relatively inefficient pK_a of 6.1. This system alone accounts for 90 percent or more of the buffering capacity of the plasma for metabolic acid loads. The bicarbonate-carbonic acid system is important for four major reasons:

1. HCO_3^- is present in high concentration (24 mEq/liter).
2. The lungs control the level of carbon dioxide in the blood by controlling the rate of ventilation (1.2 mmole CO_2/liter is present as dissolved gas).
3. The kidneys govern HCO_3^- levels by governing the rates of H^+ secretion and bicarbonate reabsorption.
4. The formation of carbon dioxide by oxidative metabolism is the major source of acid production in the healthy state (producing 12,500 mEq/day of H^+).

The Henderson-Hasselbalch equation for the HCO_3^--H_2CO_3 system is also unique. The general equation

$$pH = pK_a + \log [HCO_3^-]/[H_2CO_3]$$

while accurate is misleading. As the equation

$$CO_2 + H_2O \overset{CA}{\Longleftrightarrow} H_2CO_3 \Longleftrightarrow HCO_3^- + H^+$$

indicates, CO_2 is an **acid anhydride** and the total acid available for reaction is not equal to the $[H_2CO_3]$ but to the sum of $[H_2CO_3]$ + (CO_2 dissolved). Because carbon dioxide is present at approximately 200 times the concentration of H_2CO_3, the total concentration of acid is approximately equal to the concentration of dissolved carbon dioxide. The total content of carbon dioxide in solution **(TCO_2)** is the sum of the concentrations of carbon dioxide in all its forms (as dissolved gas, carbonic acid, carbaminohemoglobin, and bicarbonate) and is approximately equal to the partial pressure of carbon dioxide times solubility of carbon dioxide in plasma, 0.03 mmole CO_2/liter/mm Hg:

$$TCO_2 = [HCO_3^-] + [\text{dissolved } CO_2] + [\text{carbamino } CO_2] + [H_2CO_3] \approx PCO_2 \times 0.03 \text{ mmole } CO_2\text{/liter/mm Hg}$$

Substituting this into the Henderson-Hasselbalch equation:

$$pH = 6.1 + \log \frac{[HCO_3^-]}{PCO_2 \times 0.03}$$

Clinically, pH and PCO_2 are measured directly and HCO_3^- is calculated using this equation. However, during rapid changes in acid-base status, the equilibrium conditions required by this equation do not hold true. $[HCO_3^-]$ should then be calculated by determining the TCO_2 using the following equation:

$$TCO_2 = [HCO_3^-] + [\text{dissolved } CO_2] + [\text{carbamino } CO_2] + [H_2CO_3] \text{ or,}$$

$$[HCO_3^-] = TCO_2 - [\text{dissolved } CO_2] - [\text{carbamino } CO_2] - [H_2CO_3]$$

The enzyme **carbonic anhydrase (CA)** catalyzes the hydration of carbon dioxide, enabling equilibrium to be established rapidly.

$$CO_2 + H_2O \overset{CA}{\Longleftrightarrow} H_2CO_3$$

It is present in high concentration in erythrocytes and renal tubular cells as well as on the luminal aspect of proximal tubular membranes.

Proteins are weak acids at physiologic pH. They are important buffers in the plasma and red cells, and constitute the major intracellular buffer. Because its imidazole group has a pK_a of 6.0, **histidine** is the only amino acid with significant buffering capacity in the physiologic range, that is, between pH 6 and pH 8. The pK_a of cysteine's sulfhydryl group is 8.3, which is closer to normal blood pH than the pK_a of the imidazole group, but this sulfhydryl group is often tied up in disulfide linkages (such as in the dimer cystine) and is therefore unavailable to act as a buffer.

Table 5-2. Buffering of a metabolic acid load

Compartment	Major buffers (% buffering in compartment)		% of total body buffering
Plasma	Bicarbonate	(92)	13
	Plasma proteins	(8)	
	Inorganic phosphate	(<1)	
Red cell	Hemoglobin	(60)	6
	Bicarbonate	(30)	
	Inorganic phosphate and intracellular proteins	(10)	
Interstitial fluid	Bicarbonate	(>99)	30
	Inorganic phosphate	(<1)	
Intracellular fluid and bone	Proteins and inorganic phosphate		51

Phosphate is also present in plasma but due to its low concentration (1–2 mEq/liter) is not an effective buffer there. Table 5-2 illustrates the relative contributions of the plasma buffers in response to a metabolic acid load.

Red Cell Buffers

The protein **hemoglobin** is found in extremely high concentrations in red cells (15 g/dl). Although hemoglobin is technically an intracellular buffer, the red cell membrane is so permeable to H^+ that hemoglobin is functionally a plasma buffer.

The H^+ formed by the hydration of carbon dioxide in erythrocytes and by the formation of carbaminohemoglobin (Fig. 5-3) is buffered primarily by hemoglobin.

Oxygenated hemoglobin (HbO_2^{n-}, pK_a 6.68) is a stronger acid than deoxygenated hemoglobin (Hb^{n-}, pK_a 7.93). Both H^+ and carbon dioxide decrease the affinity of hemoglobin for oxygen. The reaction $HbO_2^{n-} \rightarrow Hb^{n-} + O_2$ would cause a tremendous rise in pH in the venous blood if carbon dioxide and therefore H^+ were not added at the same time, because Hb^{n-} is a stronger base than HbO_2^{n-}. Release of carbon dioxide by the tissues results in the production of approximately 1.68 mmole per liter of H^+ as carbon dioxide dissociates to form H^+ and HCO_3^-. Because pH depends only on the concentration of hydrogen ions, this increase in H^+ would be expected to cause a precipitous drop in pH. Instead, the added H^+ is buffered by deoxygenated hemoglobin, thus minimizing the pH change. Hemoglobin is the most abundant nonbicarbonate extracellular buffer and is more powerful in the short run than the bicarbonate system.

Every liter of venous blood carries 1.68 mmole of extra carbon dioxide (see above). This additional carbon dioxide is carried as follows (Fig. 5-3): 65 percent as HCO_3^-, 27 percent as carbaminohemoglobin (in the red cell), and 8 percent as dissolved gas in solution.

Like the plasma proteins, hemoglobin's buffering action is due primarily to its content of the amino acid histidine. The pK_a of hemoglobin is 6.68, making hemoglobin a better buffer than a simple solution of histidine (pK_a 6.0, see above). The protonation of the imidazole group of hemoglobin is coupled to the deoxygenation of hemoglobin:

$$HbO_2 + H^+ \Longleftrightarrow HHb + O_2$$

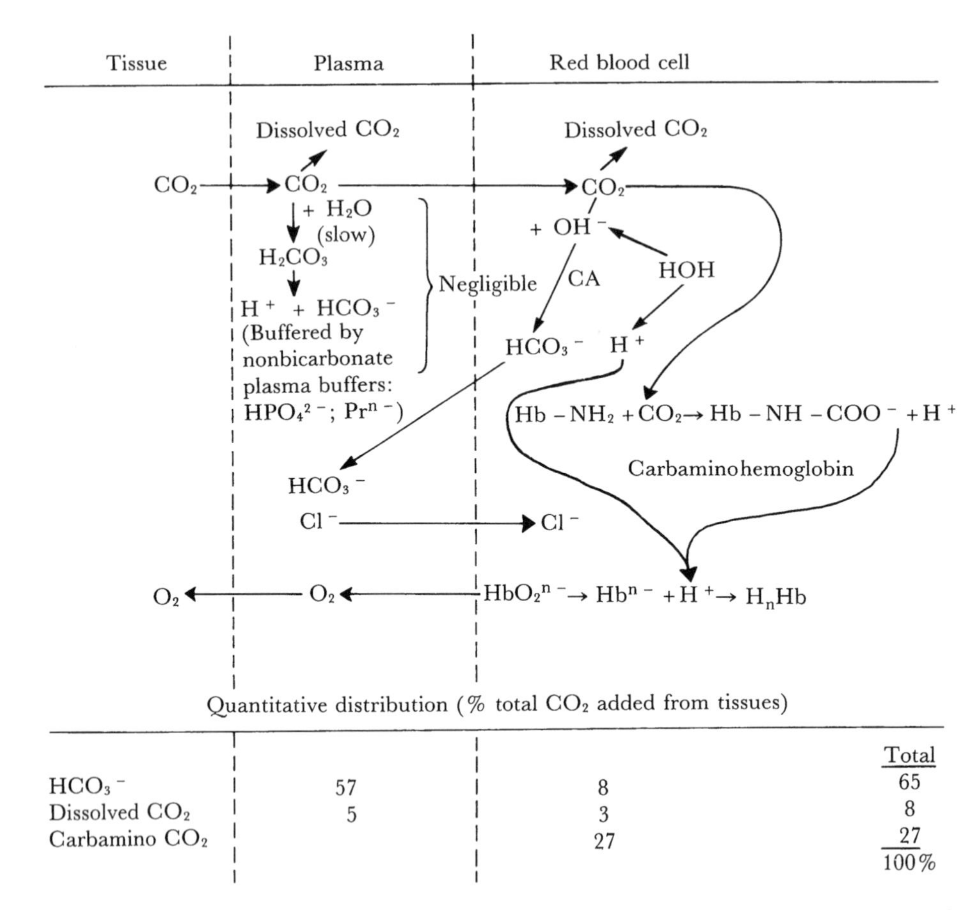

	Plasma	Red blood cell	Total
HCO_3^-	57	8	65
Dissolved CO_2	5	3	8
Carbamino CO_2		27	27
			100%

Fig. 5-3. Transport of carbon dioxide and buffering of H^+ by the blood. (Adapted from H. W. Davenport. *The ABC of Acid-Base Chemistry* [6th ed.]. Chicago, © 1974 University of Chicago Press, and E. J. Masoro and P. D. Siegel. *Acid-Base Regulation: Its Physiology, Pathophysiology and the Interpretation of Blood-Gas Analysis* [2nd ed.]. Philadelphia: Saunders, 1977.)

Both oxy- and deoxyhemoglobin may be protonated. However, because the pK_as are 6.68 and 7.93, respectively, the unloading of oxygen from hemoglobin is associated with increased affinity for hydrogen ion and therefore the net reaction is as indicated. Deoxyhemoglobin also has increased affinity for carbon dioxide. Carbon dioxide binds to the amino acid terminals of hemoglobin creating **carbaminohemoglobin:**

$$Hb\text{–}NH2 + CO_2 \Longleftrightarrow Hb\text{–}NH\text{–}COO^- + H^+$$

Ten percent or more of the total carbon dioxide content of venous blood may be carried as carbaminohemoglobin (Fig. 5-3).

Extracellular Fluid Buffers

Bicarbonate is the only significant extracellular buffer outside of the blood. Because the interstitial fluid volume is much greater than that of the plasma, its capacity to buffer metabolic acid loads is correspondingly greater. The extracellular fluid (ECF) is rapidly cleared of protein and has a very low concentration of phosphates; therefore, the ECF is unable to help buffer a carbonic acid load. Equilibration with the interstitial fluids requires several minutes, in contrast to the blood, which equilibrates quite rapidly.

Intracellular Fluid Buffers

Proteins and **phosphates** are the main intracellular buffers, with a small contribution made by intracellular bicarbonate. The pK_a of the $H_2PO_4^-$–HPO_4^{2-} system is 6.8, very close to the average pH inside cells, which is believed to be 7.0. This coupled with the high concentration of phosphate in cells accounts for its role in intracellular buffering.

$$H_2PO_4^- \Longleftrightarrow HPO_4^{2-} + H^+$$

Intracellular buffers ultimately account for 50 percent of the buffering of a metabolic acid load and 95 percent of a carbonic acid load (because the bicarbonate system is unable to help in that case). Primarily because of the lipid barrier imposed by the cell membranes, intracellular buffering requires long equilibration times (hours), in comparison to the rapid equilibration associated with extracellular buffers (seconds to minutes) in responding to a metabolic acid-base disturbance. An exception to this is the hemoglobin system, which, because of the leakiness of the red cell membrane, equilibrates in tandem with the extracellular buffers. Figure 5-4 reviews the handling of a metabolic acid load.

Respiratory Acid Loads

The discussion thus far has centered on the response of body buffers to a metabolic acid load. The major differences in the response of body buffers to a respiratory acid load are as follows.

The bicarbonate–carbonic acid system is the culprit, and must therefore be discounted in discussions of respiratory acid buffering. Because the bicarbonate system is the major extracellular buffer, ECF buffering will obviously contribute little to alleviating a respiratory acid load.

Furthermore, because carbon dioxide is lipid soluble, intracellular phosphate and protein can act to buffer respiratory disturbances without the delay associated with metabolic disturbances in which a charged molecule must generally equilibrate across the cell membranes.

Acute carbonic acid loads are therefore essentially entirely buffered by intracellular buffers with some contribution by nonbicarbonate extracellular buffers, primarily plasma proteins and inorganic phosphates. The role of hemoglobin in buffering carbonic acid loads has already been discussed. Roughly 95 percent of a carbonic acid load is buffered intracellularly: by hemoglobin and inorganic phosphates in the red cells, and by proteins and inorganic phosphates in the remainder of the ICF.

Table 5-2 illustrates the contribution of each of the various buffers to buffering a metabolic acid load while Table 5-3 reviews the differences between metabolic and respiratory acid load buffering.

Tubular Fluid Buffers

Ammonia, bicarbonate, phosphate, sulfate, and organic anions all contribute to buffering in the nephron. The three most important tubular buffers and their respective pK_as are NH_3/NH_4^+ with a pK_a of 9.2, $H_2PO_4^-/HPO_4^{2-}$ with a pK_a of 6.8, and HCO_3^-/H_2CO_3 with a pK_a of 6.1. The extent to which each is titrated depends on

1. The amount of H^+ secreted
2. The concentration of each buffer

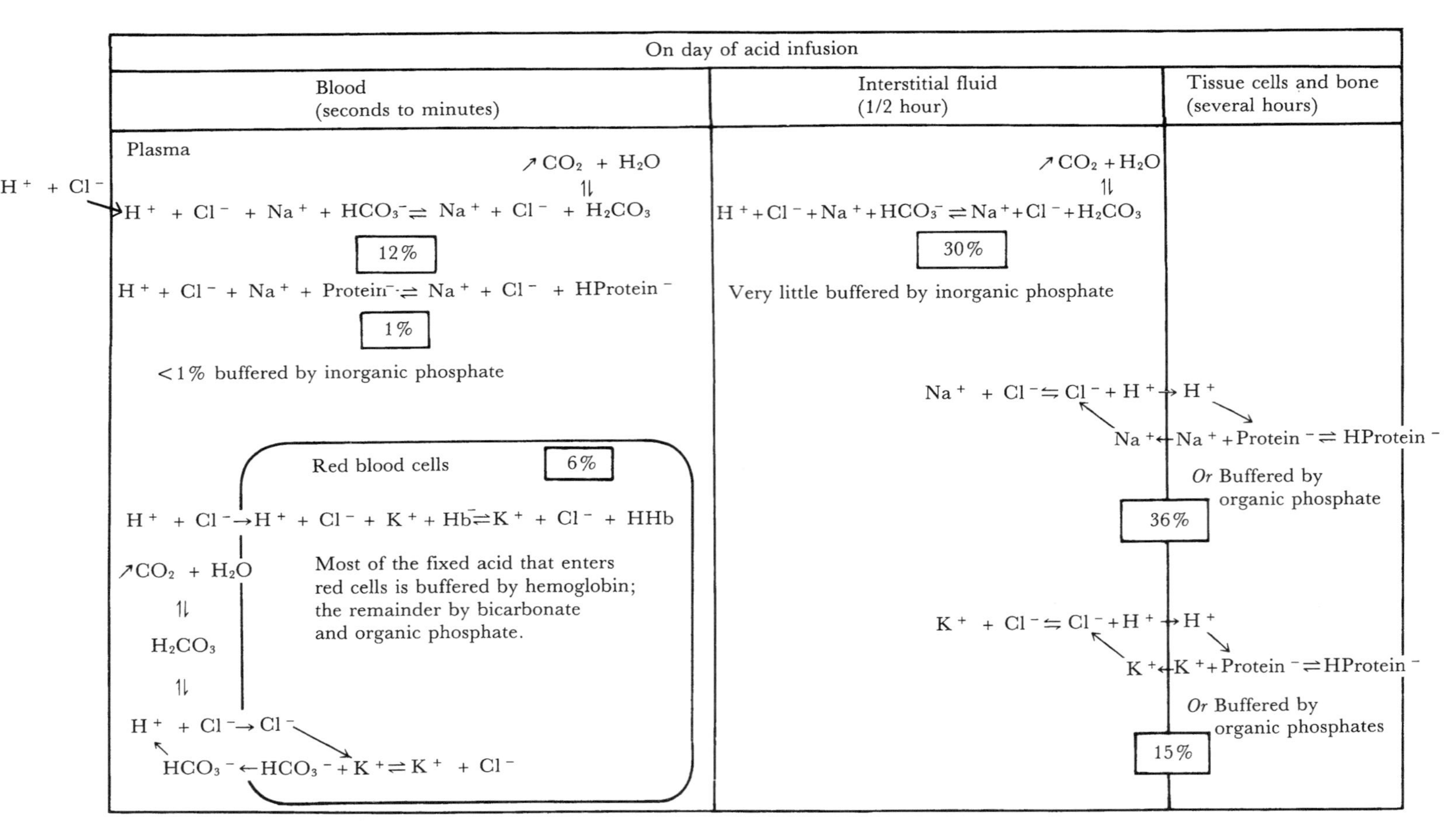

Fig. 5-4. Handling of the fixed inorganic acid, HCl, by intact dogs. For the sake of clarity, polyvalent anions such as proteins and hemoglobin have been drawn with a single negative sign. A slanted arrow next to carbon dioxide indicates that the carbon dioxide is quickly excreted through the lungs. The percentages enclosed in rectangles indicate the approximate proportion of the total acid load that is buffered by each mechanism. Twenty-four hours after infusing acid, about 25 percent of acid load has been excreted in the urine as titratable acid and NH_4^+. Extracellular pH and ionic composition are nearly normal; therefore, 75 percent of the administered acid must be sequestered and buffered in tissue cells and bone. Two to six days after infusing acid, the remaining 75 percent of the administered acid is slowly released from tissue cells and bone and excreted in the urine. (From H. Valtin. *Renal Function: Mechanisms Preserving Fluid and Solute Balance in Health* [2nd ed.]. Boston: Little, Brown, 1983. Pp. 210–211.)

Table 5-3. Metabolic acid buffering versus respiratory acid buffering

Metabolic acidosis	50% of buffering by ECF and red cells, 50% ultimately buffered by ICF Rapid buffering of acid by hemoglobin and bicarbonate in the blood (seconds) with longer equilibration times required for equilibration with the interstitial fluid (approximately ½ hour) and intracellular fluid (hours) Bicarbonate system can and does contribute
Respiratory acidosis	95% of buffering occurs intracellularly, 5% extracellularly Rapid equilibration occurs as lipid-soluble CO_2 diffuses into cell compartments to be buffered by hemoglobin, proteins, and phosphates The bicarbonate system cannot contribute

3. The pK_as of each of the buffers present
4. The pH of the tubular fluid

Buffers Summarized

In summary, all body fluids are equally good at buffering a metabolic acid load. Thus, at equilibrium, approximately 55 percent of an acid load will be buffered by the ICF and 45 percent by the ECF. Equilibration with the ECF requires approximately half an hour while equilibration with the ICF may require many hours. Due to the high concentration of hemoglobin found in red cells, red cells are somewhat more effective than plasma on a volume basis.

In respiratory acidosis, in contrast, 95 percent of the acid load is buffered by intracellular buffers and 5 percent by extracellular buffers.

Acid-Base

Renal Handling of H^+ and HCO_3^-

Role of the Kidney

The role of the kidney in acid-base regulation is to match the rate of acid excretion to the highly variable and essentially unregulated rate of dietary acid intake and metabolic production and to compensate for respiratory disorders. The kidney responds to acid-base imbalances by forming and excreting a fluid with either an excess or a deficit of acid in order to return the $[H^+]$ of the blood toward normal. The kidney is able to alter acid-base status in three basic ways:

1. Reabsorption of HCO_3^-
2. Titration of urinary buffers
3. Formation and excretion of ammonia (NH_3) as ammonium (NH_4^+)

Methods (2) and (3) result in the net excretion of H^+ while the first method results in the reabsorption of bicarbonate. It should be noted that the gain of HCO_3^- is equivalent to the loss of H^+ and that conversely the loss of HCO_3^- is equivalent to the gain of H^+ (review the Henderson-Hasselbalch equation). Furthermore, the initial source of most of the hydrogen ion secreted is via the hydration of carbon dioxide, not secretion of free H^+ by the tubular cells (Fig. 5-5A). Thus, hydrogen ion secretion (and therefore bicarbonate reabsorption) can be almost entirely abolished by carbonic anhydrase inhibitors. Some net excretion of hydrogen ion does occur in the distal tubule,

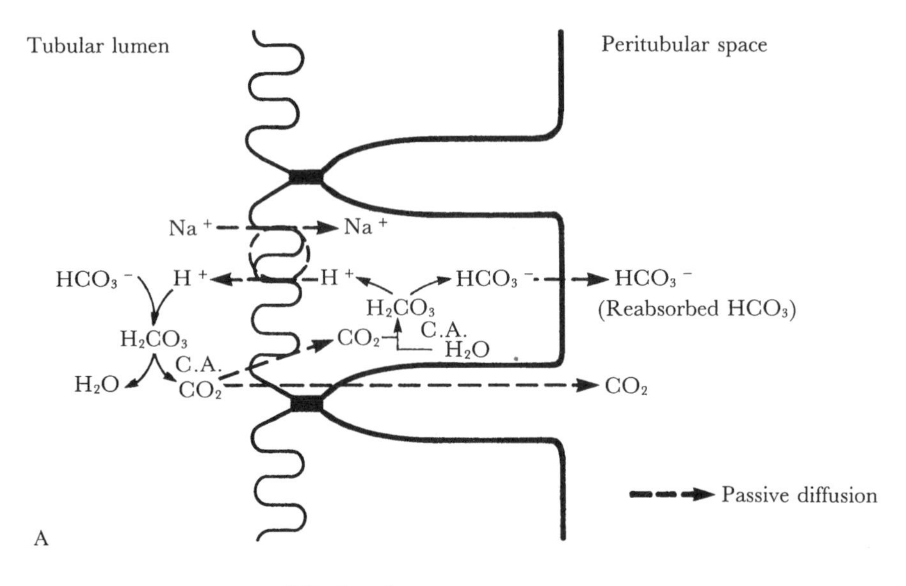

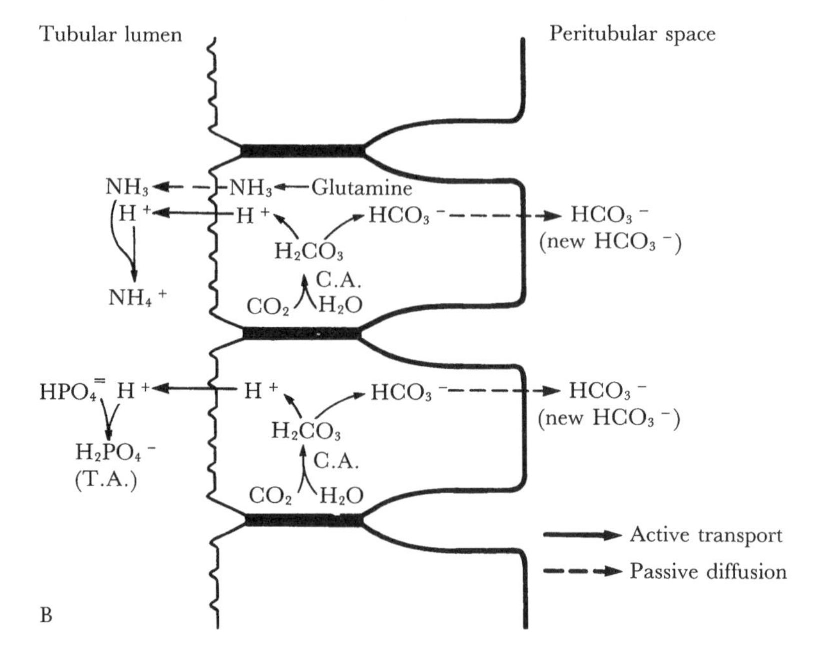

Fig. 5-5. Mechanisms for H^+ secretion in the proximal tubule and distal nephron. In both regions, carbon dioxide from metabolic processes, the peritubular capillaries, or the tubular lumen reacts with water to form H_2CO_3, a reaction catalyzed by carbonic anhydrase (CA) in the epithelial cells. The H_2CO_3 dissociates to form H^+ and HCO_3^-. The H^+ is secreted by secondary active transport via the Na^+/H^+ exchange process (antiport) in the proximal tubule (indicated by the circle) and via an active process in the distal nephron. The HCO_3^- diffuses across the basolateral surfaces into the peritubular capillaries, so that for every H^+ secreted, an HCO_3^- is returned to the systemic circulation. Filtered bicarbonate = 4500 mEq per day, reclaimed by proximal tubule = 4500 mEq per day, new bicarbonate generated by the distal tubule = 50 mEq per day, and net acid excretion = 50 mEq per day.

A. Reabsorption of filtered HCO_3^- via H^+ secretion. The secreted H^+ reacts with HCO_3^- in the tubular fluid to form H_2CO_3, which then dissociates into carbon dioxide and water; that is, an HCO_3^- ion is lost from the tubular fluid. Because the H^+ secretion process adds an HCO_3^- ion to the peritubular fluid and peritubular capillaries, however, the net result is HCO_3^- reabsorption. The proximal tubule is illustrated here, because approximately 90 percent of the filtered HCO_3^- is reabsorbed in the proximal tubule. However, similar reactions occur in more distal portions of the nephron (except for where a **low-capacity but high-gradient** H^+–Na^+ exchange pump acts to acidify the urine (Fig. 5-5B).

Reabsorption of HCO_3^-

The proximal tubule reabsorbs filtered bicarbonate by secreting H^+ and subsequently reabsorbing it as carbon dioxide (Fig. 5-5A). Greater than 75 percent of filtered bicarbonate is reabsorbed in the proximal tubule; however, the pH decreases only slightly and therefore only a relatively small amount of other buffers is titrated. The H^+ secretory pump is located in the luminal membrane and is a **high-capacity low-gradient pump.** Thus, while it is able to pump large numbers of H^+ ions, it is unable to create a pH gradient (i.e., it cannot acidify the urine). Hydrogen ions from the cell are exchanged for Na^+ from the tubule fluid, thus maintaining electroneutrality. Sodium ion moves down its concentration gradient into the cell and thus secondarily actively transports H^+ against its concentration gradient into the lumen.

The H^+ secreted into the lumen reacts with the filtered HCO_3^- to form carbonic acid. The carbonic acid is then broken down into carbon dioxide and water in the presence of carbonic anhydrase on the luminal membrane of the proximal tubule. As a result, carbon dioxide is then in greater concentration in the tubular fluid than in the cell and it therefore passively diffuses down its concentration gradient into the cells. As carbon dioxide diffuses into the cells, its increased concentration causes it to encounter intracellular carbonic anhydrase and, by **mass action,** to drive the reaction toward the formation of hydrogen ion and bicarbonate. The hydrogen ion is secreted into the tubular fluid (note that there is no net secretion of hydrogen ion in this process). The bicarbonate, now in high concentration inside the cell, is transported into the peritubular space as sodium bicarbonate in a facilitated passive transport process fueled by an active sodium pump in the peritubular basolateral membrane. Intracellular sodium is replenished as sodium from the tubular fluid enters the cells passively along the concentration gradient formed by the active transport of sodium out of the cell into the peritubular fluid.

The net result is the reabsorption of sodium bicarbonate from the tubular fluid and a slight fall in the pH of the urine. There is no change in the PCO_2 and only the minimal excretion of H^+ associated with the inadvertent titration of other buffers by hydrogen ions secreted by the proximal tubule.

The extent of hydrogen ion secretion in the proximal tubule (and in the rest of the renal tubule as well) is to a large extent governed by the availability of buffer because only by reacting with buffers can a significant amount of H^+ be excreted before the

Fig. 5-5 *(continued).* the absence of CA in the luminal surface) and are responsible for the reabsorption of most of the remaining HCO_3^- in the tubular fluid. An active Na^+-K^+ exchange pump in the basolateral membrane (not shown) actively pumps Na^+ from the cells into the peritubular fluid. This supplies the gradient for Na^+ reabsorption from the tubular fluid. (From J. B. West. *Best and Taylor's Physiological Basis of Medical Practice* [12th ed.]. P. 493. © 1990 Williams & Wilkins, Baltimore.)

B. Generation of new HCO_3^- via H^+ secretion. The HCO_3^- generated in the H^+ secretion process represents a new HCO_3^- if the secreted H^+ ion (a) reacts with NH_3 synthesized by the epithelial cells to form NH_4^+ or (b) reacts with HPO_4^{-2} and other titratable acids in the tubular fluid to form $H_2PO_4^-$ and to acidify the remaining titratable buffers. Although the distal nephron is illustrated here, similar reactions can occur in the proximal tubule (except that proximal tubular H^+ secretion is primarily coupled to Na^+ reabsorption and is gradient limited). Most of the H^+ secreted in the proximal tubule reacts with HCO_3^- in the tubular fluid and, therefore, accomplishes the reabsorption of filtered HCO_3^- rather than the generation of new HCO_3^-. The distal tubular H^+ secretory pump can create a 1000-fold $[H^+]$ gradient enabling it to acidify the urine. T.A. = titratable acid. (From J. B. West. *Best and Taylor's Physiological Basis of Medical Practice* [12th ed.]. P. 494. © 1990 Williams & Wilkins, Baltimore.)

limiting concentration gradient is reached. Even in the distal tubule the pH gradient limits urinary pH to 4.5.

The total amount of bicarbonate reabsorbed is proportional to the amount filtered and this in turn is equal to the product of the glomerular filtration rate and the serum concentration of bicarbonate. This relationship is valid until the concentration of bicarbonate in the tubular fluid exceeds the proximal tubule's ability to reabsorb it (approximately 28 mEq/liter). Normally, more than 75 percent of the bicarbonate filtered is reabsorbed in the proximal tubule. Additional bicarbonate reabsorption occurs in the loop of Henle.

Excretion of H^+: The Distal Convoluted Tubule and Collecting Ducts

The distal convoluted tubule and collecting ducts contain a luminal H^+ secretory pump (Fig. 5-5B). They are therefore able to acidify the urine and titrate the remaining urinary buffers. The gradient is limited to approximately a 1000-fold hydrogen ion gradient between the ICF and the tubular fluid. Hydrogen ion is secreted into the tubular fluid and reacts with phosphate and the other **titratable acids** (e.g., sulfate and organic acids). The relatively impermeable epithelium of the distal convoluted tubule and collecting ducts coupled with a high-gradient low-capacity H^+ secretory pump enables the kidney to form an acidic urine.

Titratable Acidity

Titratable acidity (TA) refers to the amount of acid that can be titrated by adding alkali (OH^-) to the urine until a pH of 7.4 is achieved. The amount of alkali needed to return urinary pH to pH 7.4 is equal to the amount of acid excreted bound to urinary buffers with pK_as such that the buffers will bind significant amounts of H^+ at pHs below 7.4. The major titratable acids are phosphate and sulfate (bicarbonate should not normally be present in appreciable amounts). **Organic acids** such as **creatinine** (in renal failure), and **beta-hydroxybutyric acid** and **acetoacetic acid** (in diabetes mellitus), may become important in disease states. The amount of H^+ free in solution is negligible in comparison with the amount bound to urinary buffers. Because the pK_a of ammonia is so high (9.2), addition of alkali will not titrate ammonia and thus its contribution to hydrogen ion excretion must be considered separately.

Ammonia and Ion Trapping

If renal H^+ excretion were limited to the formation of titratable acid, insufficient H^+ ion would be excreted before reaching the limiting pH of 4.5. The amount of H^+ excreted would then be limited by the amount of phosphate and other titratable acids. Because the amount of these acids present in the urine is diet dependent and not subject to regulation, total H^+ excretion would not be subject to regulation either.

Because of its high pK_a (9.2), the ammonia-ammonium system enables the kidneys to excrete H^+ as the neutral salt NH_4Cl. Measurements of TA would not reveal the contribution of this system because titration to pH 7.4 would only affect 1 to 2 percent of the ammonia buffer present.

$$NH_3 + HCl \Longleftrightarrow NH_4Cl \quad pK_a = 9.2$$

Within the renal tubule cells, carbon dioxide reacts to form H^+ and HCO_3^-. The H^+ is secreted into the tubular fluid, where it reacts with ammonia to form ammonium, while the bicarbonate is reabsorbed into the peritubular fluid.

The NH_3 in the tubular fluid is derived from the breakdown of glutamine and other amino acids by renal tubular cells (Fig. 5-5B). Being lipid soluble, the ammonia diffuses out of the cells and into the tubular fluid along its concentration gradient. In the tubular fluid, H^+ avidly combines with NH_3 to form NH_4^+. This will occur at all urinary pHs because the pK_a of ammonia is so high. However, the ratio of NH_3/NH_4^+ will decrease as the tubular fluid becomes more acidic, trapping the ammonium within the tubular fluid because ammonium is lipid insoluble. As the ammonia is trapped in the tubular fluid as ammonium, additional ammonia diffuses out of the cells to replace it. The trapped ammonium is subsequently excreted, primarily as ammonium chloride. This process is called **ion trapping** or **nonionic diffusion.**

Thus, the net function of the NH_3-NH_4^+ system is to excrete H^+ and supplement HCO_3^- stores. This probably occurs in all parts of the nephron. The mechanism by which NH_3 production and excretion is increased in acidic states is unknown. However, as the pH of the urine becomes more acidotic, more of the ammonia becomes trapped in the urine as ionized ammonium. This in turn increases the gradient favoring diffusion of the uncharged ammonia molecule out of the cells into the tubular fluid and may in some unknown fashion trigger increased ammonia production by the cells.

The final determinant of ammonia excretion is the ratio of peritubular blood flow to tubular fluid flow. When urine pH is equal to plasma pH, more ammonia would diffuse into blood than into the tubular fluid because the rapid blood flow relative to urine flow would dilute the ammonia and therefore create a greater concentration gradient (a larger **sink effect**). Normally the urine pH is acidic, however, so most of the NH_3 produced by the renal tubular cells enters into the tubular fluid by nonionic diffusion. Roughly three quarters of the daily endogenous load of nonvolatile acid is excreted as ammonium; the remainder is excreted as titratable acids.

Regulation of Renal Acid and Hydrogen Ion Excretion

Renal acid excretion is dependent on

1. The concentration of tubular fluid buffers. If the urine is already maximally acidic, then hydrogen ion excretion is directly proportional to the amount of buffer available in the urine.
2. The glomerular filtration rate (GFR). Changes in GFR alter the filtered load of bicarbonate and therefore the rate of bicarbonate reabsorption as well as the filtered load of titratable acids.
3. Carbonic anhydrase activity.
4. The partial pressure of carbon dioxide in the blood and the blood pH. Increased PCO_2 and decreased pH favor proton secretion by the renal tubular cells, which results in increased bicarbonate reabsorption, increased ammonia production and excretion, and increased titratable acidity.
5. ECF volume. Volume contraction favors secretion of hydrogen ion **(contraction alkalosis)** while volume expansion does not.
6. Potassium balance. Hydrogen ion secretory capacity is inversely proportional to body potassium stores. If either H^+ or K^+ is decreased then the other is forced to move excessively. Decreased plasma levels of potassium (hypokalemia) cause potassium to move out of cells. In order to maintain electroneutrality, H^+ then moves into cells, simultaneously causing an intracellular acidosis and an extracellular metabolic alkalosis determined by arterial blood gases. Hypokalemia also causes decreased aldosterone secretion and increased renin secretion, causing ECF volume to be decreased.

7. Aldosterone. Aldosterone causes K^+ or H^+ to be exchanged for sodium ion, resulting in salt water accumulation and the excretion of H^+ and K^+. If either H^+ or K^+ is present in less than normal amounts, the other cation will be forced to move excessively.
8. Parathyroid hormone. Parathyroid hormone causes increased bicarbonate excretion and decreased titratable acid excretion.

Regulation of Respiration

High $PaCO_2$, high $[H^+]$, and low PaO_2 all stimulate the **carotid body chemoreceptors,** which results in a reflex increase in alveolar ventilation and a corresponding increase in carbon dioxide excretion. Similarly, the **medullary respiratory center** is sensitive to $[H^+]$ (and to $PaCO_2$ to the extent that it dissociates to form H^+). A decrease in pH or an increase in $PaCO_2$ may result in an increase in respiration to three to seven times normal levels. The response to increasing the pH is less dramatic, generally resulting at most in a reduction in respiration to 50 to 75 percent of normal levels. Because the PCO_2 is directly proportional to the rate and depth of respiration, the respiratory system is able to alter pH by rapidly altering respiration. The other major determinant of body fluid PCO_2 is the rate of metabolism in the cells. If the metabolic rate increases, then carbon dioxide production increases proportionately. This leads to both an increase in PCO_2 and an acute decrease in pH. Note that fluctuations in serum pH may lead to similar or paradoxical changes in the CSF and ICF depending on the etiology of the disturbance. A discrepancy may occur as a result of the high lipid solubility of carbon dioxide and the low lipid solubility of H^+, resulting in extremely different equilibration times.

If the primary acid-base disturbance is metabolic in origin, then the lung's response cannot return the hydrogen ion concentration to normal. It can, however, partially compensate for the abnormality. A complete response is not possible because as the pH returns toward normal the driving force behind the lung's response decreases. Generally, the respiratory response to a metabolic abnormality will result in correction of 50 to 75 percent of the abnormality.

Overall, the ability of the respiratory system to respond to a metabolic acid-base disturbance is approximately one to two times that of all the chemical buffers combined.

Arterial PaO_2, $PaCO_2$, and $[H^+]$ all affect the rate and depth of alveolar ventilation. Changes in any one of these may enhance or antagonize the response to changes in another. Thus, although PaO_2 normally has little effect on the rate and depth of ventilation, when arterial PO_2 decreases below 60 mm Hg (corresponding to a hemoglobin saturation of 89%), hypoxia may drive respiration. This may result in a $PaCO_2$ as low as 20 mm Hg. In **chronic obstructive pulmonary disease (COPD),** patients frequently retain carbon dioxide and may become narcotized by the high carbon dioxide levels reached if their hypoxic drive is abolished (for example by increasing their FIO_2, thus eliminating the chronically low arterial PO_2 levels that have been driving their respirations).

In summary, then, a rise in arterial PCO_2 or hydrogen ion concentration or a drop in arterial PO_2 increases the level of respiratory center activity while changes in the opposite direction have a slight inhibitory effect. Chemoreceptors in the medulla and in the carotid and aortic bodies initiate the impulses that stimulate the respiratory center. This in turn results in a compensatory increase or decrease in alveolar ventilation.

Primary Acid-Base Disorders

There are four primary acid-base disorders:

1. Alveolar hypoventilation resulting in a primary respiratory acidosis
2. Alveolar hyperventilation resulting in a primary respiratory alkalosis
3. Metabolic acidosis
4. Metabolic alkalosis

Respiratory Acidosis

Respiratory acidosis may result from neurologic dysfunction (e.g., a decrease in central ventilatory drive or spinal cord damage), airway obstruction (e.g., COPD, reactive airway disease, etc.), muscular dysfunction (e.g., myasthenia gravis), decreased surface area for respiratory exchange (e.g., emphysema), V/Q mismatch (e.g., aspiration), decreased lung compliance (e.g., pneumonia, pleural effusion), trauma (e.g., pneumothorax, chylothorax), or physical malformation (e.g., scoliosis). Any factor that interferes with respiratory gas exchange can lead to carbon dioxide retention and respiratory acidosis. By prodigious effort, one can hold one's breath and induce a respiratory acidosis. As the pH falls, however, one will begin to lose consciousness and breathing will resume.

Respiratory Alkalosis

Respiratory alkalosis may occur as a result of an iatrogenic episode or as a response to a metabolic disturbance. Occasionally, psychoneurosis will result in overbreathing and respiratory alkalosis. Acutely, moving to a high altitude with its concomitant low partial pressure of oxygen will result in increased respiration and respiratory alkalosis.

Metabolic Acidosis

Metabolic acidosis refers to all abnormalities of acid-base imbalance other than respiratory acidosis that result in a decrease in pH below normal. The sine qua non of a metabolic acidosis is an acid pH coupled with an abnormally low bicarbonate concentration (resulting from buffering of the acid responsible for the acidosis by bicarbonate). The PCO_2 may be normal (in the absence of respiratory compensation) or decreased (if either a normal or abnormal respiratory response is present) but should not be elevated unless a mixed disorder is present. Metabolic acidosis may result from

1. Failure of the kidneys to excrete endogenous acids
2. Increased formation of endogenous acids (as in diabetes mellitus)
3. Administration of exogenous acid
4. Loss of alkali from the body (from GI or renal losses for example)

Metabolic Acidosis and the Anion Gap

Metabolic acidosis is often subdivided into those with a normal anion gap and those with an increased anion gap. The anion gap is calculated by subtracting the sum of the chloride and bicarbonate concentrations from the sodium concentration. Because sodium is the major extracellular cation, and chloride and bicarbonate are the major extracellular anions (and because the remaining cations and anions fortuitously cancel out), this difference is usually 0 to 15. Many causes of metabolic acidosis involve strong acids that contribute their conjugate bases to the anion pool, thus increasing the anion gap to greater than normal values. In contrast, administration of ammonium chloride would not result in an increased anion gap because chloride is one of the anions measured in determining anion gap.

$$\text{Anion gap} = [Na^+] - ([Cl^-] + [HCO_3^-])$$

Metabolic Alkalosis

Metabolic alkalosis is relatively rare in comparison to metabolic acidosis. Administration of diuretics (with the exception of carbonic anhydrase inhibitors) and alkaline drugs such as sodium bicarbonate may result in a metabolic alkalosis. Prolonged vomiting of gastric contents can result in a **hypochloremic metabolic alkalosis** as hydrogen ions and chloride ions are lost in the vomitus. Hydrogen ions are secreted by gastric cells in much the same fashion as renal tubular cells with concomitant reabsorption of sodium bicarbonate. Furthermore, because bicarbonate and chloride are the two major anions in the body, it is a general principle that loss of one of these anions is counterbalanced by avid retention of the other in order to maintain electroneutrality. Increased aldosterone levels may precipitate a metabolic alkalosis.

Compensation of Acid-Base Disturbances

The principle governing compensation of a disturbance in the acid-base system is that the unaffected system will act in such a fashion as to minimize the fluctuation in pH. Thus, if the disturbance is metabolic, the lungs will immediately respond by hypo- or hyperventilation as appropriate. If the disturbance is primarily that of the respiratory system, then the kidneys will respond (albeit more slowly) by holding onto or by dumping bicarbonate. This is called the **slow renal response** because, in contrast to chemical buffering and respiratory compensation, it is the slowest, taking hours to days to achieve full effect. Renal response is usually divided into acute and chronic stages. The acute renal or **metabolic response** for a respiratory disorder is actually not a response at all, but simply a reflection of the amount of carbon dioxide present in all forms including HCO_3^-. The chronic response requires that the kidneys have time to retain or discard HCO_3^- as required.

Compensation by one system for an inadequacy in the other is not 100 percent effective. It will result in moving the pH 50 to 75 percent of the way back to normal. This works relatively well for respiratory disturbances; however, the compensatory response for metabolic disturbances is neither predictable nor dependable. Table 5-4 gives the normal compensatory responses (acute and chronic) for each acid-base dis-

Table 5-4. Normal compensation for acid-base disturbances

Respiratory	Acute		Chronic	
$PaCO_2$ (mm Hg)	pH	HCO_3^- (mEq/liter plasma)	pH	HCO_3 (mEq/liter plasma)
↑10	↓0.08	↑1	↓0.04	↑3
↓10	↑0.08	↓2	↑0.04	↓5

Metabolic acidosis: For every 1 mEq fall in HCO_3^-, a compensating 1.0 mm Hg fall in $PaCO_2$ occurs unless ventilation is compromised (down to $PaCO_2$ of 5–10 mm Hg.) (From T. Raffin and J. McGee. Stanford Respiratory Therapy Department Handout, 1980.)

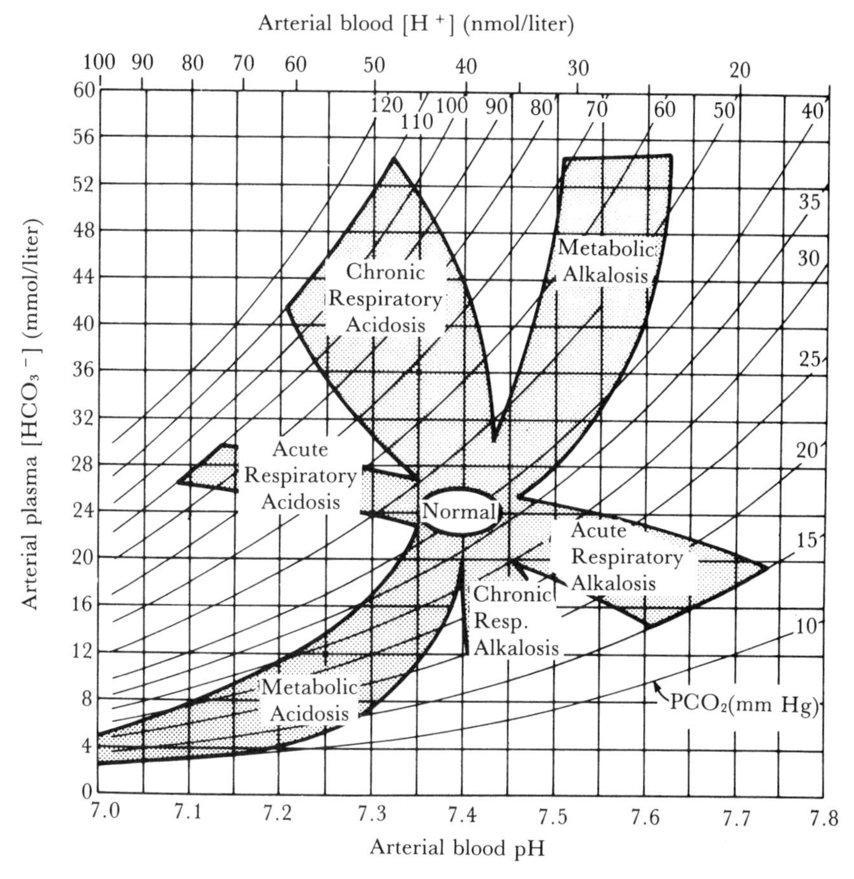

Fig. 5-6. Acid-base nomogram. Shown are the 95 percent confidence limits of the normal respiratory and metabolic compensations for primary acid-base disturbances. (From B. M. Brenner and F. C. Rector. *The Kidney* (3rd ed.). Philadelphia: Saunders, 1986. P. 473.)

order. Alternatively, an acid-base nomogram such as that pictured in Fig. 5-6 can be used to evaluate the arterial blood gas values.

An acid-base disorder is **mixed** when both the renal and respiratory systems are contributing to the problem. Thus, if the PCO_2 is elevated and the bicarbonate is low, a mixed respiratory and metabolic acidosis is present. Similarly, if a mixed problem is not present then appropriate compensation by the uninvolved system should be apparent. When plotted on an acid-base nomogram, such a mixed disturbance would yield values between those expected for each of the primary disorders present.

The following is a four-step method for determining acid-base status on the basis of arterial blood gases:

1. What is the pH? If the pH is alkaline (pH > 7.42), then the problem is an alkalosis. If the pH is acidotic (pH < 7.38), then the problem is an acidosis.
2. Is bicarbonate or is carbon dioxide the problem? If bicarbonate is the problem, then the problem is metabolic. If carbon dioxide is the problem (i.e., abnormal), then the lungs are the problem.
3. Check the compensatory response; is it appropriate? Has a mixed disorder been overlooked? If the problem is a respiratory one, is the renal response consistent with an acute or a chronic response? Is this consistent with the history?
4. Calculate the anion gap.

Clinical Correlates

The major effect of **acidosis** is to depress the CNS, which may culminate in coma and death. A metabolic acidosis will result in increased rate and depth of respiration while in respiratory acidosis it is the depressed rate and depth of respiration that are in fact causing the problem.

Alkalosis causes overexcitability of the nervous system, which may result in death due to tetany and convulsions. Both the central and peripheral nervous systems are affected. The peripheral nervous system is usually more sensitive. Tetany can frequently be demonstrated in the forearms and in the muscles of the face before the onset of CNS symptoms or tetany of the respiratory muscles.

Questions

Directions: Each of the numbered items or incomplete statements in this section is followed by answers or by completions of the statement. Select the one lettered answer or completion that is best in each case.

1. The $[H^+]$ at a pH of 7.1 is
   A. 80 μmolar
   B. 8 μmolar
   C. 80 nmolar
   D. 8 nmolar
   E. 80 picomolar
2. Carbon dioxide in the blood is carried overwhelmingly as
   A. Carbaminohemoglobin
   B. Dissolved CO_2
   C. HCO_3^-
   D. H_2CO_3
   E. CO_3^{2-}
3. The pH of one molar HCl is
   A. 0
   B. 1
   C. 2.3
   D. 3.3
   E. 4.7
4. The pH of one molar acetic acid ($pK_a = 4.7$) is
   A. 0
   B. 1
   C. 2.3
   D. 3.3
   E. 4.7
5. Lactic acid is
   A. Excreted by the kidneys
   B. Exhaled by the lungs
   C. Metabolized to HCO_3^- by the kidneys, liver, and muscle
   D. An unusual by-product of oxidative phosphorylation
   E. The usual product of aerobic metabolism
6. Beta-hydroxybutyric acid and acetoacetic acid are
   A. Important tubular fluid buffers present in high concentrations in diabetic ketoacidosis

B. Metabolized to bicarbonate by the liver when present in low concentrations but eliminated in the urine when present in high concentrations
C. Major by-products of anaerobic metabolism
D. Strong acids that may accumulate in certain disease states
E. Normal constituents of plasma and contribute to buffering of carbonic acid loads

7. The region of the titration curve corresponding to the effective buffering range is characterized by a
A. Steep slope
B. Width corresponding to plus or minus one pH unit from the pK_a of the buffer
C. Both of the above
D. Neither of the above

8.
$$K_{a1}\frac{0.03\ PCO_2}{[HCO_3^-]} = K_{a2}\frac{[H_2SO_4]}{[HSO_4^-]} = K_{a3}\frac{[HA]}{[A^-]}$$

is a statement of
A. The law of mass action
B. The Henderson-Hasselbalch equation
C. The isohydric principle
D. Ohm's law
E. Stoichiometry

9. At a given pH, which of the following blood samples would contain the most bicarbonate?
A. Arterial blood
B. Pulmonary capillary blood
C. Pulmonary venous blood
D. Systemic capillary blood
E. Venous blood

Questions 10–12 refer to the phenomenon of acclimatization to high altitude.

10. A Stanford medical student (a sea-level scutboy) joins a Himalayan expedition and climbs (or is carried) rapidly to 4 miles above sea level. Which of the following best describes his blood gases on Himalayan room air?
A. Acute metabolic acidosis
B. Acute metabolic alkalosis
C. Acute respiratory acidosis
D. Acute respiratory alkalosis
E. Mixed respiratory and metabolic alkalosis

11. The sea-level scutboy remains in the Himalayas for 2 weeks. What will his blood gases look like after acclimatization?
A. Chronic respiratory acidosis with renal compensation
B. Chronic respiratory alkalosis with renal compensation
C. Chronic metabolic acidosis with renal compensation
D. Chronic metabolic alkalosis with renal compensation
E. No acid-base disturbance, a normal ABG

12. On returning to Stanford, another blood gas is drawn. What acid-base abnormality is found?
A. Acute metabolic acidosis
B. Acute metabolic alkalosis
C. Acute respiratory acidosis

D. Acute respiratory alkalosis
E. Mixed respiratory and metabolic alkalosis

13. V/Q mismatch will most likely result in
A. Metabolic alkalosis
B. Respiratory alkalosis
C. Metabolic acidosis
D. Respiratory acidosis
E. Mixed disorder

14. Which of the following could result in a paradoxical intracellular acidosis?
A. Decreased respiratory drive resulting in an increase in PCO_2
B. Rebreathing carbon dioxide
C. Infusion of hydrochloric acid
D. Infusion of sodium bicarbonate
E. Infusion of carbonic acid

Directions: Each group of items in this section consists of lettered options followed by a set of numbered items. For each item, select the one lettered option that is most closely associated with it. Each lettered option may be selected once, more than once, or not at all.

A. Strong conjugate base
B. Strong conjugate acid
C. Weak conjugate base
D. Weak conjugate acid
E. Conjugate acid and conjugate base

15. Strong acid
16. Weak acid
17. Strong base
18. Weak base
19. HPO_4^{2-}

A. Nonvolatile acid
B. Volatile acid
C. Volatile acid anhydride
D. Exogenous acid

20. Carbon dioxide
21. Fixed acid
22. Nonfixed acid

The ratio of A^-/HA when

A. $pH = pK_a$
B. $pH = pK_a + 1$
C. $pH = pK_a + 2$
D. $pH = pK_a - 1$
E. $pH = pK_a - 2$

23. 0.01
24. 0.1
25. 1
26. 10
27. 100

Answers

1. C $[H^+] = 10^{-pH} = 10^{-7.1} = 7.94 \times 10^{-8}$ or approximately 80 nmolar. Blood is normally 40 nmolar, so at a pH of 7.1, twice as much H^+ is present as at a pH of 7.4.
2. C Bicarbonate is present in blood at concentrations of 24 mEq per liter, dissolved carbon dioxide at 1.2 µmole per liter, and carbonic acid at 0.006 µmole per liter. The ratio of bicarbonate to dissolved carbon dioxide to carbonic acid is therefore 4000 : 200 : 1.
3. A Hydrochloric acid is a very strong acid. It is therefore essentially entirely dissociated into hydrogen ion and chloride ion. One molar hydrogen ion corresponds to a pH of zero ($pH = -\log [H^+] = -\log [1] = 0$).
4. C Acetic acid is a weak acid and will therefore only partially dissociate. Given that the $pK_a = 4.7$ and $pK_a = -\log K_a$, then $K_a = 10^{-pKa} = 1.74 \times 10^{-5}$. K_a by definition is equal to the products divided by the reactants or $K_a = [H^+][RCOO^-]/[RCOOH] = 1.74 - 10^{-5}$. If acetic acid is a weak acid, then only a small fraction will dissociate. In that case, [RCOOH] will not change significantly and will remain approximately 1 mole per liter. Furthermore, $[H^+]$ will be approximately equal to $[RCOO^-]$ since acetic acid is the only significant source of H^+ present (the hydrogen ion in water will contribute only 10^{-7} mole/liter of H^+). Hence,

$$[H^+]^2 = K[RCOOH] = 1.74 \times 10^{-5},$$

or

$$[H^+] = 4.16 \times 10^{-3}$$
$$pH = -\log [H^+] = 2.3$$

The assumptions made that the H+ already in aqueous solution would not contribute significantly to the pH of the acetic acid and that the amount of RCOOH undergoing reaction would also be insignificant relative to the amount present prove reasonable, thus confirming the result.
5. C Lactic acid is metabolized to bicarbonate in the kidneys, liver, and muscle.
6. B When present in low concentrations, these substances are metabolized by the liver to bicarbonate. However, when their concentrations in the blood exceed the ability of the kidney to reabsorb them, they are eliminated in the urine (a process called ketonuria). In disease states such as diabetes mellitus, these substances may accumulate even beyond the ability of kidneys to excrete them sufficiently rapidly. This state is called diabetic ketoacidosis. They are not important tubular fluid buffers, products of anaerobic metabolism, or normal constituents of the plasma that contribute to carbonic acid buffering, so answers A, C, and E are incorrect. Answer D is incorrect because although beta-hydroxybutyric acid and acetoacetic acid accumulate in the disease state known as diabetic ketoacidosis, they are weak acids.
7. C The most effective buffering range is within one pH unit of the pK_a and corresponds to the steep portion of the sigmoidally shaped titration curve.
8. C The isohydric principle states that all buffers in a common solution must be in equilibrium with each other because they are all in equilibrium with the same $[H^+]$.
9. E The partial pressure of carbon dioxide is higher in venous blood and, because carbon dioxide is in equilibrium with HCO_3^-, more bicarbonate is present as well.

More base is also available in the form of deoxyhemoglobin, and this serves to buffer the H^+ released as the additional carbon dioxide undergoes hydration. Lastly, just the addition of base is sufficient to account for some increase in bicarbonate because by the isohydric principle, all bases in the same solution are in equilibrium. As the additional base equilibrates, H^+ ion will come off of the other buffers, thus increasing the concentrations of their conjugate bases.

10. D Initially, our scutboy will hyperventilate as pulmonary baroreceptors tell his brain that he is not getting enough air. Thus, he will experience an acute respiratory alkalosis.

11. B Over the next 2 weeks, his kidneys will adjust to the situation by excreting bicarbonate in an attempt to normalize his blood gases.

12. A On returning to Stanford level, he will suddenly experience an acute metabolic acidosis as his respiratory drive returns to normal while he continues to have below-normal levels of bicarbonate on board. This situation will normalize over the next few days as his kidneys act to conserve bicarbonate.

13. D V/Q mismatch can result in an increase in PCO_2, which by definition is a respiratory acidosis.

14. D Infusion of sodium bicarbonate can lead to paradoxical intracellular acidosis because the bicarbonate infused cannot cross the blood-brain barrier, but the carbon dioxide produced by the reaction of bicarbonate with hydrogen ion can cross the barrier freely. Once crossed, the carbon dioxide can react to form more hydrogen ion and bicarbonate, driving up the hydrogen ion concentration and thus driving down the pH inside the cells.

15. C A strong acid dissociates to form a weak conjugate base. This makes sense because if the acid has a strong propensity to dissociate into hydrogen ion (H^+) and its conjugate base, then the conjugate base must be weak; otherwise, the acid would not tend to dissociate in the first place.

16. A A weak acid dissociates to form a strong conjugate base. Again, if the acid is weak, then the conjugate base must be relatively strong. The strength of a conjugate acid or base is inversely related to its conjugate base or acid.

17. D A strong base is associated with a weak conjugate acid.

18. B A weak base is associated with a strong conjugate acid.

19. E HPO_4^{2-} is both an acid and a base. It is a weak acid and therefore has a strong conjugate base (PO_4^{3-}), but it is also a base with an associated conjugate acid, in this case $H_2PO_4^-$.

20. C Carbon dioxide is a volatile acid anhydride.

21. A A fixed acid is a nonvolatile acid; that is, it does not exist in equilibrium with a volatile product or reactant, and thus cannot evolve into a gas at usual temperatures and pressures.

22. B A nonfixed acid, such as carbonic acid, exists in equilibrium with a volatile substance, for example, carbon dioxide in this case. Because the volatile component can exist in more than one phase, its concentration on one phase (e.g., liquid) is subject to fluctuation based on changes in its concentration in another phase (e.g., gaseous).

23. E $pH = pK_a + \log [base]/[acid]$. If $pH = pK_a - 2$, then $pH - pK_a = -2 = \log [A^-]/[HA]$ and $10^{-2} = 0.01$.

24. D $pH = pK_a + \log [base]/[acid]$. If $pH = pK_a - 1$, then $pH - pK_a = -2 = \log [A^-]/[HA]$ and $10^{-1} = 0.1$.

25. A $pH = pK_a + \log [base]/[acid]$. If $pH = pK_a$, then $pH - pK_a = 0 = \log [A^-]/[HA]$ and $10^0 = 1 = [A^-]/[HA]$ or $[A^-] = [HA]$.

26. B $pH = pK_a + \log [base]/[acid]$. If $pH = pK_a + 1$, then $pH - pK_a = 1 = \log [A^-]/[HA]$ and $10^1 = 10$.

27. C $pH = pK_a + \log [base]/[acid]$. If $pH = pK_a + 2$, then $pH - pK_a = 2 = \log [A^-]/[HA]$ and $10^2 = 100$.

Bibliography

Berne, R. M., and Levy, M. N. *Physiology* (3rd ed.). St. Louis: Mosby–Year Book, 1993.

Ganong, W. F. *Review of Medical Physiology* (15th ed.). Los Altos, CA: Lange, 1991.

Guyton, A. C. *Textbook of Medical Physiology* (8th ed.). Philadelphia: Saunders, 1991.

West, J. B. *Best and Taylor's Physiological Basis of Medical Practice* (12th ed.). Baltimore: Williams & Wilkins, 1990.

West, J. B. *Respiratory Physiology: The Essentials* (4th ed.). Baltimore: Williams & Wilkins, 1990.

6

Gastrointestinal Physiology

Pauline Terebuh

The gastrointestinal (GI) system is composed of the alimentary canal (mouth, pharynx, esophagus, stomach, small intestine, large intestine, and anus) and associated glands that empty their secretions into the alimentary canal (salivary glands, pancreas, and hepatobiliary system).

This system provides a physical and immunologic barrier that separates the external environment from the internal environment while facilitating the breakdown of nutrients into small absorbable molecules **(digestion)** and the subsequent uptake of these molecules across the GI mucosa into the blood or lymphatic circulation **(absorption).** The gut is a dynamic and highly coordinated environment that secretes approximately 7 liters of various ions, organic molecules, and enzymes to the 2 liters that is typically ingested on a given day. This 7 liters of fluids is secreted by GI **exocrine glands** and mucosal cells into the gut lumen to lubricate, digest, transport, and maintain osmotic balance. Gastrointestinal motility, generated by contractions of the smooth muscle cells in the gut wall, mixes and propels the luminal contents down the GI tract.

Histology

In a cross section of the GI tract (Fig. 6-1), there are four histologic subdivisions: mucosa, submucosa, muscularis, and adventitia. The characteristics of the mucosa vary together with its function along the length of the tract: protective, secretory, and absorptive. The submucosa is a connective tissue layer. The muscularis is composed of smooth muscle cells arranged in an inner circular and an outer longitudinal layer. The spindle-shaped smooth muscle cells in each layer are electrically coupled by means of gap junctions, thus forming a functional syncytium. The exceptions to this general organization of smooth muscle can be found in the

1. Esophagus. The upper one third is striated muscle; the middle one third is mixed striated and smooth muscle.
2. Stomach. The muscularis is reinforced with an innermost oblique layer.
3. Colon. The outer longitudinal muscle layer concentrates into three bundles called the teniae coli, which fuse along the rectosigmoid, forming a uniform layer in the rectum.
4. Sphincters
   a. The upper esophageal sphincter (cricopharyngeus muscle) is striated.
   b. The lower esophageal (gastroesophageal) sphincter is physiologically but not anatomically distinct from the circular muscle layer.

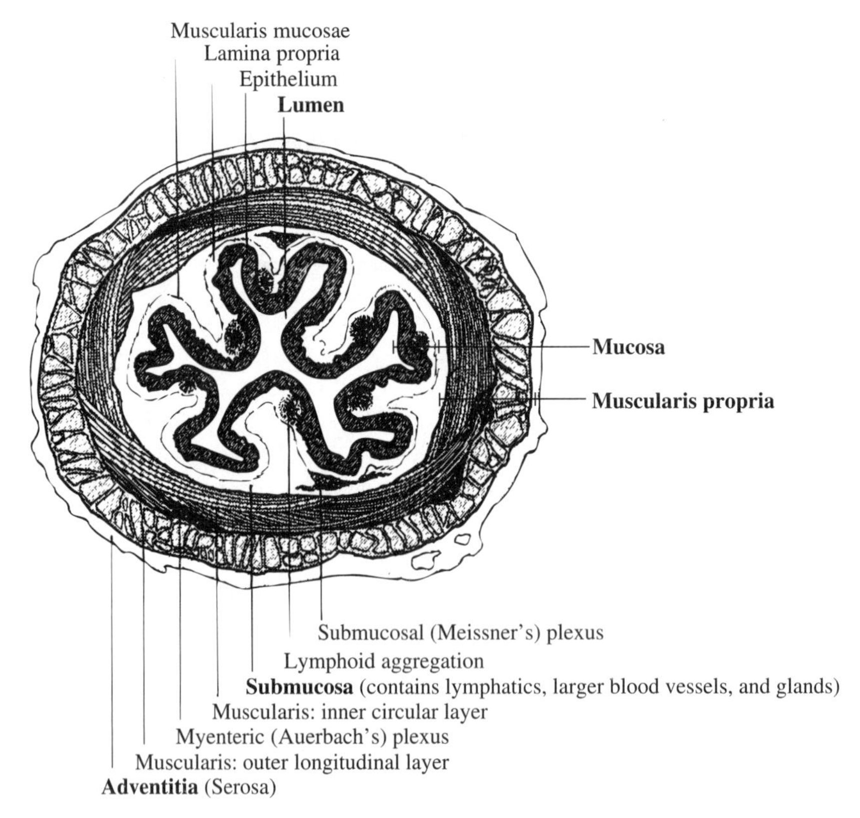

Fig. 6-1. Cross section of GI tract. (Adapted from P. R. Wheater and H. G. Burkitt. *Functional Histology: A Text and Colour Atlas* [2nd ed.]. New York: Churchill Livingstone, 1987. P. 204.)

c. The pyloric sphincter, the ileocecal sphincter, and the internal anal sphincter are thickenings of the circular muscle layer.
d. The external anal sphincter is striated muscle receiving somatic innervation.

Regulation

Almost all aspects of GI function fall under neural and hormonal regulation. This chapter will first describe the specific hormones and innervation that orchestrate the activity of the GI system and then describe the specific processes that they regulate. These processes include

1. Blood flow
2. Motility
3. Secretion
4. Absorption

Gastrointestinal Hormones and Peptides

Gastrointestinal hormones are proteins synthesized and released primarily by enteroendocrine cells that line the gut. They are released into the bloodstream and exert their effect at some distant site, usually another organ in the GI system. Additionally, other proteins classified merely as GI peptides exert their effect in a paracrine or neurocrine fashion (see Chap. 7). While some of these hormones and peptides are also synthesized in other organs such as the brain, their function there is largely undetermined and will not be discussed. Gastrointestinal hormones and peptides bind receptors in their target organs and activate these cells via two possible signal transduction mechanisms: G-proteins or activation of Ca^{2+} channels (see Chap. 1). Based primarily on sequence homology, many of the GI hormones and peptides are organized into two families: the **cholecystokinin (CCK)/gastrin family** and the **secretin family.** The major functions and sites of action of the GI hormones are reviewed in Fig. 6-2.

Cholecystokinin and gastrin share an identical carboxy-terminal sequence, but unique amino-terminal sequences. **Gastrin** is the only major hormone secreted in the stomach, hence its name. Preprogastrin, the precursor, is synthesized by **G cells** in the gastric antrum, a distal region of the stomach, and following cleavage to gastrin, it exerts its effect largely on other gastric mucosal cells. Its major action is to stimulate the stomach's parietal cells to secrete HCl. Gastrin's stimulatory effect on parietal cell secretion of gastric acid is significantly potentiated by **histamine** and **acetylcholine** (Ach). Secondly, gastrin also stimulates these same **parietal cells** to secrete intrinsic factor **(IF).** Intrinsic factor then binds **vitamin B_{12}** in the stomach and the complex is absorbed downstream in the ileum. Thirdly, gastrin stimulates **chief cells** in the stomach to secrete **pepsinogen,** the first of the many proteolytic enzymes to attack the bolus of food. The release of gastrin is stimulated by the presence of amino acids and small peptides in the stomach, distention of the stomach, and neural input (via the vagus) such as thinking about food. Gastrin release is inhibited once the lumen reaches a pH of less than 3.

Cholecystokinin is synthesized most abundantly by **I cells** that line the duodenum and jejunum. It is the gatekeeper of protein and fat digestion since it has two major effects. Cholecystokinin stimulates secretion of the pancreatic digestive enzymes and contraction of the gallbladder. The pancreatic enzymes and bile are essential for luminal digestion and the emulsification of fats. Cholecystokinin also sends a chemical signal from the small intestine that inhibits gastric emptying. Not surprisingly, it is the digestive products of proteins and fats and acidity of the duodenum that stimulate the release of CCK.

The secretin family includes secretin, vasoactive intestinal peptide **(VIP),** gastric inhibitory peptide **(GIP),** as well as lesser hormones with as yet unspecified physiologic roles (enteroglucagon, growth hormone releasing factor). Their membership in the family is based on sequence homology to secretin. **Secretin** is synthesized and secreted by **S cells** of duodenum and jejunum. It is responsible for the bulk of fluids that enter the small bowel since it stimulates H_2O and HCO_3^- secretion by the pancreas and bile ducts. Binding of cholera toxin to the secretin receptor mimics its effects producing a secretory diarrhea. Secretin supplements the effect of CCK by also stimulating, but to a much lesser extent, the secretion of pancreatic digestive enzymes. The overall effect is to raise the pH of the chyme as it emerges from the stomach; secretin secretion is predictably stimulated by acidification of the duodenum and also by products of fat digestion.

Vasoactive intestinal peptide is produced by neurons in the enteric system and elsewhere in the body. Its actions, while not firmly established, are likely to include secretion of ions, H_2O, and HCO_3^- from the intestine and pancreas; increase in intestinal blood flow; and relaxation of GI smooth muscle. GIP is synthesized and secreted by **K cells** in the small intestine. Its homology to glucagon confers its ability to stimulate insulin secretion in response to oral glucose. It also inhibits water and electrolyte absorption in the small intestine.

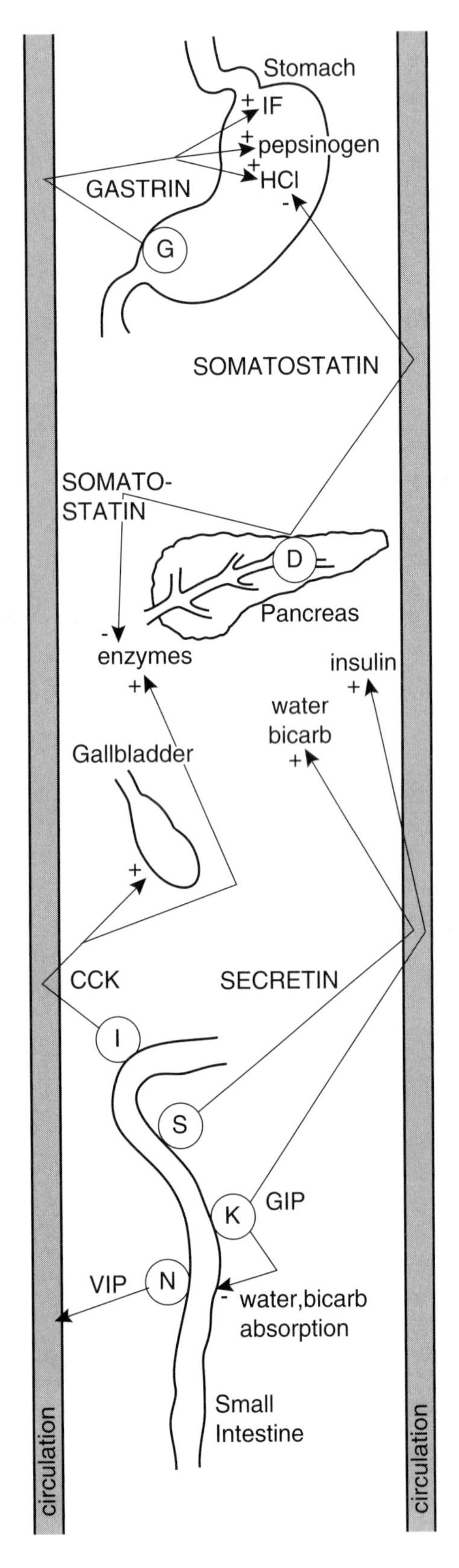

	Action	Stimulus	Pathway
Gastrin/CCK Family			
GASTRIN G cells gastric antrum preprogastrin	- HCL secretion pepsinogen secretion (chief cells) - IF secretion (parietal cells)	- aa peptides - vagal discharge - stomach distention	G-protein ↑$[Ca^{2+}]$ via IP_3
CCK I cells duodenum jejunum preproCCK	- pancreatic enzyme secretion - gallbladder contraction - inhibits gastric emptying	- protein digests - fat digests - acid lumen	G-protein ↑$[Ca^{2+}]$ via IP_3
Secretin Family			
SECRETIN S cells duodenum jejunum	- HCO_3^-, H_2O secretion (pancreas, bile ducts) - potentiates CCK	- acid lumen - fat digests	G-protein cAMP
VIP Neurons enteric, others	SPECULATED - intestinal secretion HCO_3^- (pancreas) - vasodilation - GI SM relaxation	- unknown	G-protein cAMP
GIP K cells duodenum jejunum	- insulin secretion - inhibits intestinal H_2O, HCO_3^- absorption	- oral glucose, fats, amino acids	?
Other Hormones			
SOMATO- STATIN D cells pancreas neurons enteric, others	- inhibits HCl secretion (stomach) - inhibits enzyme secretion (pancreas)		G-protein inactivates adenylate cyclase via $G_{\alpha i}$
MOTILIN ?cells, intestine	- peristaltic contraction		

Minor GI Hormones and Proteins

Enteroglucagon - synthesized in the intestine; processed to glucagon, action is undetermined.

Neurotensin - synthesized by N cells in the ileum; speculated actions include inhibition of acid secretion in response to fat, stimulation of pancreatic secretion, stimulation of intestinal fluid secretion, increased mucosal blood flow.

Pancreatic Polypeptide - synthesized in pancreas; action is inhibition of pancreatic exocrine secretion; stimulated by proteins in intestinal lumen.

Fig. 6-2. Major GI hormones and their actions.

Outside the two main GI hormone families, there exist several other heterogeneous hormones. Two of these are somatostatin and motilin. **Somatostatin** is synthesized by many GI-associated cells, including enteric neurons, endocrine cells in the stomach, and **D cells** in the pancreas. Most of somatostatin's actions are inhibitory, for example, the inhibition of HCl secretion in the stomach and the inhibition of pancreatic enzyme secretion. **Motilin** is synthesized in the intestine and induces peristaltic contractions.

Nerve Supply of the Gastrointestinal Tract

The GI system has its own enteric nervous system, that is, neurons that have their cell bodies within the wall of the alimentary canal (Fig. 6-3). These intrinsic neurons also receive input from extrinsic neurons forming an extensively integrated network. There are two major plexuses: (1) the **submucosal (Meissner's) plexus,** which is within the submucosa and governs secretion, and (2) the **myenteric (Auerbach's) plexus,** which is between the circular and longitudinal muscle layers and governs motility.

These neurons receive afferent signals from the GI mucosa and send efferent signals back, thereby mediating local reflexes. Acetylcholine is the excitatory transmitter, while adenosine triphosphate (ATP) and VIP appear to be the principal inhibitory transmitters.

The extrinsic neurons that also synapse onto these plexuses include the

1. Voluntary nervous system. The fibers that innervate the extreme ends of the GI system are alpha motor neurons traveling via the vagus to the esophagus and via the pudendal to the external anal sphincter, allowing voluntary control of swallowing and defecation.
2. Sympathetic nervous system. Postganglionic sympathetic nerve fibers innervate all portions of the GI tract. The generally inhibitory effect is mediated through the release of norepinephrine (NE). Its inhibitory action on smooth muscle and glan-

Fig. 6-3. Neural control of the gut wall showing (1) the myenteric and submucosal plexuses, (2) extrinsic control of these plexuses by the sympathetic and parasympathetic nervous systems, and (3) sensory fibers passing from the luminal epithelium and gut wall to the enteric plexuses and from there to the prevertebral ganglia, spinal cord, and brainstem. (Adapted from A. C. Guyton. *Textbook of Medical Physiology* [8th ed.]. Philadelphia: Saunders, 1991. P. 691.)

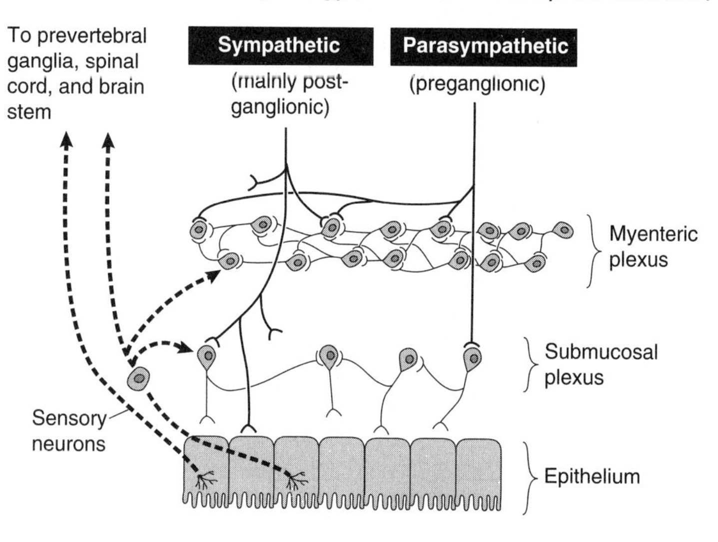

dular cells is both direct and indirect by binding to the presynaptic (alpha-2) receptors on cholinergic fibers, causing a decreased parasympathetic output.

3. Parasympathetic nervous system. Preganglionic parasympathetic fibers travel via the vagus and the sacral outflows. The postganglionic neurons are located in the submucosal and myenteric plexuses. The general effect is stimulatory, although some of the enteric neurons that they influence are inhibitory (e.g., inhibit tone of most sphincters with the notable exception of the lower esophageal sphincter, which they stimulate).
4. Sensory pathways. Feedback from the mucosal receptors that sense the lumen's chemical environment, mucosal irritation, and gut distention travels to enteric neurons, prevertebral sympathetic ganglia, or all the way to the CNS. This allows for multiple levels of control of GI function.

Blood Flow

The splanchnic circulation supplies blood to the components of the GI tract. Branches off the aorta (celiac, superior mesenteric, inferior mesenteric) supply the stomach, intestines, pancreas, spleen, and liver with arterial blood. Venous blood from these organs reunites in the portal vein en route to the liver. This blood carries numerous constituents, including nutrients absorbed from the intestines, bilirubin from the spleen and gut, and hormones from the gut and pancreas. After percolating through the hepatic sinusoids, the blood flows to the inferior vena cava via the hepatic vein. Blood flow to the gut is under neural and hormonal regulation such that it can be profoundly reduced by sympathetic discharge and increased by the neurocrine mediator VIP, which is liberated by enteric neurons.

Motility

Physiologic Properties of Gastrointestinal Smooth Muscle

Smooth muscle cells of the GI tract are generally connected to adjacent cells by gap junctions. This arrangement allows ions and currents to flow through many interconnected cells so that large groups of muscle cells can act like a **functional syncytium.** The smooth muscle cells differ from skeletal muscle cells in several respects. Rather than maintaining a constant resting potential, the resting potential of the intestinal smooth muscle cells spontaneously oscillates 5 to 15 mV in cycles that occur at a frequency of 3 to 12 times per minute. This pattern is described as **"slow waves,"** resulting from the flow of Na^+ ions. These waves generally do not result in contraction.

A second type of electrical activity in intestinal smooth muscle cells is called **"spikes."** These membrane potential phenomena are more like the typical all-or-nothing action potentials in neurons and result in contraction of the smooth muscle fiber. The stimuli for the spikes include stretch, Ach, parasympathetic discharge, and certain GI hormones. The spikes are generated by an influx of Ca^{2+} through voltage-gated Ca^{2+} channels. The increase in intracellular Ca^{2+} in the muscle fiber causes contraction via the calmodulin pathway. Since these ion channels are much slower than the Na^+ channels in skeletal muscle, the spike may last much longer than the potential in skeletal muscle. The timing of the spike potentials is influenced by the background of slow waves such that they tend to superimpose themselves during the depolarizing phase of the slow wave cycle (Fig. 6-4). The slow waves thereby serve as a sort of pacesetter potential, the superimposed rhythmic contractions of peristalsis. The GI smooth muscle is also capable of tonic contractions that last from min-

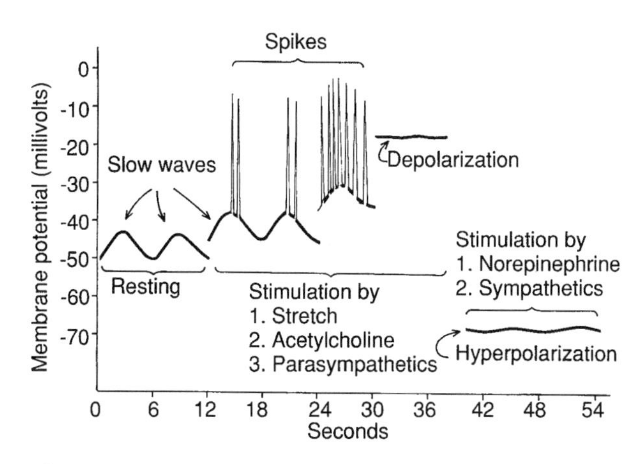

Fig. 6-4. Membrane potentials in intestinal smooth muscle. Note the slow waves, the spike potentials, total depolarization, and hyperpolarization, all of which occur under different physiologic conditions of the intestine. (From A. C. Guyton. *Textbook of Medical Physiology* [8th ed.]. Philadelphia: Saunders, 1991. P. 689.)

utes to hours. These are not dependent on the pacesetting of the slow waves, but are a pattern more typical of sphincter smooth muscle. Various stimuli, including repeated spike potentials and hormones, are capable of inducing this type of contraction.

Finally, NE and sympathetic stimulation hyperpolarize the smooth muscle membrane, inhibiting contraction of the muscle fiber. Figure 6-4 summarizes the sequential changes in intestinal smooth muscle potential (1) at rest with only slow waves, (2) stimulated with slow waves and spikes, and (3) inhibited with little fluctuation.

The peristaltic wave of contraction of the gut proceeds in an oral to anal direction. It is a reflex governed entirely within the enteric nervous system that helps to propel the bolus of food unidirectionally analward. A single ganglion of the myenteric plexus innervates a circumferential ring of intestine. Stimulation of a ring just proximal to the bolus is coupled with inhibition a few centimeters downstream. The end result is a contraction of a ring of intestine simultaneous with receptive relaxation distally.

Secretions

As the roughly 2 liters of ingested food travels through the proximal half of the gut, it is broken down into progressively smaller particles. The additional 7 liters of secretions from the GI organs lubricates this passage, digests and solubilizes the particles, and maintains osmotic equilibrium. Since the number of particles increases along the way, the increase in fluid volume is necessary to maintain an isoosmotic lumen.

Salivary Secretions

The salivary glands secrete approximately 1500 ml saliva per day. Although there are many small buccal salivary glands, more than 90 percent of saliva is secreted by the paired parotid, submandibular, and sublingual salivary glands. The salivary secretory unit is composed of acini and ducts. The **acinar cells** elaborate a primary secretion that contains enzymes, mucoproteins, and electrolytes at concentrations similar to those of plasma. The cells that line the proximal salivary ducts, called **striated ducts,** modify the electrolyte composition of this primary acinar secretion as it flows out the

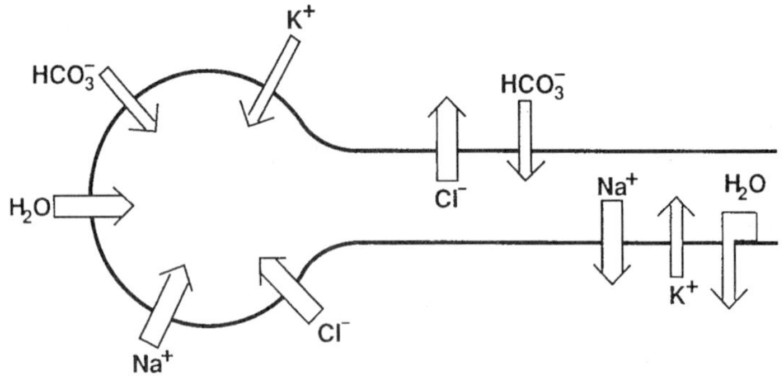

Fig. 6-5. Movements of ions and water in the acinus and duct of the salivon. The acinus secretes electrolytes at concentrations similar to those of plasma. At slow flow rates, the duct has sufficient time to modify the concentrations to closer to that of interstitial fluid. (From L. R. Johnson. *Gastrointestinal Physiology* [4th ed.]. St. Louis: Mosby, 1991. P. 63.)

duct by actively reabsorbing Na^+, passively reabsorbing Cl^-, and secreting K^+ and HCO_3^- (Fig. 6-5). The final composition of saliva is dependent on the salivary flow rate.

At slow rates, when there is ample time for active transport processes of the striated ducts, the ratio of Na^+ to K^+ concentration in saliva is similar to that of intracellular fluids, which are relatively K^+ rich. At faster rates the ratio of Na^+ to K^+ concentration remains closer to that of extracellular fluids, which are relatively Na^+ rich. Aldosterone influences Na^+-K^+ exchange in the striated ducts in the same way that it does in the distal convoluted tubule of the nephron: It increases Na^+ reabsorption and K^+ secretion.

There are two distinguishable types of salivary acini. **Serous acini** produce a watery secretion rich in an alpha-amylase called ptyalin. **Mucinous acini** elaborate a more viscous secretion that contains a variety of mucoproteins. The parotid glands are almost entirely composed of serous acini, while the sublingual glands are predominantly composed of mucinous acini. The submandibular glands are mixed salivary glands. They contain both serous and mucinous acini.

The most important function of saliva is the role that it plays in oral hygiene. Retained food particles between teeth serve as nutrients for bacteria. The acid that is liberated from bacterial metabolism can dissolve the enamel of teeth. The flow of saliva protects the enamel by buffering and washing away this acid and also by removing retained food particles. Saliva contains immunoglobulin A (IgA), antimicrobial proteins such as lysozyme, and thiocyanate ions, all of which act to protect the oral cavity against bacteria.

Saliva also plays a role in digestion. Ptyalin, an alpha-amylase capable of breaking internal alpha-1,4 glycosidic linkages, is secreted in saliva. The breakdown of carbohydrates is initiated in the mouth by salivary ptyalin and is continued until inactivation of ptyalin by gastric acid. Although ptyalin is important in carbohydrate digestion, it is a nonessential enzyme because the pancreas secretes amylases with the same function as ptyalin. Salivary secretion is under neural regulation. Salivary glands are innervated by both sympathetic and parasympathetic nerves. Signals such as taste, sight, smell, or thought of food are conveyed via afferent nerves to the salivatory nuclei in the brainstem, which in turn causes efferent parasympathetic nerves to discharge. Parasympathetic stimulation can increase the rate of secretion by as much as 20-fold over baseline. Unlike other physiologic functions under autonomic regulation, in which sympathetic and parasympathetic nerves have antagonistic effects, in the regulation of salivary secretion, sympathetic stimulation is complementary to parasympathetic stimulation. The intracellular effects of autonomic stimulation on secretory cells are mediated by the second messengers cyclic adenosine monophosphate (cAMP), Ca^{2+}, and diacylglycerol (DAG). Although both branches of the autonomic nervous system increase salivary secretion, the parasympathetic system is the predominant regulator of salivary glands.

Gastric Secretions

The stomach secretes about 2500 ml gastric juices per day. Anatomically, the stomach is divided into two major regions: (1) the fundus and body (the proximal 80%) and (2) the antrum (the distal 20%). The secretory glands of the stomach localize according to this same geography.

1. Fundus and body (oxyntic) glands
   a. Mucus cells—secrete mucus
   b. Parietal cells—secrete HCl, IF
   c. Chief cells—secrete pepsinogen
2. Antral glands
   a. Mucus cells—secrete mucus
   b. G cells—secrete gastrin

The pH of gastric secretions can be as low as 1.0. Parietal cells achieve this tremendous H^+ gradient of over 2,000,000 : 1 by using their unique H^+-K^+ adenosine triphosphatase (ATPase) pumps. A proposed mechanism of this process is outlined (Fig. 6-6).

1. The H^+-K^+ ATPase pump is a transmembrane protein located within the luminal surface of the parietal cell. It is similar to the basolaterally located Na^+-K^+ ATPase in other mucosal cells with an alpha and beta catalytic subunit, powering the expulsion of each proton out of the cell by hydrolyzing an ATP molecule. Unlike the Na^+-K^+ ATPase, the exchange is 1 : 1 and, as such, not electrogenic. The absence of electrical gradient allows for the generation of the large pH gradient.

Fig. 6-6. Proposed mechanism for parietal cell HCl secretion and cellular homeostasis.

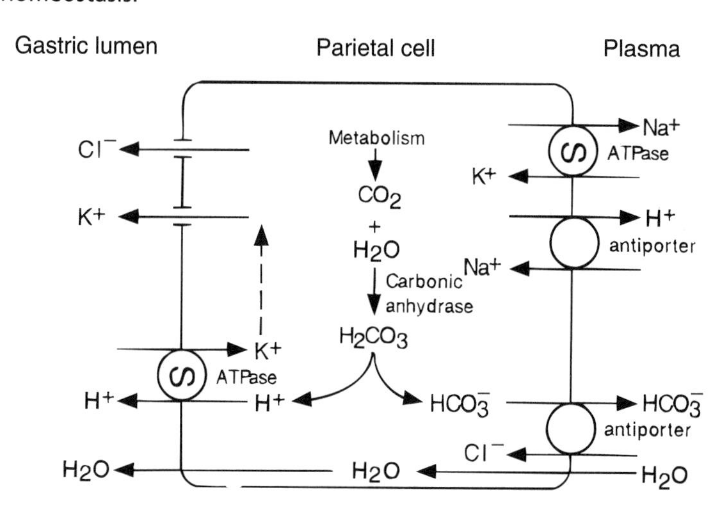

2. K^+ must passively recirculate to the lumen to regain access to the pump; Cl^- follows to maintain electroneutrality. These ions also draw water into the lumen such that the cell's secretions are isotonic. Cl^- channels in the apical membrane have been detected, and their conductance is regulated by intracellular cAMP. The search for K^+ channels is still on.
3. Regulation of proton pumping can occur on two levels: the number of pumps in the membrane and the activity of pumps. Vesicles of plasma membrane containing the pumps can endocytose into the cytosol, decreasing the number and availability of the pumps, until they receive a signal to exocytose and fuse with the apical membrane, increasing the availability of the pumps. A drop in cAMP decreases Cl^- channel conductance, which limits the availability of K^+ to the pump thereby limiting its activity.
4. Channels and pumps in the basolateral membrane have the responsibility of maintaining cytosolic homeostasis. For each H^+ that is pumped into the lumen, an HCO_3^- is generated by catalysis of carbonic anhydrase. The HCO_3^- molecule is transported out to the plasma side by a Cl^--HCO_3^- antiporter. Along with the Na^+-K^+ ATPase and the Na^+-H^+ antiporter, they maintain intracellular pH and act as a source of K^+ and Cl^-.

The relative concentrations of electrolytes in gastric secretions change with flow rate. While the concentrations of K^+ and Cl^- in gastric secretions are relatively constant with increasing flow rate, H^+ concentrations increase and Na^+ concentrations decrease because the H^+-pumping parietal cells make a larger contribution to the total gastric secretions.

Pepsinogen is an inactive **zymogen** secreted by gastric chief cells. Once in the gastric lumen, pepsinogen is converted to pepsin by the action of gastric acid. Pepsin can, in turn, activate other pepsinogen molecules. As a proteolytic enzyme capable of cleaving internal peptide linkages, pepsin initiates the digestion of proteins in the stomach.

Intrinsic factor is a peptide secreted by parietal cells. Its function is to bind vitamin B_{12}, forming a complex that is then absorbed downstream in the distal ileum. In the absence of IF, vitamin B_{12} cannot be absorbed.

In addition to the typical soluble mucus elaborated throughout the GI tract, the mucus cells of the stomach secrete a viscous insoluble mucus that coats the surface epithelium of the stomach, conferring a crucial protective barrier to the gastric mucosa.

The function of acid in the stomach is to kill ingested bacteria, activate pepsinogen, and promote the binding of IF to vitamin B_{12}. Regulation of its secretion is controlled by the interplay of multiple factors. The three activators are gastrin, Ach, and histamine; each potentiates the effects of the other and, therefore, functions synergistically in stimulating acid secretion. The inhibitory factors include low luminal pH, CCK, and neural input. All of these factors work cooperatively to influence gastric acid secretion during its three phases: cephalic, gastric, and intestinal.

During the cephalic phase, before the entrance of food into the stomach, signals such as taste, smell, sight, or thought of food can activate the vagal nuclei in the medulla, causing vagal stimulation of the stomach. This enhances gastric secretion in two ways. First, Ach released from vagal nerve endings directly stimulates parietal and chief cells to increase their secretion. Second, vagal stimulation indirectly enhances acid secretion by stimulating release of gastrin from G cells in the antral mucosa. Gastrin, as mentioned earlier, is also a potent stimulus for acid secretion.

During the gastric phase, the presence of food in the stomach is sensed by mechanoreceptors that are activated by distention and chemoreceptors that are activated primarily by proteins and products of protein digestion. Activation of these receptors results in increased gastric secretion through the following mechanisms:

1. Vagovagal reflexes. Afferent fibers from gastric mechanoreceptors are coupled with efferent fibers in the vagus, increasing vagal output.
2. Local reflexes. Mechanical distention can also activate local reflexes that have both their afferent and efferent fibers located within the gastric wall.
3. Mechanical distention in the antrum stimulates G cells directly to increase the release of gastrin.
4. Some of the chemical constituents of food, principally proteins and products of their digestion, stimulate G cells to release gastrin.
5. Other food constituents such as caffeine can directly stimulate parietal cells.
6. While the source and regulation of histamine release is unclear, histamine binds to the H2 receptor on parietal cells, which produces an increase in intracellular cAMP. Histamine potentiates the secretion of acid in response to Ach and gastrin.

During the intestinal phase, presence of food in the intestine is a minor stimulant of gastric secretion. The effect is mediated through release of gastrin from endocrine cells in the mucosa of the proximal duodenum.

Several factors inhibit gastric secretion of acid. These include

1. Low luminal pH. When the lumen reaches pH 3 or less, release of gastrin secretion is inhibited.
2. Enterogastric reflex. Mechanical distention of the small intestine decreases vagal output to the stomach.
3. Cholecystokinin. Triggered by the presence of fat, protein, or acid in the small intestine, CCK inhibits acid secretion.

While the stomach actively initiates some processes of digestion, its secretions are not essential for digestion. The stomach's *critical function is the storage of ingested food.* The compliance of the stomach is mediated by the smooth muscle relaxant nitric oxide. Peristaltic contractions mechanically break down food particles to an average size of a couple of millimeters before they exit the pylorus. Regulation of gastric emptying is the domain of both the stomach and the small intestine. In general, gastric distention and gastrin promote gastric emptying. Signals from the small intestine that communicate the unreadiness of the small intestine to accept chyme for definitive digestion are largely inhibitory. These inhibitory signals include CCK and the enterogastric reflex, which likewise inhibit gastric acid secretion.

Pancreatic Secretions

The pancreas secretes approximately 1500 ml pancreatic juice into the duodenum per day. The basic secretory unit of exocrine pancreas, like that of the salivary gland, is composed of acini and ducts. The acinar cells secrete digestive enzymes while the centroacinar cells and the cells that line the small ductules secrete HCO_3^- and water.

The acinar cells secrete enzymes for the digestion of proteins, carbohydrates, fats, and nucleic acids. Unlike the digestive enzymes of saliva and the stomach, the pancreatic enzymes are essential to digestion and absorption of nutrients. To prevent autodigestion of pancreatic tissues, the potent proteolytic enzymes of the pancreas are synthesized, stored, and secreted as inactive zymogens. These include trypsinogen, chymotrypsinogen, proelastase, and procarboxypeptidase. Once in the duodenum, trypsinogen is converted to the active enzyme trypsin by the action of **enterokinase** (misnamed a kinase rather than a peptidase), which is an enzyme present in the duodenal brush border. Trypsin can itself activate trypsinogen, thus setting up a positive feedback loop for activating trypsinogen. Trypsin also activates all the other proteolytic enzymes. To prevent premature activation of trypsinogen in the pancreatic ducts, the acinar cells secrete a substance called trypsin inhibitor.

In addition to proteolytic enzymes, the pancreatic acini secrete ribonuclease and deoxyribonuclease, which digest nucleic acids. Other enzymes found in pancreatic secretions include those engaged in fat digestion (lipases, phospholipases, and cholesterol esterase), as well as pancreatic amylase, which is involved in carbohydrate digestion.

Pancreatic secretions are basic; therefore, they (1) neutralize gastric acid and thereby protect the duodenal mucosa from injury by acid, and (2) raise the pH of duodenal fluids to a range that is optimal for the activity of pancreatic digestive enzymes. Figure 6-7 diagrams a proposed mechanism for HCO_3^- secretion. Carbonic acid, formed from carbon dioxide and water by action of carbonic anhydrase, dissociates into H^+ and HCO_3^-. On the luminal surface of the centroacinar and duct cells, HCO_3^- is exchanged for Cl^-. On the basolateral surface Na^+ is exchanged for H^+. The energy for these exchanges is the gradient generated by the Na^+-K^+ ATPase in the basolateral membrane. The electrolyte composition of the pancreatic juice varies with increasing flow rate such that the concentration of HCO_3^- increases as the concentration of Cl^- decreases due to the 1 : 1 exchange. The concentrations of Na^+ and K^+ in the juice remain equivalent to their concentrations in plasma, independent of flow rate.

Pancreatic secretions are under both neural and hormonal regulation. During the cephalic phase, before entry of chyme into the duodenum, vagal stimulation increases enzyme secretion by acinar cells. Later, as chyme enters the duodenum, the hormone CCK is secreted by I cells of the duodenal and jejunal mucosa. Peptides, amino acids, and fatty acids are particularly potent signals to the secretion of CCK. Cholecystokinin stimulates the acinar cells to secrete the digestive enzymes that, following activation, will digest the molecules that promoted their secretion. Presence of acid in the duodenum is a potent signal for the release of secretin from the S cells of the duodenal and jejunal mucosa. Secretin stimulates the ductule cells to secrete a watery, HCO_3^--rich secretion.

Fig. 6-7. Proposed mechanism for pancreatic ductule cell secretion of electrolytes. With increasing flow rates, the concentration of HCO_3^- increases as the concentration of Cl^- decreases. The concentrations of Na^+ and K^+ are independent of flow rate and remain approximately equal to their concentrations in plasma.

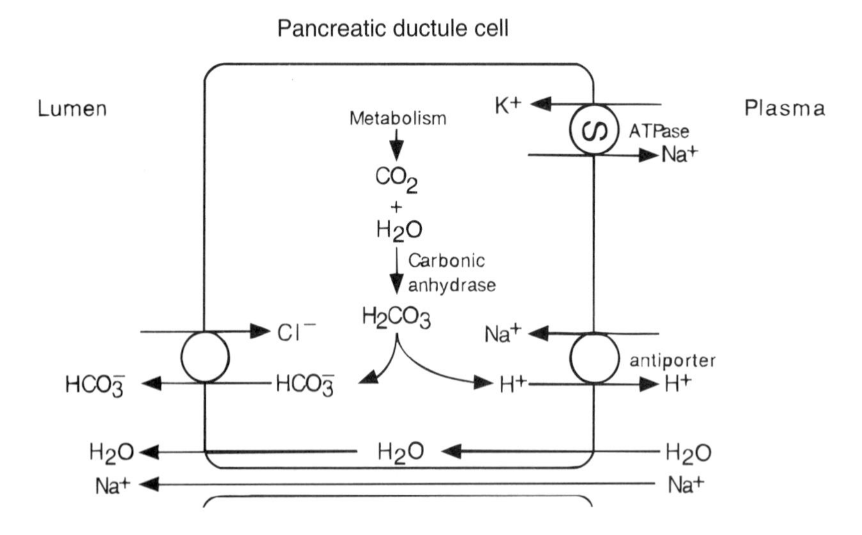

General Physiology of Bile

Approximately 500 ml bile is expelled into the duodenum daily. Bile is synthesized and secreted continuously by hepatocytes into the bile canaliculi of the liver. The canaliculi flow into ductules and eventually feed into the hepatic duct. The bile then follows the path of least resistance to either the cystic duct en route to the gallbladder or directly down the common bile duct to the first part of the duodenum. Between meals, the sphincter of Oddi is contracted, so bile travels to the gallbladder for temporary storage. While in the gallbladder, bile is concentrated 3- to 15-fold by active transport of Na^+ with either Cl^- or HCO_3^- to maintain electroneutrality; water passively follows. Following a meal, the sphincter of Oddi relaxes, bile flows from the liver into the small intestine directly, and the gallbladder contracts so that bile flows out of the gallbladder and into the small intestine.

Bile plays an essential role in both excretion and in digestion/absorption. Components of bile aid in the digestion and emulsification of fats. Bile secretion into the intestine is also a route for elimination of lipid-soluble waste products from both endogenous sources (cholesterol, bilirubin) and exogenous sources (lipophilic drugs, toxins). Constituents of bile include

Water (84%)
Bile salts (11%)
Lecithin (3%)
Cholesterol (0.5%)
Bilirubin, protein, inorganic ions (1%)
Electrolytes at concentrations similar to those in plasma except more HCO_3^- and less Cl^- (pH 7.5–9.0)

Bile Acids

Bile acids are **amphipathic** molecules; that is, they contain both hydrophilic and hydrophobic domains, making them capable of forming an interface between lipophilic molecules and aqueous solution. In the aqueous lumen of organs such as the gallbladder and small intestine, they form thermodynamically stable spheres called **mixed micelles.** This arrangement allows the bile acids to incorporate other amphipathic molecules such as lecithin into the micellar shell or to sequester insoluble lipids within the core, thereby suspending these lipids in aqueous solution. Micelles provide these hydrophobic molecules, that is, excess body cholesterol, bilirubin, drugs, and toxins, with a means of exiting the body. As bile is concentrated in the lumen of the gallbladder, the lipid components spontaneously assemble into mixed micelles at the **critical micellar concentration,** the minimum concentration of bile salts at which micelles form. If the proportions of bile acids, lecithin, and cholesterol are not in balance, cholesterol may precipitate out of solution, contributing to the formation of a gallstone. Once in the lumen of the small intestine, bile acids behave similarly by emulsifying fat. Mixed micelles reform, this time incorporating dietary lipids.

Only a small fraction of the bile acids released into the duodenum is lost in the feces. The remainder is reabsorbed from the intestine, taken up by the liver, and resecreted in bile. This is called the **enterohepatic circulation of bile.** The total body pool of bile acids is 2 to 4 g. This pool undergoes 6 to 10 cycles of secretion and reabsorption per day. Only 0.2 to 0.6 g of bile acids (10%) is lost in feces each day. In order to maintain steady state for the total body pool of bile acids, the rate of hepatic synthesis must be equal to the rate of fecal loss.

Bile acids are derivatives of cholesterol synthesized by hepatocytes. The synthesis of the bile acids begins with the catabolism of cholesterol via two separate pathways to either cholic acid or chenodeoxycholic acid. Within the typical pH range, these acids are likely to be partially ionized, existing as bile salts. These two acids are called the **primary bile acids** and can be further modified by conjugation with either of two amino acids, glycine or taurine. The conjugated bile acids have a lower pK_a, such that they are almost completely ionized within the typical pH range. Thus, there are four **conjugated primary bile acids:** (glyco-, tauro-) cholate and (glyco-, tauro-) chenodeoxycholate. Once in the small intestine, resident bacteria can deconjugate the primary bile acids. They can also further modify the primary bile acids by a dehydroxylation to what are then called **secondary bile acids.** Cholate is converted to deoxycholate; chenodeoxycholate is converted to lithocholate. Because lithocholate is insoluble, it is excreted. Therefore, only one of the two secondary bile salts, deoxycholate, can be reabsorbed into the enterohepatic circulation and returned to the liver. In the liver, the secondary bile acid is conjugated to (glyco-, tauro-) deoxycholate and secreted with the four conjugated primary bile salts.

There are two mechanisms for intestinal reabsorption of bile acids. First, un-ionized bile acids are more lipophilic and can be reabsorbed passively anywhere along the length of the intestine. Second, ionized bile salts are reabsorbed via an active transport process in the distal ileum. The overall efficiency of the reabsorption process is approximately 95 percent. Figure 6-8 summarizes the circulation of bile acids.

Bilirubin

Heme is degraded by the reticuloendothelial system (RES) into its waste product, bilirubin. Most of the daily bilirubin production comes from the hemoglobin in senescent red blood cells, and the remainder is derived from the breakdown of other heme-containing proteins. The average daily production of bilirubin in a normal adult is 250 to 300 mg per day. There is an enormous functional reserve in the capacity of the RES to produce bilirubin. Faced with an increased load of hemoglobin degradation, for example, in hemolytic anemia, the RES can increase daily production of bilirubin to as high as 50 g per day. Bilirubin has no known useful physiologic function.

Bilirubin is transported from its site of production in the RES to the liver by blood. Because it is hydrophobic, it is transported by the plasma carrier protein albumin. Albumin-bound bilirubin is in equilibrium with a small amount of unbound bilirubin. Because of its lipid solubility, this unbound bilirubin freely crosses cell membranes to enter cells. The hepatic uptake of free bilirubin is specially boosted by several proteins within hepatocytes, called Y and Z ligands. They have a high affinity for bilirubin, which allows bilirubin to accumulate in hepatocytes. A hepatocyte-specific carrier for bilirubin, both bound and unbound, has also been hypothesized to transport bilirubin across the hepatocyte's sinusoidal membrane and into the cell.

Within the hepatocyte, bilirubin is conjugated to a hydrophilic moiety to make it more soluble for excretion. Two glucuronic acids are transferred to form bilirubin diglucuronide. At least the first transfer, and perhaps both, are catalyzed by enzymes in the endoplasmic reticulum called glucuronyl transferases from an activated donor called uridine diphosphate glucuronic acid. The conjugated product, bilirubin diglucuronide, is soluble and can be excreted in bile. Under pathologic conditions, conjugated bilirubin may be circulating in the blood. Because of its water solubility, this form of bilirubin is less tightly bound to albumin and can be filtered by the kidney and excreted in the urine. This is in contrast to unconjugated bilirubin, which is very tightly bound to albumin, is lipid soluble, and as such cannot be filtered by the kidney.

Secretion of conjugated bilirubin out of hepatocytes and into the bile canaliculus is generally the rate-limiting step of the entire process. This occurs against a concen-

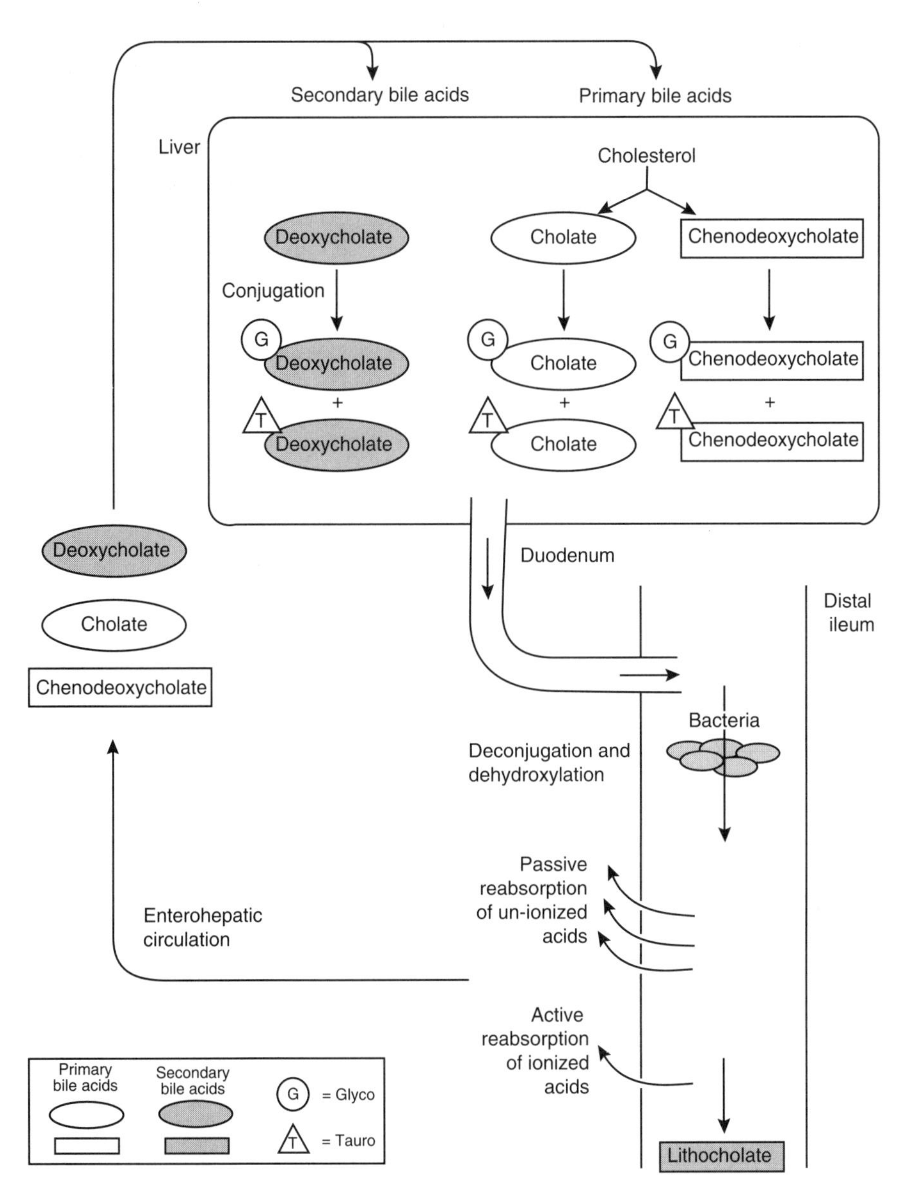

Fig. 6-8. Hepatocytes conjugate primary and secondary bile acids. They are secreted into the first part of the duodenum, where resident bacteria deconjugate and dehydroxylate them to secondary bile acids. Lithocholate is insoluble and is excreted. Un-ionized bile acids are passively reabsorbed along the length of the small intestine while ionized bile acids are actively reabsorbed in the distal ileum. They return to the liver via the enterohepatic circulation to continue the cycle.

tration gradient that requires active transport. Once in the small intestine, the fate of bilirubin has some parallels with that of bile acids. It can be deconjugated and modified by resident bacteria; some derivatives are reabsorbed into the enterohepatic circulation. Bilirubin diglucuronide is deconjugated and then metabolized to urobilinogen by bacteria. Urobilinogen can be either reabsorbed (~10%), or it can continue along in the gut to be converted to stercobilinogen, and finally oxidized to stercobilin on exposure to air as it is excreted in feces.

Stercobilinogen/stercobilin give feces its brown color. Most of the urobilinogen that is reabsorbed in the gut is taken up by hepatocytes and resecreted in bile. Some

Gastrointestinal

continues into the systemic circulation to be excreted in the urine. There it is oxidized to urobilin on exposure to air and gives urine its yellow color. This is summarized in Fig. 6-9.

Bilirubin and its derivatives can thus be found in the blood, in the urine, and in stool. In the serum, both unconjugated (indirect, includes unbound and albumin-bound) bilirubin and conjugated (direct) bilirubin can be measured. In the urine, bilirubin and urobilinogen can be measured. In the stool, only stercobilinogen (sometimes identified as urobilinogen) is measured. Jaundice, an excess of bilirubin in the serum, results from any one or combination of the following processes: (1) overproduction, (2) decreased uptake into the hepatocyte, (3) decreased conjugation, and (4) decreased excretion. Increased production of bilirubin, as in hemolytic anemia, would cause an increase primarily in unconjugated (indirect) bilirubin. A mechanical obstruction of bile flow, for example, a gallstone, would lead to decreased excretion and an accumulation of conjugated (direct) bilirubin in the plasma. An increase in bilirubin in the urine would be measured and in extreme cases, if no bilirubin could escape into the gut, there might be no urobilinogen in the urine and the stool would lack its normal brown color and appear clay colored.

Regulation of Bile Secretion and Excretion

The rate of hepatic secretion of bile is controlled by three factors: the size of the bile acid pool in the enterohepatic circulation, hepatic blood flow, and secretin. Bile acids are potent promoters of bile secretion. The greater the amount of bile acids in the enterohepatic circulation, the greater is the rate of hepatic bile secretion. Reduction of hepatic blood flow reduces the rate of bile secretion. The hormone secretin, released from the duodenal mucosa in response to the presence of acid in the duodenum, causes the secretion of a dilute but HCO_3^--rich bile, presumably by stimulating an HCO_3^- pump in the cells that line the bile ductules.

While the hepatic secretion of bile is a continuous process, the entrance of bile into the duodenum is not. Between meals, bile is stored in the gallbladder since biliary pressures are not sufficient to drive the bile through a contracted sphincter of Oddi. The presence of chyme in the duodenum, especially one rich in lipids, causes the release of the hormone CCK from the duodenal mucosa. Cholecystokinin causes the contraction of the gallbladder while it relaxes the sphincter of Oddi, resulting in the expulsion of bile into the duodenum.

Intestinal Secretion

The intestines secrete a total of about 1000 ml per day. The most proximal segment of duodenum contains submucosal glands called **Brunner's glands,** which secrete an alkaline mucus. The function of these glands is to protect the duodenal mucosa adjacent to the stomach by neutralizing gastric acid as it enters the duodenum. The daily secretion of Brunner's glands is less than 100 ml.

Throughout the rest of the small intestine, the epithelial cells located in the crypts secrete the bulk of the intestinal fluid, which is very similar in composition to that of interstitial fluid. Simultaneously, reabsorption of luminal contents is occurring into cells on adjacent villi. Secretion of water is mediated by a Cl^- channel in the apical membrane of the intestinal mucosa (Fig. 6-10), but is generally limited to the proximal small intestine. Na^+ follows the Cl^- via paracellular pathways, and water passively follows. Conductance of the channel is increased by intracellular cAMP. Cholera toxin, which causes a secretory diarrhea, increases intracellular cAMP via a G-protein. The end result is hypersecretion, which is manifested as a profuse watery diarrhea.

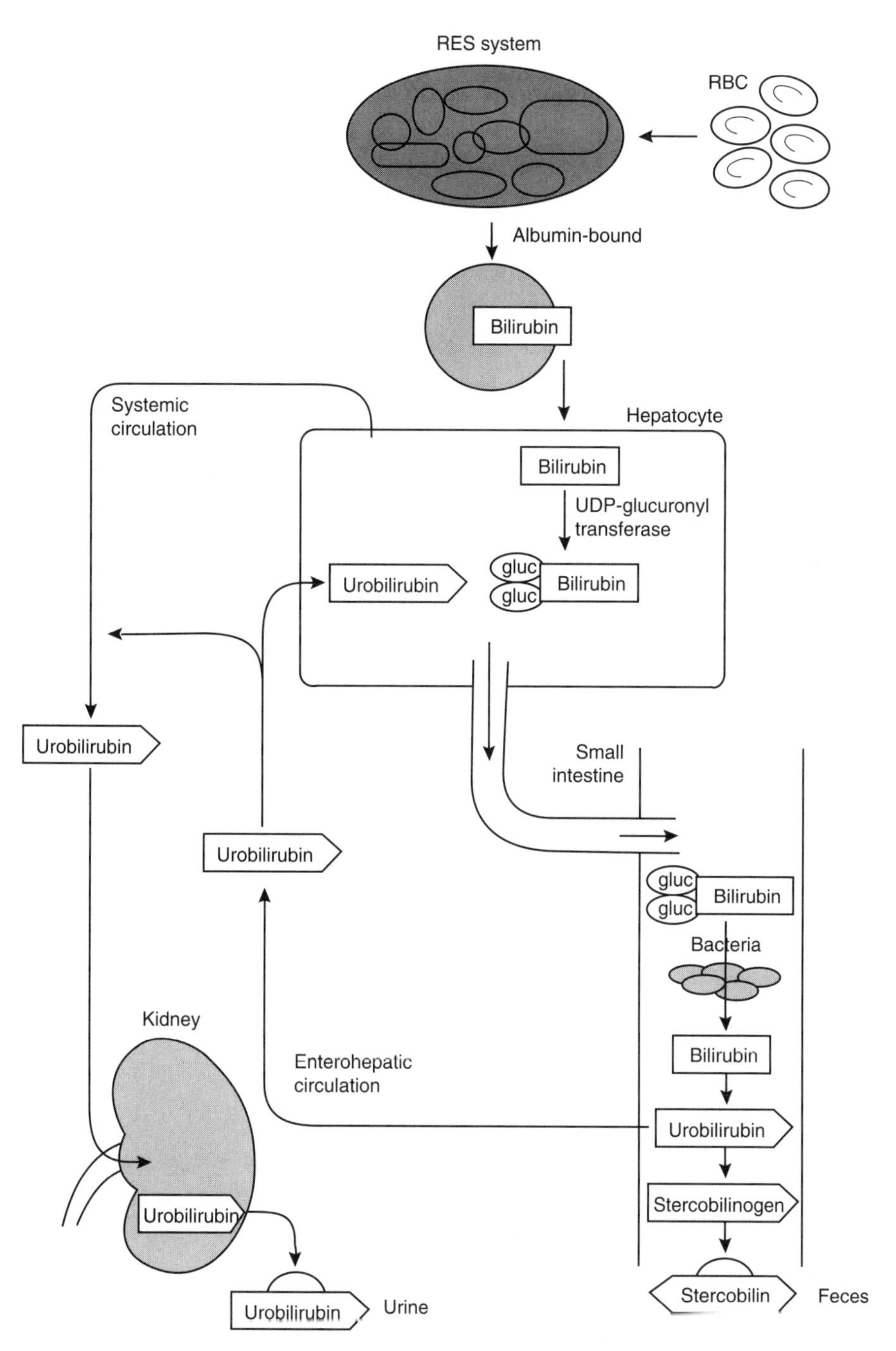

Fig. 6-9. Bilirubin, a breakdown product of heme from red blood cells, is synthesized in the RES system and travels in the plasma tightly bound to albumin. Following hepatocyte uptake of the unbound bilirubin, it is conjugated and secreted. In the small intestine, bacteria deconjugate bilirubin, which allows it to be further metabolized to urobilinogen. Urobilinogen is partially reabsorbed and is eventually either taken up by the liver or excreted by the kidney. The final metabolites are urobilin in the urine and stercobilin in the feces. UDP = uridine diphosphate.

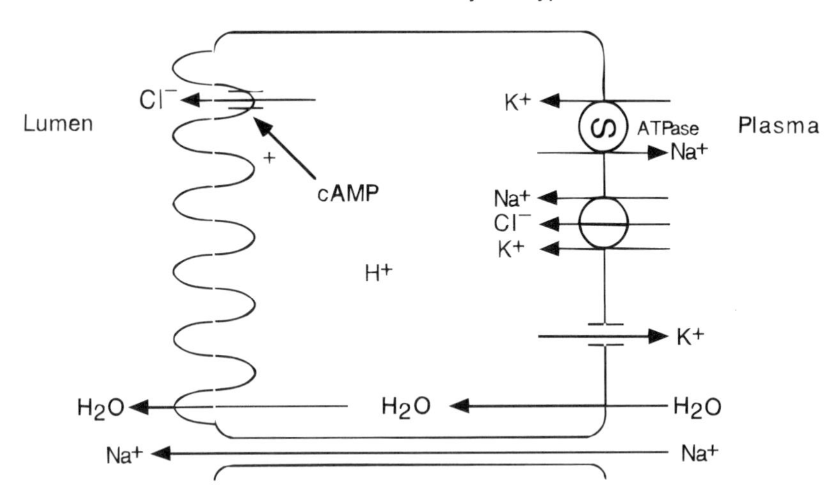

Fig. 6-10. Enterocytes located in the intestinal crypts secrete intestinal fluid mediated by the apical Cl^- channel. Na^+ passively follows via paracellular routes. The conductance of the Cl^- channel is regulated by cAMP. An increase in intracellular cAMP enhances secretion.

In addition to the fluids, there are goblet cells in the intestinal mucosa that secrete mucus. The colon also secretes mucus, which has a high concentration of K^+ and HCO_3^- ions. Intestinal secretion is stimulated by the presence of chyme in the intestine by either neural reflexes in response to tactile or irritant stimuli or by secretory hormones (secretin, CCK). Figure 6-11 summarizes the secretory functions of the GI tract.

Digestion and Absorption

Digestion, the chemical breakdown of foods to absorbable components, can be divided into luminal and membrane digestion. **Luminal digestion** is carried out by enzymes that are present in salivary, gastric, and pancreatic secretions. Membrane, or **brush border, digestion** is carried out by enzymes that are synthesized by enterocytes and are bound to the small-intestinal brush border (microvilli). Although some digestion occurs in the mouth and stomach, the vast majority of protein and carbohydrate digestion, as well as all fat digestion, takes place in the small intestine.

Absorption, the transfer of digestive products from the intestinal lumen to blood and lymph, takes place almost exclusively in the small intestine. (A few drugs and some lipid-soluble substances such as ethanol are absorbed in the stomach.) The absorptive surface area of the small intestine is increased by three superimposed levels of mucosal folding. First, there are circularly arranged folds of mucosa and submucosa called **plica circulares** that increase the absorptive surface area by three- to fourfold. Second, the plica circulares are covered by finger-like projections of the mucosa called **villi.** The villi increase the absorptive surface area by an additional 10-fold. Third, the apical membrane of the enterocytes (intestinal epithelial cells) lining the villi are thrown into numerous folds called **microvilli** that increase the absorptive area by another 20-fold. Thus, the combination of these structural adaptations increases the area of the absorptive small-intestinal mucosa by 600- to 800-fold to the size of a tennis court. Digestion and absorption are complementary and often linked processes that occur simultaneously throughout the small intestine.

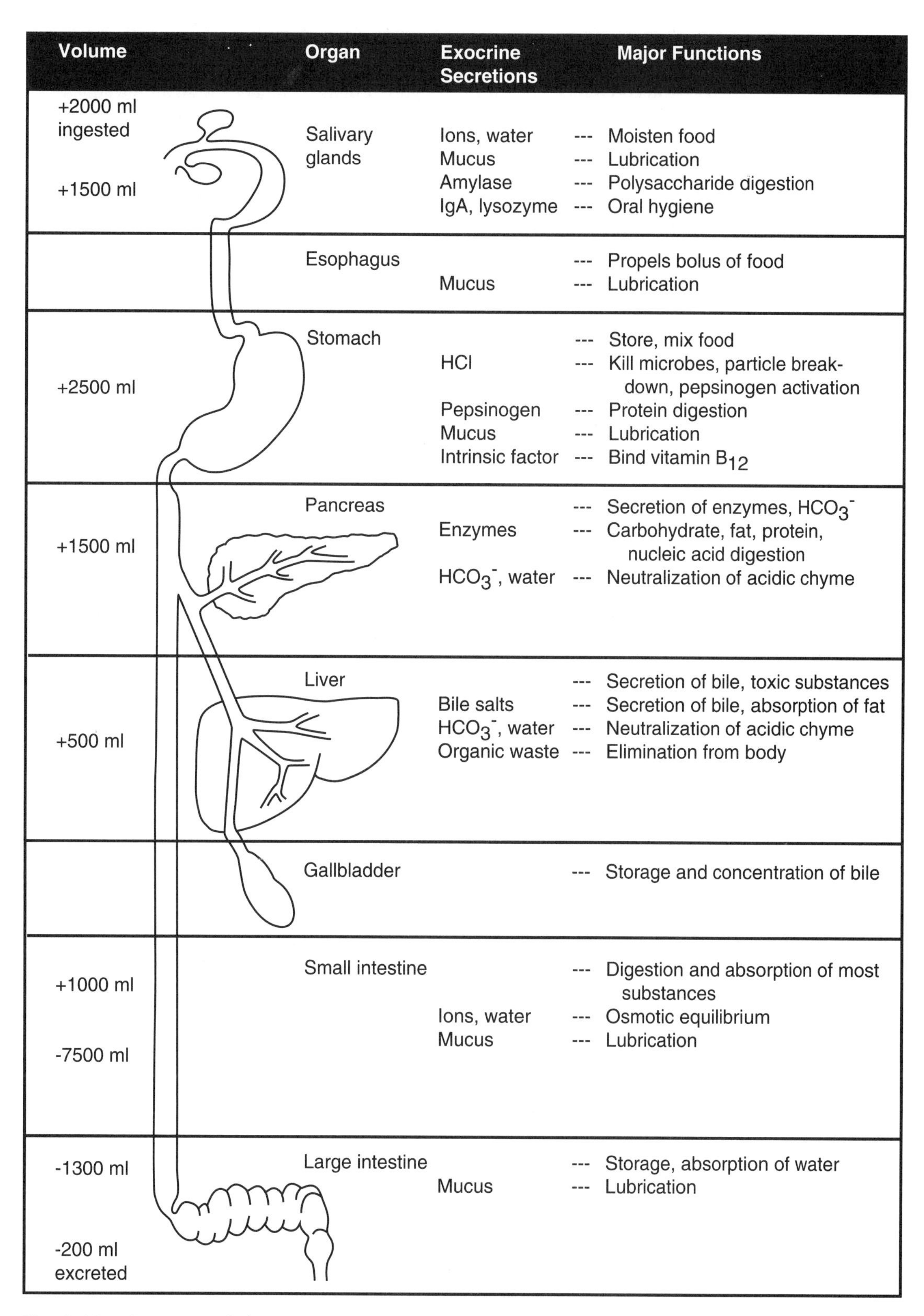

Volume	Organ	Exocrine Secretions	Major Functions
+2000 ml ingested	Salivary glands	Ions, water	Moisten food
		Mucus	Lubrication
+1500 ml		Amylase	Polysaccharide digestion
		IgA, lysozyme	Oral hygiene
	Esophagus		Propels bolus of food
		Mucus	Lubrication
	Stomach		Store, mix food
+2500 ml		HCl	Kill microbes, particle breakdown, pepsinogen activation
		Pepsinogen	Protein digestion
		Mucus	Lubrication
		Intrinsic factor	Bind vitamin B_{12}
+1500 ml	Pancreas		Secretion of enzymes, HCO_3^-
		Enzymes	Carbohydrate, fat, protein, nucleic acid digestion
		HCO_3^-, water	Neutralization of acidic chyme
+500 ml	Liver		Secretion of bile, toxic substances
		Bile salts	Secretion of bile, absorption of fat
		HCO_3^-, water	Neutralization of acidic chyme
		Organic waste	Elimination from body
	Gallbladder		Storage and concentration of bile
+1000 ml	Small intestine		Digestion and absorption of most substances
		Ions, water	Osmotic equilibrium
-7500 ml		Mucus	Lubrication
-1300 ml	Large intestine		Storage, absorption of water
-200 ml excreted		Mucus	Lubrication

Fig. 6-11. Summary of the major GI secretions and their functions. (Adapted from A. J. Vander, J. H. Sherman, and D. S. Luciano. *Human Physiology—The Mechanism of Body Function* [4th ed.]. New York: McGraw-Hill, 1994. P. 564.)

Water and Electrolyte Absorption

Approximately 90 percent of the roughly 9 liters that enter the small intestine daily is absorbed by the end of the ileum. In the colon, an additional 1 liter is absorbed so that all but 100 to 200 ml is reabsorbed. The luminal contents of the intestine are isoosmotic with plasma. As chyme is digested and broken into smaller particles, the number of osmoles increases, drawing water passively into the lumen to equilibrate the contents. Likewise, with active absorption of osmoles (e.g., Na^+, glucose, and amino acids), water follows. Chyme remains isotonic with plasma throughout the intestine. The solid nature of stool is a product of large particles that are relatively few in number and which therefore exert a very small osmotic pressure.

While water diffuses passively into and out of the intestinal lumen, electrolytes (e.g., Na^+, K^+, Cl^-, and HCO_3^-), which are charged ions, must use specific ion channels. The GI tract is lined by a functionally polarized epithelium with distinct apical and basolateral membranes that permits asymmetric transport of ions. Ions can either diffuse passively down their concentration gradient, be cotransported with glucose, or be coupled with other ions. The last two are examples of **secondary active transport.** The major driving force for solute transport is the Na^+-K^+ ATPase pump. Its location in the basolateral membrane generates an electrochemical gradient within the cell that allows other solutes to enter the cell against their electrochemical gradients.

The epithelium of the small intestine is "leaky" so that there is a significant amount of ion conductance through paracellular pathways. The Na^+ pump on the basolateral membrane causes a net transfer of Na^+ from the gut lumen to the blood. Na^+ can enter the mucosal side of the cell through two processes: coupled to glucose transport or coupled to other ions (Fig. 6-12). For the latter, there are two models. In one model, Na^+ and Cl^- enter the cell on a single cotransporter. In a second model, there are two independent antiporters where Na^+ is exchanged for H^+ and Cl^- is exchanged for HCO_3^-. K^+ travels from the lumen to the blood largely by passive diffusion down its

Fig. 6-12. Proposed mechanism for enterocyte secretion. Enterocytes located on intestinal villi absorb Na^+ by both nutrient-dependent (cotransport with glucose) and nutrient-independent (Na^+/Cl^- cotransport, or Na^+/H^+ antiporter coupled with an HCO_3^-/Cl^- antiporter). Intracellular cAMP inhibits absorption of NaCl while it enhances its secretion (see Fig. 6-10). In the ileum, HCO_3^- is secreted while Cl^- is reciprocally absorbed. In the jejunum, both are absorbed.

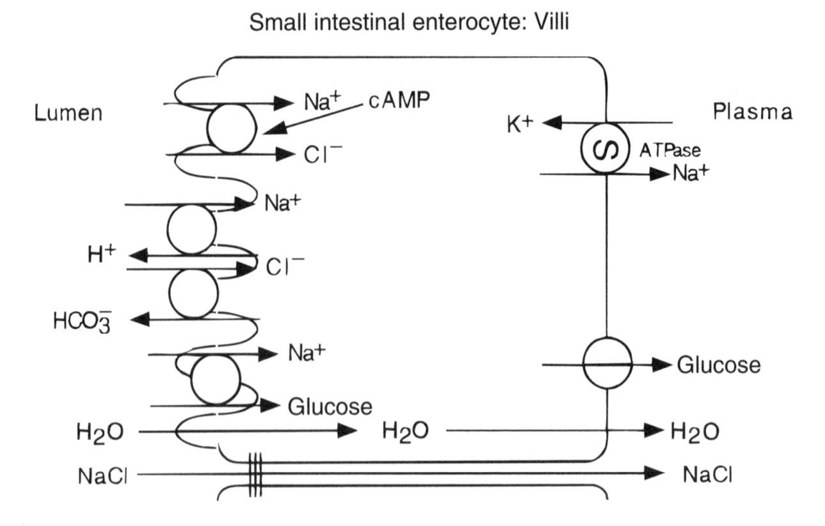

concentration gradient. The mechanism of transport for Cl^- and HCO_3^- is uncertain and may be interrelated. There is evidence for both active and passive transport. Both Cl^- and HCO_3^- are absorbed in the jejunum. In the ileum Cl^- is absorbed, but HCO_3^- is secreted. The mechanism of HCO_3^- "absorption" in the jejunum may not be absorption of the HCO_3^- but may be due to secretion of H^+ into the lumen, which titrates the HCO_3^- to water and carbon dioxide.

Transport in the colon differs from that in the small intestine in three ways. First, it is a "tight" epithelium with low paracellular conductance such that absorption of Na^+ is electrogenic. Consequently, the lumen is negative (15–50 mV) with respect to the blood side. Second, there is no cotransport of ions with glucose or other nutrients because most nutrients have been absorbed in the small intestine. Third, colonic ion transport is responsive to mineralocorticoids. These three differences significantly impact ion absorption in the colon. K^+ is secreted rather than absorbed in response to the negative lumen. As in the ileum, there is an electroneutral exchange of HCO_3^- for Cl^-. In some cases of diarrhea, with a large volume of stool presented to the colon, the increased colonic absorption can result in hyperchloremic hypokalemic metabolic acidosis as determined by the direction of ion flow. Aldosterone, released in response to volume depletion, augments Na^+ conductance in the apical membrane and may also stimulate Na^+-K^+ ATPase activity. Antidiuretic hormone stimulates coupled Na^+ and Cl^- absorption and may also inhibit Cl^- secretion.

Carbohydrate Digestion and Absorption

Digestible dietary carbohydrates take the form of polysaccharides (e.g., starch, glycogen), disaccharides (e.g., sucrose, lactose), and monosaccharides (e.g., glucose, fructose). Cellulose, a carbohydrate, is indigestible and contributes to dietary fiber. Carbohydrate assimilation involves three steps: **luminal digestion, surface digestion, and cell transport** (Fig. 6-13).

1. Luminal digestion is initiated by **ptyalin** in the salivary secretions (which are inactivated by gastric acid) and completed by **alpha-amylases** from the pancreas in

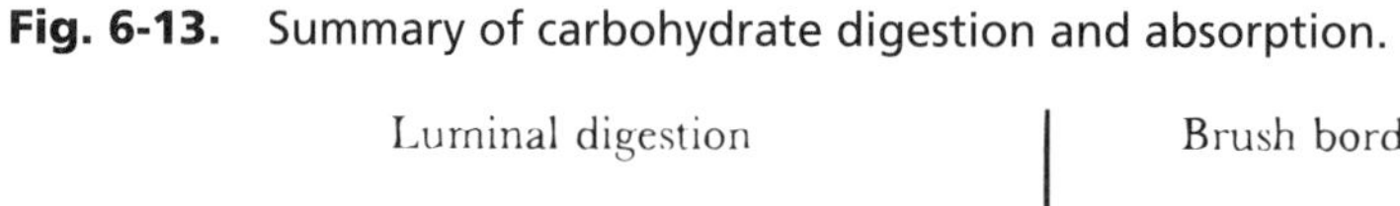

Fig. 6-13. Summary of carbohydrate digestion and absorption.

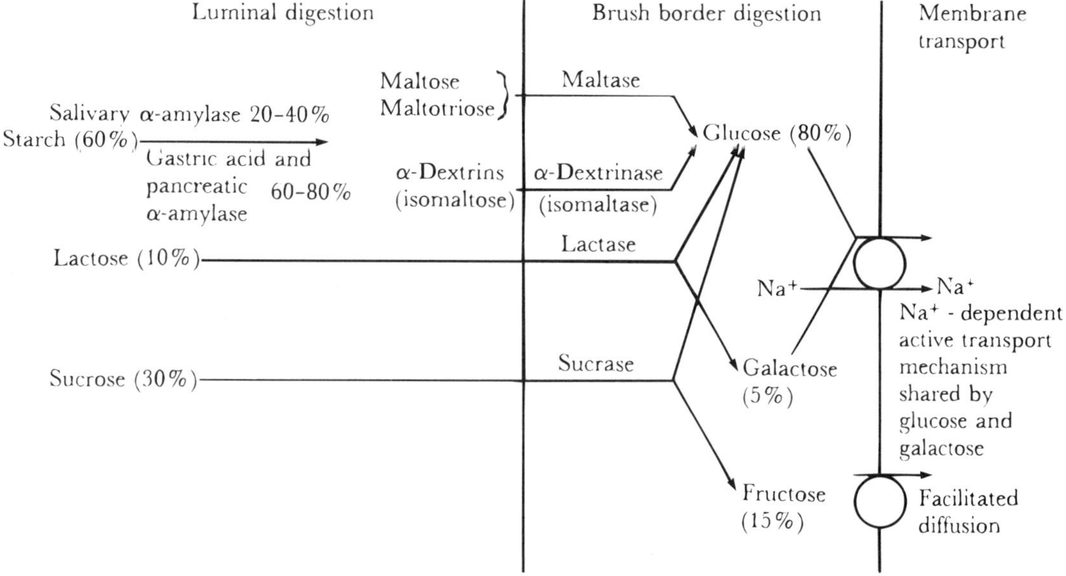

the lumen of the small intestine. These enzymes break down polysaccharides to oligo- and disaccharides.

2. Surface digestion is catalyzed by the brush border enzymes that are synthesized in enterocytes and incorporated in the microvillous membranes of these cells. The brush border enzymes hydrolyze the products of luminal digestion into their component monosaccharides, which can then be absorbed.
3. The rate-limiting step in carbohydrate assimilation is generally the transport of digestive products across the brush border membrane. Several different transport processes have been identified. Glucose and galactose are actively transported by a common Na^+-dependent carrier system as described for Na^+ absorption. The energy for this active transport process is provided by the Na^+-K^+ ATPase on the basolateral membrane of these intestinal cells. Fructose is transported by facilitated diffusion. Gastrin has been shown to greatly enhance carbohydrate absorption.

Once absorbed, the monosaccharides enter the portal circulation en route to the liver. Digestion and absorption of carbohydrates are efficient and usually completed in the proximal jejunum.

In the case of lactose, the rate-limiting step in the process of carbohydrate assimilation is not transport, but rather cleavage to glucose and galactose by the brush border enzyme lactase. Deficiency of this enzyme further retards the process, which leads to an intolerance of milk products. The inability to assimilate the lactose increases the number of osmoles remaining in the lumen, which results in an osmotic diarrhea following ingestion of a milk product. Deficiencies of other digestive enzymes may have similar consequences.

Fat Digestion and Absorption

Triglycerides account for over 90 percent of dietary fat. Phospholipids, cholesterol, cholesterol esters, and **fat-soluble vitamins (A, D, E, K)** make up the rest. While there is some mixing and emulsion in the stomach, digestion of fats is begun largely in the duodenum. The presence of chyme (especially one that is rich in lipids and proteins) and gastric acid in the duodenum causes the release of two peptide hormones, CCK and secretin from the upper small-intestinal mucosa. The combined effect of these hormones results in pancreatic secretion of HCO_3^- and digestive enzymes as well as contraction of the gallbladder and flow of bile into the duodenum. Through the action of these hormones, the presence of lipids in the duodenum brings about the conditions that are necessary for fat emulsification, digestion, and absorption.

Because lipolytic enzymes are water soluble and therefore function only at the lipid-water interface, dietary lipids must be emulsified before being digested. **Emulsification** is the process of breaking large fat globules into smaller particles, which greatly increases their total surface area. It takes place in the duodenum and requires two conditions: a neutral pH and the presence of bile salts. Pancreatic HCO_3^- brings the chyme from the stomach up to pH 6. Bile salts reduce the surface tension of the lipid-aqueous interface so that, on agitation, the fat is emulsified. In addition, pancreatic secretions also contain several enzymes involved in lipid digestion. These include

1. **Pancreatic lipase.** This enzyme is secreted in its active form and catalyzes the hydrolysis of ester linkages in the 1 and 3 positions of triglycerides, resulting in the formation of fatty acids and 2-monoglycerides. For optimum function, pancreatic lipase requires the presence of a peptide called colipase, which is also secreted by the pancreas.
2. **Colipase.** This enzyme has two functions. First, the optimum pH for lipase is changed from 8 to 6 by binding to colipase; therefore, lipase is made more active

at the physiologic pH of the small intestine. Second, colipase facilitates the action of lipase by attaching to the lipid-water interface of emulsified lipids, allowing fixation of lipase to this interface. The formation of lipase-colipase–bile salt–lipid complex is the rate-limiting step in fat digestion and absorption.

3. Phospholipase. Several phospholipases with specificities for different bonds in the phospholipid molecule are present in pancreatic secretions. These enzymes convert lecithin (a membrane lipid) to lysolecithin.
4. Cholesterol esterase. This enzyme hydrolyzes the esters of cholesterol as well as esters of the vitamins A, C, and E.

As digestion proceeds, most of the products of lipid digestion are incorporated into tiny packages called **micelles.** Bile salts line the outside of micelles and sequester the lipid digests in the centers. To be absorbed, the lipids must be "ferried" to the brush border of enterocytes. One model for intestinal absorption hypothesizes the presence of an **"unstirred water layer" (UWL)** adjacent to the brush border and glycocalyx of the enterocytes. This layer is not mixed with luminal chyme regardless of peristaltic activity, and therefore movement of solute molecules through it is by diffusion. With the exception of short- and medium-chain fatty acids, the products of lipid digestion are insoluble in the UWL and must be "ferried" through by micelles. These micelles are not themselves absorbed by intestinal cells. In the UWL and near the brush border membrane of intestinal cells an equilibrium exists between free lipid digests and lipid digests in micelles. It is the free form of lipid digests that penetrates the intestinal cell membrane by passive diffusion. The bile salts continue their function in the lumen until most of them are reabsorbed in the distal ileum.

Once inside the small-intestinal cell, the products of lipid digestion are reconstituted into triglycerides, phospholipids, and cholesterol esters by reesterification in the smooth endoplasmic reticulum. They are then incorporated into **chylomicrons,** which, like micelles, serve to solubilize them in aqueous solution. Chylomicrons are composed of triglycerides (84%) and cholesterol (2%), which form the core, and phospholipids (13%) and lipoproteins (e.g., Apo-B-48; 1%) synthesized in the endoplasmic reticulum, which form the outer coat. These chylomicrons then leave the basolateral membrane of intestinal cells by exocytosis and enter lymph through the lacteals.

Short- and medium-chain triglycerides have a unique path of absorption. They are more water soluble and as such are not digested by lipases and are not incorporated into micelles. Instead, they diffuse through the UWL and are absorbed intact. Once inside the cell, they are hydrolyzed by esterase. The short- and medium-chain fatty acids diffuse across the basolateral membrane and enter the portal circulation. Triglyceride absorption is summarized in Fig. 6-14.

Protein Digestion and Absorption

In addition to the proteins ingested in the diet, there are several endogenous proteins that enter the GI tract and are thus subject to the same digestive and absorptive processes as dietary proteins. These include proteins in GI secretions, proteins in desquamated mucosal cells of the GI tract, and plasma proteins. Like carbohydrate digestion, protein digestion involves two steps: luminal digestion and surface digestion.

Protein digestion begins in the stomach by the action of pepsin and gastric acid (Fig. 6-15). Both of these, however, are dispensable, and in their absence proteins are readily digested and absorbed. The principal enzymes involved in protein digestion are secreted by the pancreas as inactive **zymogens.** These include trypsinogen, chymotrypsinogen, proelastase, and procarboxypeptidases A and B. Trypsinogen is ac-

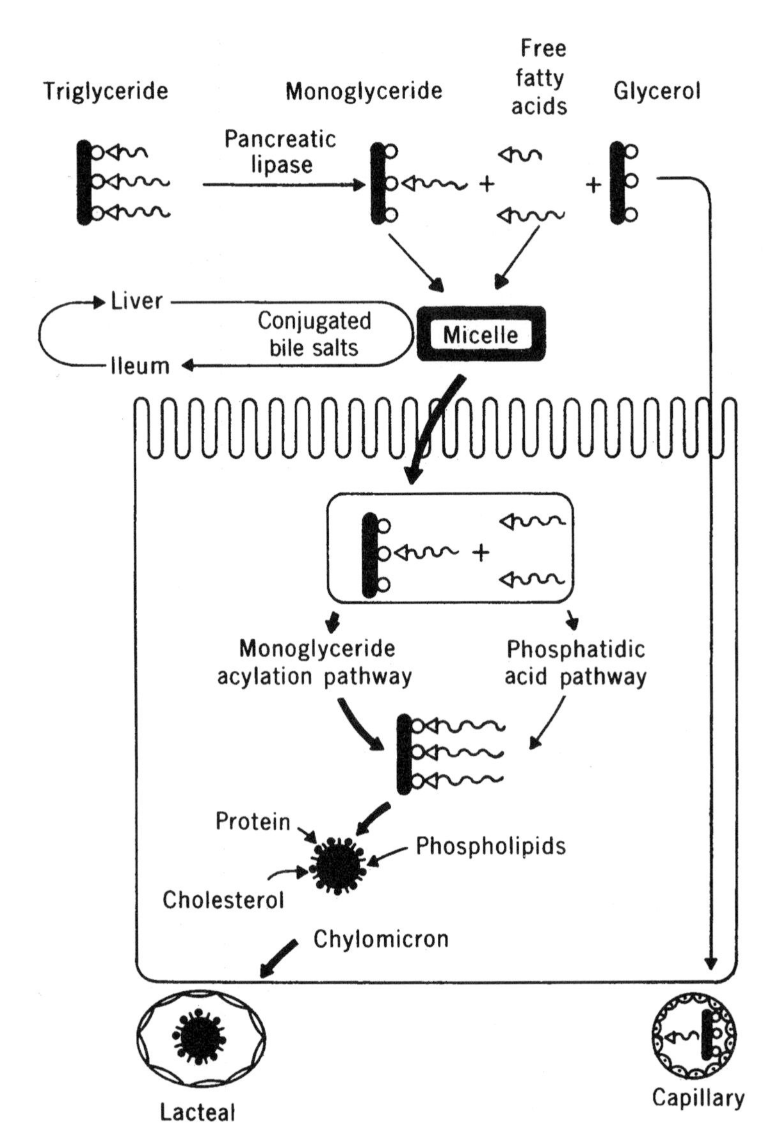

Fig. 6-14. Summary of the digestion and absorption of triglycerides. Monoglyceride and long-chain fatty acids enter the cells after first being incorporated into micelles. Glycerol and short-chain and medium-chain acids, because of their solubility in the aqueous unstirred layer, enter without micelle solubilization. (From L. R. Johnson. *Gastrointestinal Physiology* [4th ed.]. St. Louis: Mosby, 1991. P. 126.)

tivated to trypsin by a duodenal brush border peptidase called enterokinase. Trypsin, in turn, activates other pancreatic proteases. Luminal digestion of proteins by these proteases results in free amino acids and small peptides. The small peptides are either transported intact or further broken down by brush border peptidases. The various peptidases have specificities for different di- and oligopeptidases.

Absorption of amino acids by enterocytes is dependent on several active transport processes, all of which are specific for L-amino acids. Separate carrier systems are known to exist for neutral, basic, and acidic amino acids. The amino acids proline and hydroxyproline are also known to have a separate active transport carrier. According to the current model, all of these active transport systems cotransport the amino acid with a proton through the apical membrane. An antiporter in the apical membrane ex-

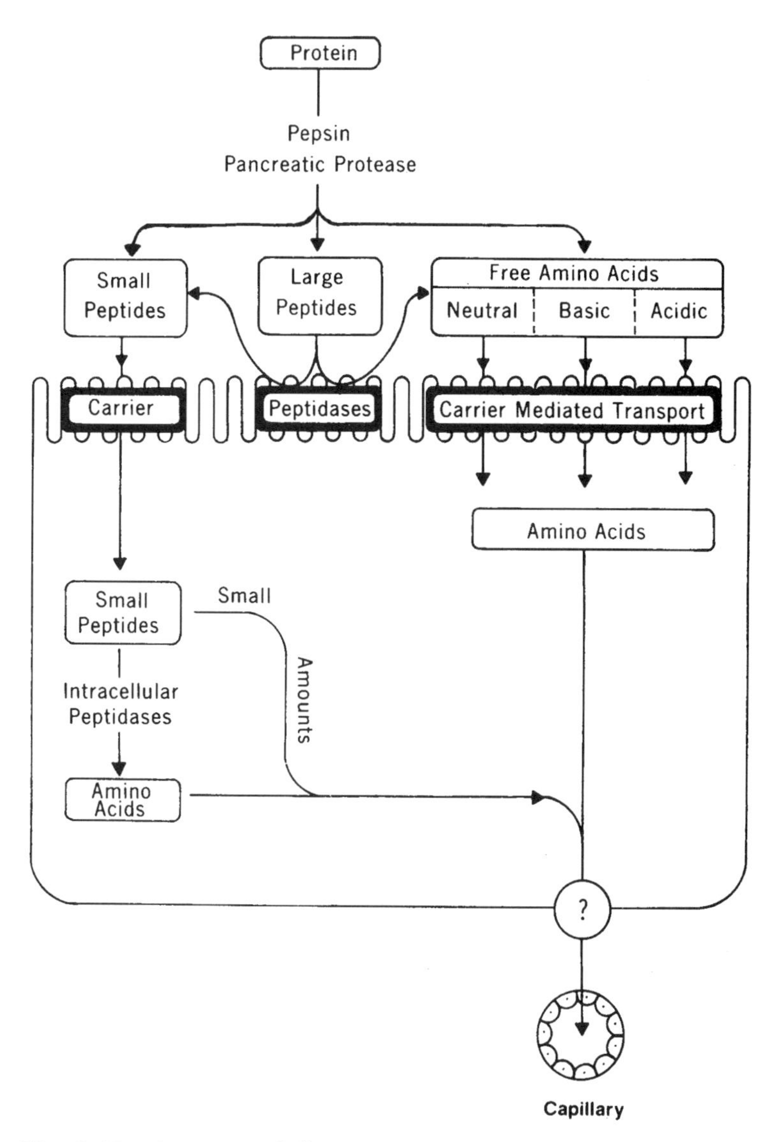

Fig. 6-15. Summary of digestion and absorption of dietary protein. Approximately one third of the total amino acids is absorbed in free form after luminal digestion. The remaining is absorbed as oligopeptides or dipeptide and tripeptide following membrane processes. (From L. R. Johnson. *Gastrointestinal Physiology* [4th ed.]. St. Louis: Mosby, 1991. P. 120.)

changes H^+ for Na^+. Ultimately, the energy for this secondary active transport of amino acids into the cell against a concentration gradient is provided by the Na^+-K^+ ATPase in the basolateral membrane.

In addition to active transport of amino acids, there are specific carriers for the active transport of some di- and tripeptides. Alternatively, they can be broken into their constituent amino acids and absorbed as described above. The multiple complementary events at each stage of digestion and transport lead to efficient assimilation of proteins. Within the cell, the di- and tripeptides are hydrolyzed by cytosolic peptidases. Ultimately, only amino acids are delivered to the portal circulation. Figure 6-16 summarizes the major digestive enzymes, and Fig. 6-17 summarizes the intestinal distribution of absorption.

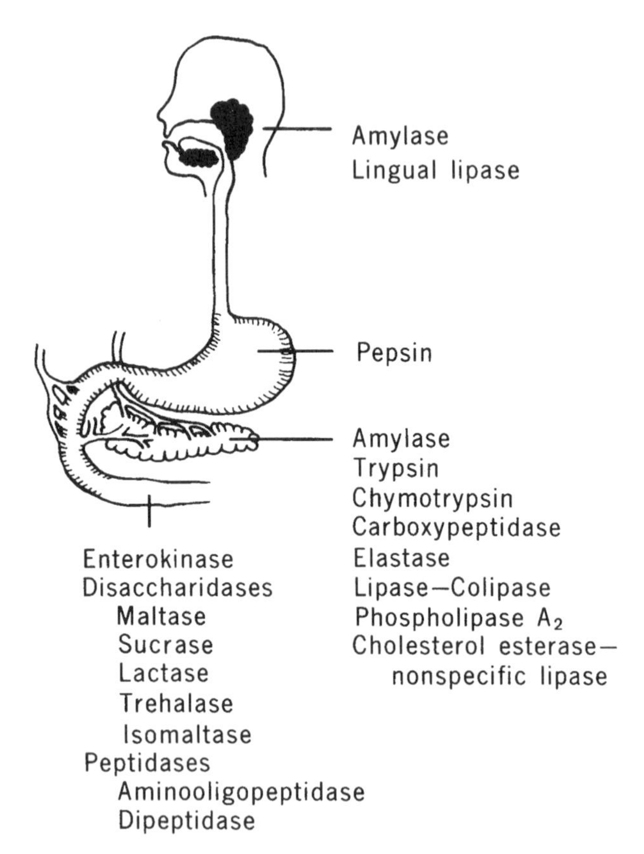

Fig. 6-16. Source of the principal luminal and membrane-bound digestive enzymes. (From L. R. Johnson. *Gastrointestinal Physiology* [4th ed.]. St. Louis: Mosby, 1991. P. 109.)

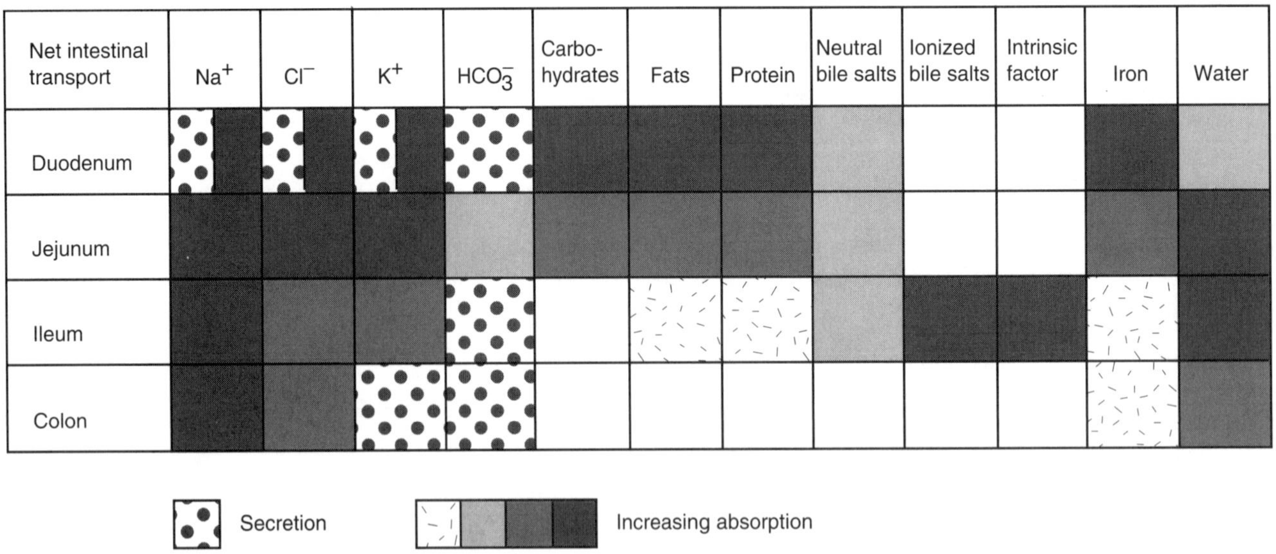

Net intestinal transport	Na^+	Cl^-	K^+	HCO_3^-	Carbo-hydrates	Fats	Protein	Neutral bile salts	Ionized bile salts	Intrinsic factor	Iron	Water
Duodenum												
Jejunum												
Ileum												
Colon												

Secretion   Increasing absorption

Fig. 6-17. Anatomic location of absorption of select nutrients.

Questions

Directions: Each group of items in this section consists of lettered options followed by a set of numbered items. For each item, select the one lettered option that is most closely associated with it. Each lettered option may be selected once, more than once, or not at all.

A. Inhibits gallbladder contraction
B. Inhibits gastric emptying
C. Inhibits pancreatic secretion
D. Promotes insulin secretion
E. Promotes secretion of pepsinogen

1. Cholecystokinin (CCK)
2. Somatostatin
3. Gastric inhibitory peptide (GIP)

Directions: Each of the numbered items or incomplete statements in this section is followed by answers or by completions of the statement. Select the one lettered answer or completion that is best in each case.

4. Which of the following characteristics do both CCK and secretin share?
   A. They are secreted by mucosal I cells
   B. In both, signal transduction occurs via cAMP
   C. They are stimulated by acidic intestinal lumen
   D. They are synthesized as prepropeptides
5. Regarding GI hormones, it is **true** to say that they
   A. Are grouped into families based on function
   B. Are released primarily by enteroendocrine cells that line the gut
   C. Bind transmembrane receptors that are generally serine kinases
   D. Have receptors only within the GI tract and associated organs
   E. Reach their target organ by traversing the gut lumen
6. Vagotomy would produce an increase in which of the following?
   A. Gastric acid secretion
   B. Gastrin secretion
   C. Motility
   D. Pancreatic enzyme secretion
   E. Parietal cell threshold to histamine or gastrin
7. The submucosal plexus
   A. Employs VIP as an excitatory neurotransmitter
   B. Is also known as Auerbach's plexus
   C. Is generally stimulated by the sympathetic nervous system
   D. Is involved primarily in motility
   E. Is stimulated by mucosal irritation
8. Regarding GI smooth muscle, which of the following is **false?**
   A. The resting potential oscillates
   B. The slow wave is caused by Na^+ flow
   C. Some GI smooth muscle can maintain a tonic contraction for hours
   D. The spike is caused by Ca^{2+} flow
   E. Stretch hyperpolarizes the cell
9. Peristaltic waves of contraction in the small intestine
   A. Are induced by motilin
   B. Cannot occur in the absence of parasympathetic innervation

C. Involve receptive relaxation proximal to the bolus of food
D. Occur primarily in response to chemical stimuli from digested foodstuffs
E. Proceed in both directions from their site of origin

10. Regarding the histology of the GI tract, which of the following statements is **false?**
    A. Both enteric plexuses are located within the submucosa
    B. The outer longitudinal muscle layer of the colon concentrates into three bundles called teniae coli
    C. The mucosa and submucosa of the small intestine are arranged into folds called plica circulares
    D. The four major subdivisions include the mucosa, submucosa, muscularis, and adventitia
    E. Villi are not found in the colon
11. A wrestler trying to "make weight" finds that he is a few grams over. In an attempt to lose the extra grams of weight, he sucks on a hard candy to induce copious salivation. The saliva that he spits up
    A. Contains IgD
    B. Contains a peptidase that initiates protein digestion
    C. Is regulated primarily by sympathetic stimulation
    D. Is relatively rich in Na^+ and poor in K^+
    E. None of the above is correct
12. Gastric secretion of H^+
    A. Decreases in concentration with increasing flow rate
    B. Is by chief cells
    C. Is enhanced by carbonic anhydrase inhibitors
    D. Is in exchange for luminal Na^+
    E. Is regulated in part by the number of H^+ pumps
13. The parietal cell H^+-K^+ ATPase
    A. Exchanges H^+ for K^+ in a 1 : 1 ratio
    B. Generates an electrical potential of approximately −15 mV
    C. Is located on the basolateral surface of the parietal cell
    D. Requires 3-ATP for the expulsion of each proton
    E. Requires a mole of K^+ to be absorbed into the blood for each mole pumped
14. For patients who do not respond to medical interventions for peptic ulcer disease, surgical interventions may be considered. A complete antrectomy (resection of the gastric antrum) would decrease gastric acid production by
    A. Reducing pepsin production by approximately one-half
    B. Reducing the number of parietal cells by approximately one-half
    C. Virtually eliminating the source of Ach
    D. Virtually eliminating the source of gastrin
    E. Virtually eliminating the source of histamine
15. Gastric emptying time is reduced by
    A. Decreased gastrin production
    B. Increased CCK production
    C. Increased discharge of nitric oxide
    D. Increased gastric distention
    E. Neural discharge via the enterogastric reflex
16. Which of the following is **true** regarding pancreatic secretion?
    A. Enzymes involved in fat digestion are stored as inactive zymogens
    B. The process is stimulated by hormones liberated by the intestinal mucosa
    C. The process is not under neural control
    D. With increasing flow rate, HCO_3^- decreases in relative concentration
    E. With increasing flow rate, the concentration of Na^+ increases as the concentration of K^+ decreases

17. Secondary bile acids
    A. Are a derivative of bilirubin
    B. Are formed by conjugation of a bile acid with glycine or taurine
    C. Are formed by deconjugation of a bile acid within the gut
    D. Are incorporated into mixed micelles in the gallbladder
    E. Do not reenter the enterohepatic circulation
18. Which of the following would decrease the rate of bile acid synthesis?
    A. Chelation in the intestine
    B. Enhanced conversion of cholate to lithocholate
    C. Enhanced intestinal reabsorption
    D. Increased hepatic cholesterol load
    E. Resection of the distal ileum
19. VIPoma, a pancreatic islet cell tumor, produces an explosive watery diarrhea that results from the induction of hypersecretion by mucosal cells of the small intestine. Which of the following is most likely associated with this tumor?
    A. Alkalosis
    B. Decreased mucosal cell cAMP
    C. Enhanced Cl^- conductance
    D. Hypernatremia

Directions: Each group of items in this section consists of lettered options followed by a set of numbered items. For each item, select the one lettered option that is most closely associated with it. Each lettered option may be selected once, more than once, or not at all.

20–21.
    A. Cholestasis
    B. Crigler-Najjar syndrome (decreased hepatocellular conjugation)
    C. Hemolytic anemia
    D. Sepsis causing decreased hepatocellular uptake
20. A patient is found to have extremely elevated urine bilirubin. To which disorder can this finding be attributed?
21. A patient with jaundice has the following laboratory abnormalities:

    Serum bilirubin
        Indirect: increased
        Direct: normal
    Urine bilirubin: absent
    Urobilinogen
        Stool: increased
        Urine: increased

Which disorder is the most likely cause of the jaundice?

22–25.
    A. Salivary secretion
    B. Gastric secretion
    C. Pancreatic secretion
    D. Hepatic secretion
    E. Intestinal secretion
22. In which process is the concentration of both Na^+ and Cl^- constant regardless of flow rate?
23. Which process is increased in response to sympathetic stimulation?

24. Which process is essential for digestion?
25. Which process serves in part an excretory function?

Directions: Each of the numbered items or incomplete statements in this section is followed by answers or by completions of the statement. Select the one lettered answer or completion that is best in each case.

26. A 62-year-old man suffering from Crohn's disease had a section of bowel removed therapeutically. Removal of ileum would interfere most with absorption of which of the following?
    A. Carbohydrates
    B. Iron
    C. Un-ionized bile salts
    D. Vitamin B_{12}
    E. Water
27. In moderate diarrhea, where a large volume of stool is presented to the colon, the most likely electrolyte disturbances are
    A. Acidemia, hyperchloremia, hyperkalemia
    B. Acidemia, hyperchloremia, hypokalemia
    C. Acidemia, hypochloremia, hyperkalemia
    D. Alkalosis, hyperchloremia, hypokalemia
    E. Alkalosis, hypochloremia, hyperkalemia
28. Which of the following is **true** concerning electrolyte absorption?
    A. Aldosterone enhances K^+ absorption in the colon
    B. GIP enhances H_2O and HCO_3^- absorption
    C. Nutrient-independent Na^+ absorption occurs throughout the intestine
    D. Na^+-glucose cotransport occurs primarily in the colon
    E. Sixty percent of the luminal electrolytes is absorbed in the colon
29. Which of the following is essential for digestion and absorption of glucose from complex carbohydrates?
    A. Salivary amylase
    B. Enterokinase
    C. Na^+-K^+ ATPase
    D. Secretin
    E. Sucrase
30. Malabsorption of fat could be due to all of the following **except**
    A. Celiac sprue (blunting of intestinal villi)
    B. Cholestasis
    C. Chronic pancreatitis
    D. Low-pH Zollinger-Ellison syndrome (acid hypersecretion)
    E. Resection of the proximal colon
31. Which of the following is **true** regarding protein digestion and absorption?
    A. Absorbed dipeptides are transported on albumin
    B. Amino acids are absorbed into the lymphatics
    C. Each amino acid has its own specific carrier
    D. Enteropeptidase activates the brush border enzymes
    E. Some dipeptides have specific carriers
32. Vomiting is a symptom commonly seen in intestinal obstruction. Based on your understanding of gastrointestinal physiology, which of the following would be associated with the highest volume of vomiting?
    A. Esophageal obstruction
    B. Gastric outlet obstruction
    C. Obstruction in the first part of the duodenum

D. Obstruction in the third part of the duodenum
E. Colonic obstruction

Directions: Each group of items in this section consists of lettered options followed by a set of numbered items. For each item, select the one lettered option that is most closely associated with it. Each lettered option may be selected once, more than once, or not at all.

A. Electrolyte absorption
B. Carbohydrate digestion/absorption
C. Fat digestion/absorption
D. Protein digestion/absorption
E. None of the above

33. A process that does **not** involve secondary active transport (i.e., basolateral Na^+-K^+ ATPase)
34. A process that does **not** require pancreatic secretion
35. A process that occurs primarily in the intestinal crypts
36. A process that requires neutral pH

Answers

1. B Cholecystokinin's main actions are to stimulate gallbladder contraction and pancreatic secretion, but it also inhibits gastric emptying, which allows the chyme in the duodenum additional time to digest before receiving more chyme from the stomach. CCK has no effect on insulin or pepsinogen secretion.

2. C Somatostatin is synthesized in the pancreas, acts in a paracrine fashion, and has a generally inhibitory effect. It therefore inhibits pancreatic endocrine and exocrine function by inhibiting insulin secretion and pancreatic exocrine secretions. While it also inhibits stomach acid secretion, it does not play an important role in gastric emptying, gallbladder contraction, or pepsinogen secretion.

3. D Gastric inhibitory peptide promotes insulin secretion in response to oral glucose. It does not play a role in any of the other functions.

4. C Cholecystokinin and secretin are both stimulated by acidic chyme that enters the intestinal lumen. While CCK is secreted by I cells, secretin is secreted by S cells. CCK uses IP_3 as a second messenger while secretin uses cAMP. CCK is synthesized as a prepropeptide while secretin is synthesized in its active form.

5. B Gastrointestinal hormones are released primarily from enteroendocrine cells in the gut's mucosa. While some have paracrine effects, they are largely released into the bloodstream, not into the lumen, and travel in the circulation before reaching their target organs. While there are abundant receptors in the gut and associated organs, there are also receptors (and hormone) elsewhere in the body, for example, the brain. The receptors activate G-proteins and Ca^{2+} channels rather than serine kinase receptors. The family membership is based on sequence homology.

6. E Vagal discharge directly promotes gastrin secretion, motility, and pancreatic enzyme secretion. It potentiates gastric acid secretion by lowering the parietal cell threshold to its other stimulants: histamine and gastrin. Vagotomy would decrease these functions and increase the threshold.

7. E The submucosal plexus (Meissner's plexus) is primarily involved in secretion. VIP is an inhibitory transmitter while the sympathetic nervous system is also primarily inhibitory. Secretion can be stimulated by mucosal irritation.

8. E All of the statements are true except the last one. Stretch depolarizes the cell and can result in a spike potential.

9. A The peristaltic wave of contraction is a reflex that occurs within the enteric nervous system. While parasympathetic innervation can induce contraction, a piece of gut void of any extrinsic innervation will undergo this characteristic reflex. It involves a unidirectional wave of contraction (not bi-), where contraction begins proximal to the bolus of food with distal (not proximal) receptive relaxation. Stimulators for contraction include mechanical distention and the hormone motilin.

10. A All of the statements are true except that while the submucosal plexus is located in the submucosa, the other enteric plexus is the myenteric plexus, which is located in the muscularis.

11. D The composition of the saliva will have ion concentrations typical of a fast flow rate. Since it is flowing quickly, the composition will be similar to the composition secreted by the acinus since the ductules have little time to modify it. This concentration is similar to that of plasma, which is rich in Na^+ and poor in K^+. Furthermore, saliva has IgA and amylase, which is involved in carbohydrate digestion, and is stimulated by both sympathetic and parasympathetic stimulation, but primarily parasympathetic.

12. E The rate of gastric acid secretion is regulated by the activity of the H^+ pumps, but also by their numbers. Stimulation of secretion causes small vesicles containing H^+ pumps to fuse with the plasma membrane, thereby increasing the total number of pumps. Gastric acid is secreted by parietal cells. Its concentration increases with increasing flow rate because the HCl-rich secretion then represents a greater percentage of the total than it does at basal flow rates. H^+ is exchanged for K^+. Carbonic anhydrase catalyzes the production of H^+ and HCO_3^- from carbon dioxide and water. An inhibitor would interfere with the production of H^+ (see Fig. 6-6).

13. A The H^+-K^+ ATPase is located in the luminal surface of the parietal cell. There is a 1 : 1 exchange of the two ions, requiring 1 ATP per proton expulsion. The pump is electroneutral and does not generate a charge gradient. This makes it possible to generate a very large chemical gradient of H^+ ions so that the extraordinarily low pH can be achieved. A mole-for-mole supply of K^+ need not be absorbed by the blood because the K^+ passively recirculates to the lumen of the stomach to regain access to the pump. The overall electroneutrality is still maintained by Cl^- accompanying the K^+ into the lumen (see Fig. 6-6).

14. D The two major subdivisions of the stomach are the body and the antrum. The parietal cells that produce gastric acid are located primarily in the body. G cells that produce gastrin are located primarily within the antrum. The three major stimulants of gastric acid production are gastrin, Ach, and histamine. The vagus innervates both the body and the antrum. Its effect on the antrum is to enhance gastrin production. The source of histamine, while not entirely clear, is not localized to a single subdivision and can reach the stomach by blood. Therefore, the most salient effect of antrectomy would be to remove virtually all G cells and therefore also the source of gastrin. Vagal input to the body of the stomach would still be present. Pepsin plays no role in gastric acid production or induction.

15. D Gastric emptying is stimulated by (the time is reduced by) gastric distention. Gastrin is also stimulatory, whereas nitric oxide which enhances compliance would increase the emptying time. Signals from the small intestine further inhibit emptying, including CCK and neural feedback, known as the enterogastric reflex.

16. B Pancreatic secretion is stimulated by CCK and secretin, which are secreted by intestinal mucosal I and S cells, respectively. The process is also stimulated by vagal discharge. Peptidases, not lipases, are stored as inactive zymogens. HCO_3^-

concentration increases with flow rate. The concentrations of Na^+ and K^+, however, are constant.

17. D Secondary bile acids can be either conjugated or unconjugated. They are formed from primary bile acids, derivatives of cholesterol, by entrance into the gut and dehydroxylation by enteric bacteria. After entering the enterohepatic circulation, they can be resecreted. In the gallbladder, they are incorporated into mixed micelles at the critical micellar concentration (see Fig. 6-8).
18. C The total bile acid pool in circulation stays relatively constant. Therefore, the rate of synthesis must equal the rate of excretion. Enhanced intestinal absorption would decrease the rate of excretion and therefore also the rate of synthesis. Chelation in the intestine would decrease absorption and increase synthesis. Increased lithocholate, the bile acid that is insoluble and excreted, would also decrease absorption. The distal ileum is the site of absorption of ionized bile acids. Resection would decrease absorption. Increased cholesterol, the precursor of bile acids, would not decrease their rate of synthesis.
19. C Secretion from enterocytes is mediated by the Cl^- channel (see Fig. 6-10). Enhanced conductance, triggered by an increase in intracellular cAMP, would increase secretion. Na^+ follows Cl^-; increased secretion would not be expected to result in hypernatremia. Bicarbonate secretion would also be expected to increase and result in an acidosis.
20. A Only conjugated bilirubin is found in the urine because it is water soluble and can be filtered by the kidney. Unconjugated bilirubin is not water soluble and remains tightly bound to albumin. Cholestasis is the only disorder that would cause an increase in primarily conjugated bilirubin. Bilirubin enters the cell and is conjugated, but cannot be excreted. Therefore, reflux of conjugated bilirubin into the serum can find its way into urine. The other three disorders cause an increase in primarily unconjugated bilirubin because steps before conjugation are inhibited or overwhelmed.
21. C Hemolytic anemia presents a greater load of red blood cells to the RES system. The increase is primarily in unconjugated serum bilirubin. There is also an increase in stool and urine urobilinogen because the rate of bilirubin excretion is increased even though the enhanced excretion rate cannot entirely compensate for the increased load. Cholestasis or decreased hepatocellular conjugation could not result in an increase in stool and urine urobilinogen.
22. C In pancreatic secretions, Na^+ and K^+ remain relatively constant, similar to their concentrations in plasma. In the other secretions, they vary with flow rate.
23. A Salivary secretion is stimulated by both parasympathetic and sympathetic innervation. In every other case, sympathetic innervation is inhibitory.
24. C While salivary and gastric secretions contribute enzymes that aid in digestion, they are relatively expendable. The digestive enzymes in pancreatic secretion are, however, essential.
25. D All of the secretions aid in digestion and absorption by contributing vital enzymes or other factors, or by simply maintaining osmotic equilibrium. Most of these secretions are reabsorbed. Hepatic secretions, however, also include lipid-soluble waste products such as bilirubin and drugs destined for excretion.
26. D The intrinsic factor–vitamin B_{12} complex is actively absorbed in the distal ileum. Carbohydrates, iron, and un-ionized bile acids are absorbed in the duodenum and jejunum (see Fig. 6-17). Water is absorbed throughout.
27. B In the colon, as elsewhere, water absorption is coupled to electrolyte absorption. An increased volume load would lead to enhanced absorption of electrolytes. In the colon, Na^+ and Cl^- absorption are coupled to HCO_3^- and K^+ secretion. The net result of increased absorption is acidemia, hyperchloremia, and hypokalemia.
28. C Na^+ absorption can occur by two different mechanisms: cotransport with glucose (nutrient-dependent), or coupled to other electrolytes (nutrient-

independent). Nutrient-dependent absorption occurs where glucose is absorbed, that is, in the small intestine, but nutrient-independent absorption occurs throughout the intestine. The vast majority of total absorption occurs in the small intestine. Aldosterone enhances Na^+ absorption in the colon, which is coupled to K^+ secretion. GIP decreases H_2O and HCO_3^- absorption.

29. C The Na^+-K^+ ATPase provides the electrochemical gradient for Na^+-glucose cotransport. While salivary amylase plays a role in digestion, pancreatic amylases can compensate. Enterokinase is involved in protein digestion, not carbohydrate digestion. While secretin potentiates the effects of CCK, it is CCK that is the primary stimulus for pancreatic enzyme secretion. Secretin is primarily responsible for pancreatic water and bicarbonate secretion. While sucrase is an important brush border enzyme that converts sucrose to glucose and fructose, lactase, maltase, and isomaltase can also yield glucose from other carbohydrates.

30. E Resection of the colon would not affect fat absorption because fat is absorbed in the small intestine. Blunting of intestinal villi would decrease the surface area for absorption. Cholestasis would interfere with bile secretion, which is essential for solubilization of all but a few fats. Pancreatitis would interfere with the secretion of lipases. Low pH would interfere with the action of the lipases and with micelle formation.

31. E Amino acids are absorbed into the blood as mono- amino acids only, but some di- and tripeptides can be absorbed into the enterocyte. These di- and tripeptides have specific carriers, but groups of amino acids share common carriers, for example, the basic amino acids. Enteropeptidase is responsible for activating the luminal peptidase, trypsinogen.

32. D Obstruction at this level is distal to the ampulla of Vater, the point at which pancreatic and biliary secretions enter the duodenum. Salivary, gastric, pancreatic, and biliary secretion would, therefore, accumulate proximal to the obstruction, leading to increased volume of vomitus. In colonic obstruction, upper GI secretions can be absorbed in the small intestine, and vomiting is not a major part of the clinical presentation.

33. C Absorption of fat is a passive process. With the help of bile acids, most fats are shuttled to the mucosal membrane. Since they are lipophilic, they do not require specific transporters or the expenditure of energy for absorption. For the absorption of the other constituents, specific transporters are required. In each case, these transporters function with the help of a gradient established by an energy-requiring pump in the basolateral membrane, usually the Na^+-K^+ ATPase.

34. A The pancreas contributes digestive enzymes that digest foodstuffs into constituents that can either be absorbed directly, as in the case of fats, or further digested by the brush border enzymes, as in the case of proteins and carbohydrates. Electrolytes do not require anything of the pancreas.

35. E Absorption occurs primarily on the villi. Secretion occurs primarily in the crypts.

36. C The digestion of fats requires a neutral pH because the lipolytic enzyme lipase functions only at neutral pH. Colipase allows lipase to function at a somewhat lower pH.

Bibliography

Guyton, A. C. *Textbook of Medical Physiology* (8th ed.). Philadelphia: Saunders, 1991.

Johnson, L. R. (ed.). *Gastrointestinal Physiology* (4th ed.). St. Louis: Mosby, 1991.

Lowe, A. (ed.). *Gastrointestinal Physiology Syllabus.* Lecture handout, Stanford University School of Medicine, 1993.

Vander, A. J., Sherman, J. H., and Luciano, D. S. *Human Physiology—The Mechanism of Body Function* (4th ed.). New York: McGraw-Hill, 1994.

Wheater, P. R., and Burkitt, H. G. *Functional Histology: A Text and Colour Atlas* (2nd ed.). New York: Churchill Livingstone, 1987.

7

Endocrine Physiology

Jeffrey Ustin

Hormone Classification

Hormone Types

There are three types of hormones based on chemical structure: **peptides, steroids,** and **amino acid derivatives** (Table 7-1). There are two exceptions to this classification. The **thyroid hormones** are amino acid derivatives, but they act like steroid hormones; and 1,25-vitamin D is not a steroid but is chemically similar. For the following discussion, it can be considered a steroid. The thyroid and steroid hormones are primarily hydrophobic. Being lipid soluble, these compounds must utilize complex storage schemes, which is the case with the thyroid hormones, or be released into circulation as soon as they are made, which is the case with the steroid hormones. The thyroid and steroid hormones bind intracellular receptor proteins in target cells, forming a complex that binds DNA. The complex activates transcription. The major difference in mechanism between thyroid hormones and steroid hormones is that the thyroid hormone receptors are intranuclear and the steroid hormone receptors are cytoplasmic. In contrast to the thyroid and steroid hormones, the peptide hormones and the amino acid derivatives are hydrophilic. This feature allows them to be stored in vesicles and to travel in the plasma unbound to carrier proteins. Peptides and amino acid derivatives act at the cell membrane, through **second messengers,** with a rapid onset. However, they are degraded quickly.

Signal Transduction

In general, the peptide and amino acid derivatives use second-messenger systems (G-proteins or inositol triphosphate; see Chap. 1 for more information on second messengers). Notable exceptions to this rule are insulin and the somatomedins (also called insulin-like growth factors), which bind to receptors that are tyrosine kinases themselves. As mentioned above, steroid hormones, thyroid hormones, and 1,25-vitamin D form a complex with their receptors. These complexes act as transcription factors within the nucleus.

Communication

There are four basic patterns of hormone communication in the endocrine system (Fig. 7-1):

1. Autocrine. Hormone is secreted by a cell and acts on that cell.
2. Paracrine. Hormone is secreted by a cell and acts on nearby cells.
3. Neuroendocrine. Hormone is secreted by a nerve cell into the bloodstream and acts on cells at a distance.

Table 7-1. Comparison of peptide, steroid, and amino acid derivative hormones

	Peptides	Steroids	Amino acid derivatives	
Hormones	GRH, STS, CRH, PRL, ADH, PTH, TRH, PIH, GnRH, FSH, TSH, ACTH, LH, GH, insulin, oxytocin, glucagon, somatomedins	Glucocorticoids, mineralocorticoids, androgens, vitamin D, estrogens, progestins	(Function like steroids) Thyroxine (T^4), triiodothyronine (T^3)	(Function like peptides) Dopamine, epinephrine, norepinephrine
Source	Hypothalamus, pituitary, pancreas, parathyroid	Adrenal cortex, ovaries, testes, placenta	Thyroid gland nerves	Adrenal medulla,
Production	rER	sER	Cytoplasm	Cytoplasm
Storage	Vesicles	None	Follicles Vesicles	
Plasma form	Unbound	Protein bound	Protein bound	Unbound
Site of receptor	Cell membrane	Cytoplasm	Nucleus	Cell membrane
Onset of effects	Fast	Slow	Slow	Fast

ACTH = adrenocorticotropic hormone; ADH = antidiuretic hormone; CRH = corticotropin-releasing hormone; FSH = follicle-stimulating hormone; GH = growth hormone; GnRH = gonadotropin-releasing hormone; GRH = growth hormone–releasing hormone; LH = luteinizing hormone; PIH = prolactin-inhibiting hormone = dopamine; PRL = prolactin; PTH = parathyroid hormone; rER = rough endoplasmic reticulum; sER = smooth endoplasmic reticulum; STS = somatostatin; TRH = thyrotropin-releasing hormone; TSH = thyroid-stimulating hormone.

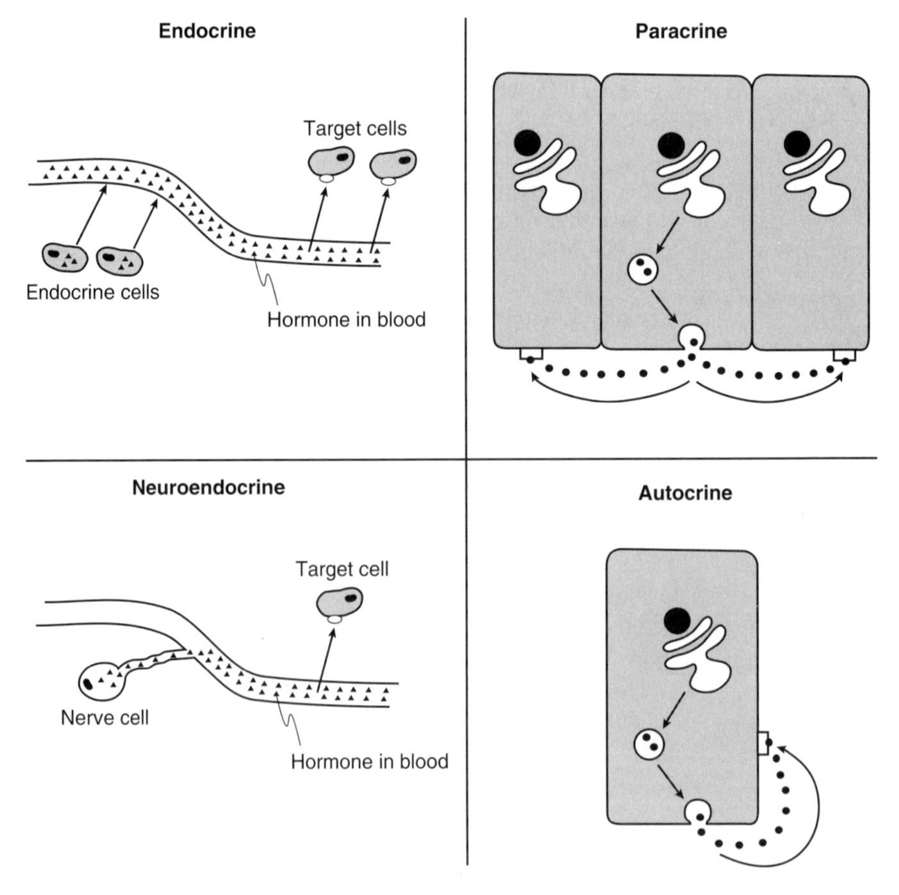

Fig. 7-1. Patterns of hormone communication. (Redrawn from S. H. Snyder. The molecular basis of communication between cells. © *Scientific American* 253(4), Oct. 1985, p. 132.)

4. Endocrine. Hormone is secreted by a glandular cell into the bloodstream and acts on cells at a distance.

Regulation of Carbohydrate Metabolism

Intermediary Metabolism

Anabolic Phase

In the anabolic phase (e.g., after digesting food) energy intake exceeds the energy requirements of the body, and under the direction of **insulin,** energy fuel is stored in the form of **glycogen,** structural protein, and fat. After a meal, the major product of carbohydrate digestion is glucose. Most of the glucose is taken up by the liver, where it is stored as glycogen or used to form **triglycerides,** which are subsequently transported to adipose tissue. Glucose is also taken up by skeletal muscle, where it is incorporated into glycogen or is oxidized.

Catabolic Phase

The catabolic phase (e.g., fasting state) occurs when the body's energy needs are met only by **endogenous** sources. Mediated largely by **glucagon,** catabolism involves the breakdown of glycogen, structural protein, and triglyceride stores to maintain a constant energy supply to the body. The brain can function on only two fuel sources, glucose and ketone bodies. During the first 12 to 24 hours of fasting, glycogen breakdown in the liver provides sufficient glucose for the brain. When glycogen stores become depleted, **gluconeogenesis** takes place, in which noncarbohydrate molecules are converted to glucose. Gluconeogenic substrates include amino acids, glycerol, and lactate. Fatty acids cannot be converted to glucose because the oxidation of pyruvate to acetyl coenzyme A (CoA) is irreversible. Glycerol gives only half a glucose. Thus, the breakdown of triglycerides cannot be a quantitatively significant source of glucose, because each triglyceride is composed of glycerol and fatty acids. Amino acids derived from the breakdown of skeletal muscle structural protein are the major gluconeogenic substrates. As a consequence, muscle wasting is a price paid to maintain brain function during the first few days of fasting. Amino acids are sent from muscle to the liver, where the bulk of gluconeogenesis takes place. The other site is the kidney.

The body's fat stores are critical to survival during prolonged starvation. Although lipolysis of triglyceride from adipose tissue cannot supply much glucose to the brain, it does supply the rest of the body with free fatty acids. Free fatty acids are also converted to ketone bodies in the liver. After 2 to 4 days of fasting, most tissues of the body use free fatty acids and ketone bodies for energy, allowing only the central nervous system (CNS) the privilege of consuming glucose. Protein loss from muscle is thus minimized. When the fasting period extends into weeks, the brain eventually makes use of ketones as its major fuel source. After approximately 4 to 6 weeks, fat stores are exhausted and rapid protein catabolism is resumed.

The pancreatic hormones, insulin and glucagon, regulate not only carbohydrate metabolism but protein and fat metabolism.

Insulin

The Molecule and its Effects

Insulin is produced by the beta cells of the pancreatic islets. It is a polypeptide hormone made up of an A chain and a B chain that are connected by disulfide bridges.

The action of this hormone is anabolic, increasing the storage of glucose, fatty acid, and amino acids. Insulin increases the efficiency of glucose transport into cells of most tissues except brain, kidney tubules, intestinal mucosa, and red blood cells. In the liver, insulin increases glycogenesis by stimulating glucokinase and glycogen synthetase, and inhibits gluconeogenesis and glycogenolysis by inhibiting fructose 1,6-bisphosphatase and glycogen phosphorylase (Fig. 7-2). The incorporation of glucose into glycogen and the decrease in gluconeogenesis result in net glucose entry into hepatocytes. Hence, insulin facilitates glucose entry into liver cells by an indirect mechanism, while it does so in other tissues by a direct action on cell membranes. Insulin also increases lipid and protein synthesis in the liver.

In adipose tissue, insulin increases formation of glycerol and long-chain fatty acids from glucose and increases esterification to form triglycerides. Insulin also inhibits the hormone-sensitive lipase in adipocytes and, thereby, decreases lipolysis. In skeletal muscle, insulin increases amino acid uptake as well as glucose entry, glycogen synthesis, and protein synthesis.

Figure 7-3 summarizes the major metabolic effects of insulin.

Mechanism of Action

The mechanism of insulin action begins with binding to highly specific receptors in the cell membrane. Internalization of the receptor-hormone complex via endocytosis may be part of the mechanism of action. The insulin receptor is a protein kinase that phosphorylates tyrosine residues on the receptor itself **(autophosphorylation)** and on other proteins. From this point, the exact mechanism leading to insulin's ultimate biologic effects remains unclear. One known effect is the translocation of glucose transporters from vesicles within the cell to the cell membrane.

Fig. 7-2. Major steps and enzymes involved with glucose metabolism.

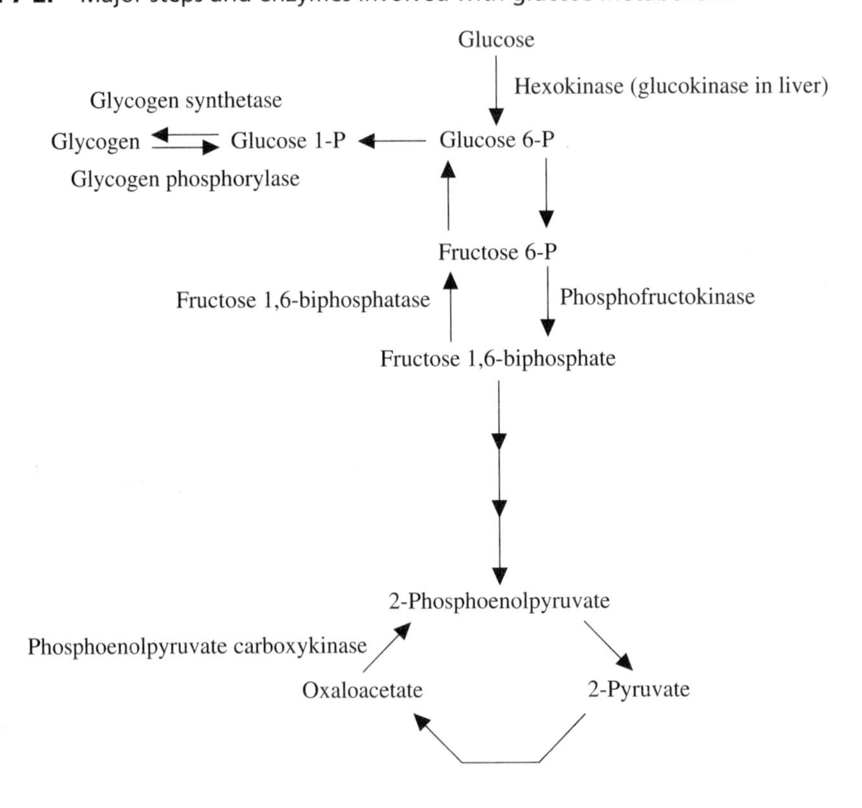

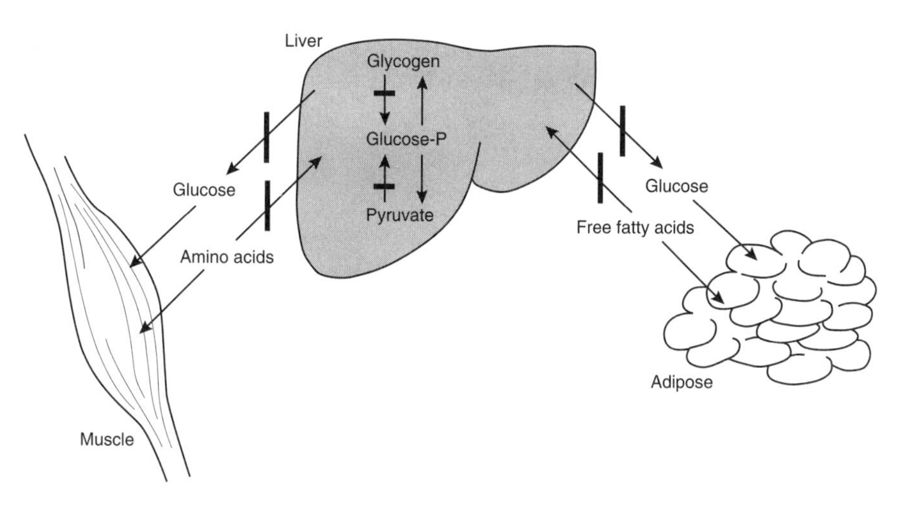

Fig. 7-3. Major metabolic effects of insulin. Lines with bars through them indicate that insulin inhibits that action. Lines without bars indicate that insulin causes that action. (Redrawn from R. M. Berne and M. N. Levy. *Physiology* [3rd ed.]. P. 863. © 1993 Mosby–Year Book, St. Louis.)

Insulin receptors show a phenomenon called **"down-regulation,"** in which high levels of insulin cause a decrease in receptor number. Down-regulation may be associated with the insulin resistance in obesity and non–insulin-dependent diabetes mellitus.

Regulation

The most important stimulator of insulin secretion is a rise in blood glucose. Amino acids from protein digestion are potent stimulators too. A more minor role is played by intestinal hormones such as gastric inhibitory protein. Fat ingestion does not affect release. Strong inhibitors include sympathetic stimulation and catecholamines. Somatostatin plays a lesser role.

Stimulators of Release	**Inhibitors of Release**
Elevated blood glucose	Sympathetic stimulation
Amino acids	Catecholamines
Intestinal hormones	Somatostatin

Insufficiency

Insulin insufficiency leads to several metabolic derangements that constitute **diabetes mellitus.**

1. **Hyperglycemia.** Caused by decreased glucose entry into tissues, increased gluconeogenesis, and increased glucose release by the liver
2. **Osmotic diuresis.** Caused by the kidney's failure to reabsorb the increased filtered glucose load; free water follows the glucose osmoles into the tubules
3. **Polyuria.** Caused by diuresis; results in urinary loss of sodium and potassium, hypovolemic hypotension, and dehydration
4. **Polydipsia.** Caused by dehydration secondary to diuresis
5. **Protein catabolism.** Leads to muscle wasting
6. **Intracellular glucose deficiency.** Caused by lack of glucose entry into cells
7. **Fat catabolism.** Free fatty acids in the bloodstream are converted to acetyl-CoA in the liver, which enters the tricarboxylic acid cycle

8. **Polyphagia.** Caused by cells in the hypothalamic ventromedial nuclei that sense an intracellular glucose deficiency and increase appetite
9. **Ketoacidosis.** Caused by excess acetyl-CoA being converted to ketone bodies (acetone, acetoacetate, beta-hydroxybutyrate)

Excess

Insulin excess, seen most commonly in exogenous insulin overdose, leads to hypoglycemia, the effects of which are most pronounced in the CNS (confusion, convulsions, coma). The body compensates by releasing glucagon and epinephrine, both of which raise the blood glucose concentration. Glucocorticoids and growth hormone also elevate the blood glucose concentration, but their role is of less importance.

Glucagon

Glucagon is a polypeptide hormone secreted by the alpha cells of pancreatic islets. Its actions are catabolic, generally opposing insulin's actions. It works primarily in the liver, increasing glycogenolysis (activates glycogen phosphorylase) and gluconeogenesis (activates fructose 1,6-bisphosphatase and phosphoenolpyruvate carboxykinase; see Fig. 7-2). It also increases uptake of substrate used for glucose production. Finally, it increases lipolysis and ketogenesis within the liver.

Stimulators of Release	**Inhibitors of Release**
Hypoglycemia	Elevated blood glucose
Amino acids	Somatostatin
Beta-adrenergic stimulation	

Catecholamines

Catecholamines maintain plasma glucose levels by increasing glycogenolysis and gluconeogenesis. To this extent they act like glucagon. However, catecholamines will act on fat and muscle to increase substrate availability whereas glucagon only acts in the liver.

Glucocorticoids and Growth Hormone

Glucocorticoids and growth hormone have little effect acutely, but will increase plasma glucose levels and augment glucagon and catecholamine action in long-standing hypoglycemia. Glucocorticoids mobilize protein stores whereas growth hormone mobilizes fat stores.

Glucose Homeostasis

The relative importance of the above hormones in counteracting hypoglycemia is as follows:

1. Decrease insulin levels
2. Increase glucagon levels
3. Increase catecholamines
4. Increase glucocorticoids and growth hormone

Somatostatin and Pancreatic Polypeptide

Pancreatic islet delta cells secrete somatostatin. Somatostatin inhibits secretion of the other three pancreatic islet hormones and its role may only be one of local regulation.

It is the same somatostatin as that released by the hypothalamus to inhibit growth hormone secretion. The PP cells secrete pancreatic polypeptide, the function of which is unknown.

Regulation of Calcium and Phosphate Metabolism

Calcium in the Body

The adult human body contains about 1 to 2 kg of calcium, of which over 98 percent is contained in the skeleton, and the rest in the plasma and cells. The total plasma calcium concentration is normally about 10 mg/dl, half of which is protein bound (mainly to albumin) and half of which exists as free ions. Plasma calcium is in equilibrium with bone calcium, but only 0.5 percent of bone calcium is readily exchangeable, with the remainder only slowly exchangeable. Free plasma calcium is the physiologically important component and is critical in maintaining neuromuscular and other cellular functions. It is under tight hormonal control and the body chooses to sacrifice mineral calcium stores in order to maintain serum calcium levels. Calcium enters the plasma via intestinal absorption and bone resorption and is removed from extracellular fluid (ECF) by urinary excretion, secretion into the gastrointestinal tract, and deposition in bone. The intestinal absorption of calcium is regulated by a vitamin D metabolite (calcitriol), its kidney reabsorption and excretion by parathyroid hormone (PTH) and calcitonin, and its bone resorption and deposition by PTH and calcitonin. Short-term control of serum calcium is achieved by hormonal effects on bone, while long-term control is achieved through hormonal effects on the intestinal tract and kidney. Figure 7-4 summarizes calcium metabolism.

Phosphorus in the Body

The average adult body contains about 1 kg phosphorus, 85 percent of which is in the skeleton and 15 percent in muscle and other tissues. Fasting plasma phosphorus concentrations range from 3 to 4 mg/dl, about 12 percent of which is protein bound. Un-

Fig. 7-4. Calcium metabolism. Arrows indicate movement of calcium from one compartment to another. The hormone next to an arrow indicates that hormone stimulates the movement shown by the arrow. (Adapted from A. Despopoulos and S. Silbernagl. *Color Atlas of Physiology.* P. 255. © 1986 Georg Thieme Verlag, Stuttgart.)

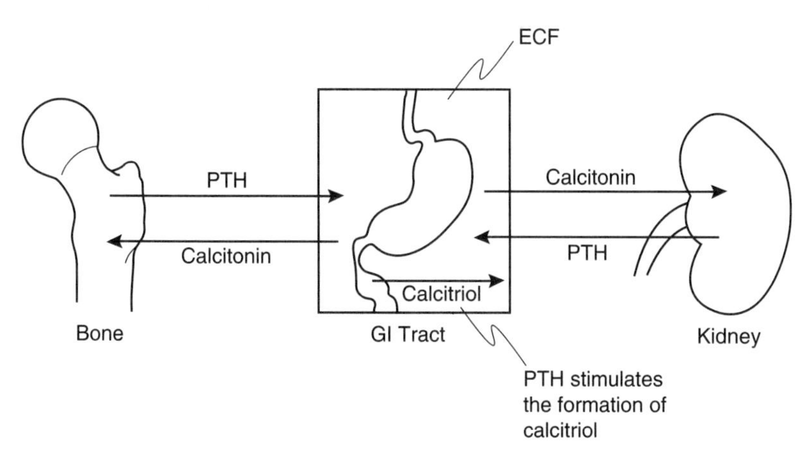

Endocrine

like calcium, serum phosphate levels can vary widely, because phosphate not only moves between ECF and bone, but also between ECF and intracellular fluid (ICF). (It is involved in almost all intracellular metabolic processes.) As well, there is no direct feedback as with calcium. Phosphate enters the ECF via the intestinal tract, ICF, and bone; it leaves the ECF via urine and movement to ICF or bone. Phosphorus absorption from the intestine is very efficient, from 70 to 90 percent. The major site of control of the body's phosphorus is the kidney, in which urinary excretion parallels dietary intake. Parathyroid hormone modifies urinary excretion of phosphate. Figure 7-5 summarizes phosphate metabolism.

Parathyroid Hormone

Synthesis and Actions

Parathyroid hormone is produced by the chief cells of the parathyroids. This peptide hormone is first synthesized as prepro-PTH, processed by cleavage to pro-PTH, and further cleaved to PTH, the active form. Parathyroid hormone has the following actions:

1. Increases bone resorption to mobilize calcium by stimulating osteoclasts and inhibiting osteoblasts
2. Increases reabsorption of calcium in the kidney's distal tubules
3. Decreases reabsorption of phosphate in the renal tubules
4. Increases the production of 1,25-dihydroxycholecalciferol, the vitamin D metabolite that enhances intestinal calcium absorption.

The net result is increased serum calcium, decreased serum phosphate, and increased urine phosphate.

Fig. 7-5. Phosphate metabolism. Arrows indicate movement of phosphate from one compartment to another. A solid arrow with a hormone next to it indicates that hormone stimulates the movement shown by the arrow. A dashed arrow indicates that the movement occurs independently of hormone stimulation or inhibition.

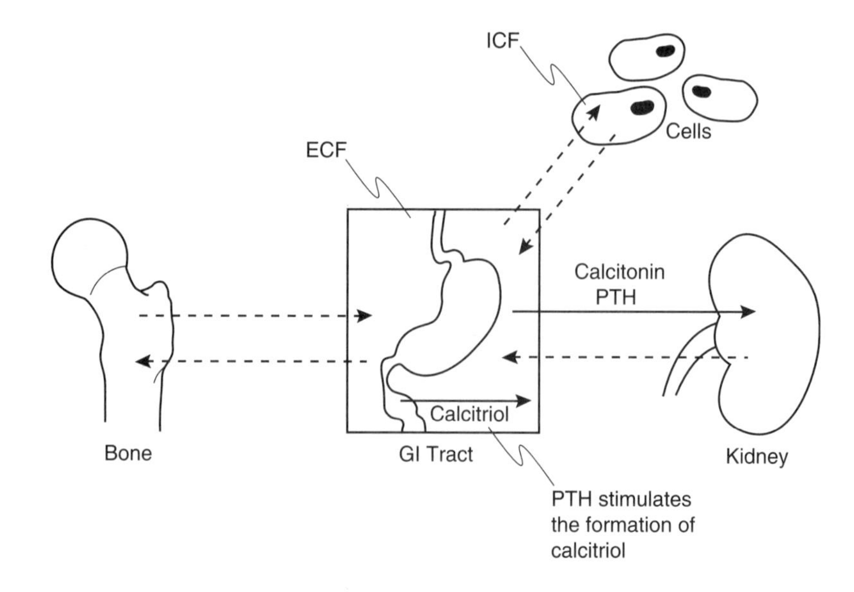

Regulation

The major regulator of PTH secretion is serum calcium. Increased serum calcium depresses PTH secretion, while decreased serum calcium elevates PTH secretion.

Excess Hormone

Primary hyperparathyroidism can be defined as the inappropriate hypersecretion of PTH that results in hypercalcemia, and **secondary hyperparathyroidism** as the hypersecretion of PTH in response to a hypocalcemic stress. Hyperparathyroidism gives rise to hypercalcemia, hypercalciuria (the kidney cannot reabsorb the entire increased calcium load even in the presence of elevated PTH), hypophosphatemia, hyperphosphaturia, and bone demineralization.

Insufficient Hormone

If all parathyroid glands are inadvertently removed during thyroid surgery, serum calcium declines and neuromuscular hyperexcitability develops. Calcification of the lens, intracranial calcifications, and abnormal cardiac conduction can be seen as well. Unless the patient is given supplemental calcium, a fatal hypocalcemic tetany may occur.

Calcitonin

Calcitonin is produced in the parafollicular (also called clear cells or C cells) of the thyroid gland. It is a peptide hormone whose principal pharmacologic actions include (1) inhibition of bone resorption by inhibiting osteoclasts and (2) increased urinary excretion of calcium and phosphate. The result is a lowered serum calcium and phosphate. Hormonal secretion is controlled by serum calcium; that is, increased serum calcium increases secretion and vice versa. The exact physiologic role of calcitonin is not entirely clear, because neither an excess of hormone (e.g., medullary carcinoma of the thyroid) nor a deficiency (e.g., thyroidectomy) leads to any defects in bone or calcium metabolism.

Vitamin D

Synthesis and Actions

Vitamin D_3, or cholecalciferol, is produced in the skin from 7-dehydrocholesterol after exposure to ultraviolet light. Vitamin D_2, or ergocalciferol (Fig. 7-6), is derived from plants and used to supplement dairy products. This dietary supplement is not required for humans who are adequately exposed to sunlight. Vitamin D_2, like other fat-soluble vitamins, does require the fat-absorbing mechanisms of the intestine to be intact. Vitamins D_2 and D_3 are equally potent biologically. Vitamins D_2 and D_3 are transported to the liver, where they are hydroxylated to produce 25-hydroxy-D_2 and 25-hydroxy-D_3, respectively. The 25-hydroxy D_3 is the most abundant form of vitamin D in the circulation and the major storage form of vitamin D. It is further hydroxylated in the kidneys to **1,25-dihydroxy-D_3, calcitriol** (Fig. 7-6), the active form of vitamin D. The major action of 1,25-dihydroxy-D_3 is to increase the absorption of dietary calcium and phosphate from the intestinal tract. It initially binds to a specific cytoplasmic receptor molecule. The resulting complex is translocated to the nucleus and causes an increased synthesis of an intestinal calcium-binding protein. It also enhances the action of PTH on the osteoclasts.

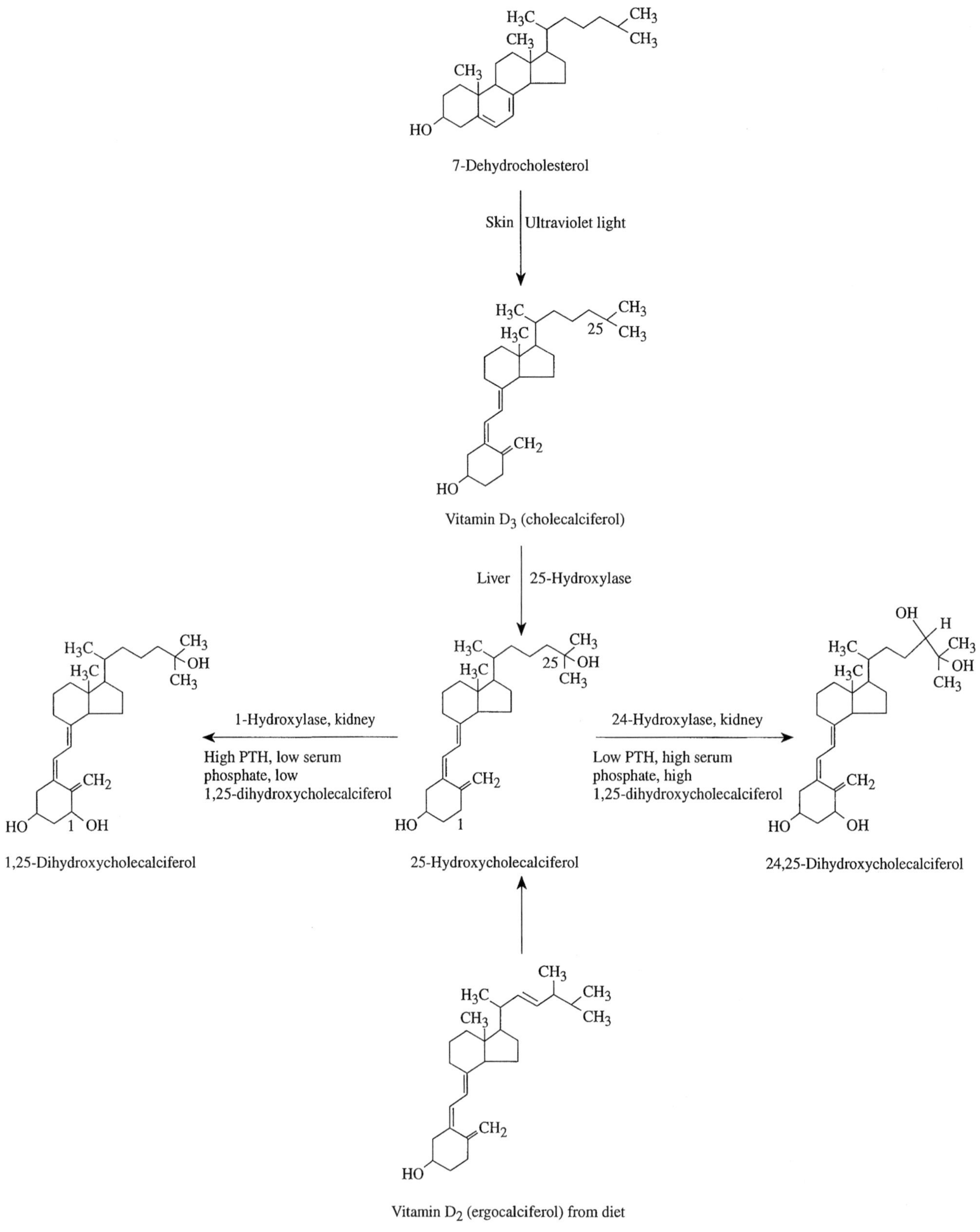

7-Dehydrocholesterol

Skin | Ultraviolet light

Vitamin D_3 (cholecalciferol)

Liver | 25-Hydroxylase

25-Hydroxycholecalciferol

1-Hydroxylase, kidney

High PTH, low serum phosphate, low 1,25-dihydroxycholecalciferol

1,25-Dihydroxycholecalciferol

24-Hydroxylase, kidney

Low PTH, high serum phosphate, high 1,25-dihydroxycholecalciferol

24,25-Dihydroxycholecalciferol

Vitamin D_2 (ergocalciferol) from diet

Fig. 7-6. Vitamin D synthesis.

Regulation

The production of 1,25-dihydroxy-D_3 in the kidney is regulated by PTH, serum phosphate, and its own level. An increase in PTH or a decrease in serum phosphate stimulates 1,25-dihydroxy-D_3 production. The hormone is also regulated by feedback inhibition. When PTH is low, serum phosphate is high, or 1,25-dihydroxy-D_3 is high, the kidney hydroxylates another position, producing the inactive metabolite 24,25-dihydroxy-D_3 instead.

Deficiency of Hormone

Vitamin D deficiency may be caused by either inadequate exposure to sunlight or fat malabsorption. Decreased vitamin D causes decreased absorption of calcium and phosphate, decreased serum calcium and phosphate, increased PTH, and increased bone resorption. Bones are not adequately calcified and in adults **osteomalacia** results. In children, vitamin D deficiency causes **rickets,** in which bone fails to mineralize, epiphyses fail to fuse, epiphyseal plates widen, and fractures occur.

Hypothalamic-Pituitary Relationships

Anterior and Posterior Pituitary

The hypothalamus communicates differently with the anterior pituitary than it does with the posterior pituitary. Neurons in the supraoptic and paraventricular nuclei of the hypothalamus send axons directly to the posterior pituitary (neurohypophysis) delivering hormones. The anterior pituitary does not receive axons. Instead it receives hormones via the portal hypophyseal vessels from the median eminence of the hypothalamus.

The hormones oxytocin and vasopressin (antidiuretic hormone [ADH]) are synthesized in the cell bodies of neurons in the supraoptic and paraventricular nuclei, transported down their axons to the posterior pituitary, and released into the bloodstream. Antidiuretic hormone increases water reabsorption in the kidney to conserve fluid volume and oxytocin stimulates milk secretion from the mammary glands.

The Hormones

The median eminence of the hypothalamus secretes the hypothalamic-releasing and inhibiting hormones: growth hormone–releasing hormone (GRH), growth hormone–inhibiting hormone (GIH), corticotropin-releasing hormone (CRH), thyrotropin-releasing hormone (TRH), gonadotropin-releasing hormone (GnRH), prolactin-releasing factor (PRF), and prolactin-inhibiting hormone (PIH). (Note: GIH and STS, somatostatin, are identical. PIH and dopamine are identical.) These hypophysiotropic hormones are polypeptides that regulate the secretion of six anterior pituitary hormones: growth hormone (GH), adrenocorticotropic hormone (ACTH), thyroid-stimulating hormone (TSH), follicle-stimulating hormone (FSH), luteinizing hormone (LH), and prolactin. GnRH stimulates the secretion of both FSH and LH.

Endocrine Function of Hypothalamic and Pituitary Hormones

Growth hormone promotes body growth via stimulation of somatomedin secretion by the liver. **ACTH** stimulates corticosteroid secretion and **TSH** stimulates thyroid secretion. In the female, **FSH** stimulates early growth of the ovarian follicle, and in the male, spermatogenesis. **LH** stimulates ovulation and luteinization of the ovarian follicle in the female and testosterone secretion in the male. **Prolactin** stimulates lactation. These hormonal relationships are summarized in Fig. 7-7 and each hormone is discussed in further detail later in this chapter.

Adrenocorticotropic hormone is derived from a larger precursor protein, called proopiomelanocortin (POMC), which is synthesized in the hypothalamus, and while en route to the anterior pituitary it is cleaved into ACTH, beta-lipotropin (beta-LPH), and gamma melanocyte-stimulating hormone (gamma-MSH). A small amount of β-LPH is cleaved into gamma-LPH and beta-endorphin. All these substances are secreted from the anterior pituitary. The physiologic function of only ACTH is known. Beta-endorphin is an endogenous opiate that binds to opioid receptors. In pharmacologic amounts, gamma-MSH stimulates melanin synthesis in melanocytes. The function of gamma-LPH is not known.

Regulation

The secretion of hypothalamic and pituitary hormones is controlled by feedback mechanisms to preserve homeostasis. For example, high plasma levels of cortisol exert a negative feedback on CRH and ACTH secretion. Superimposed on the feedback control are chronobiologic and emotional factors. ACTH displays a circadian rhythm in that pulses of secretion are more frequent in the early-morning hours and less frequent in the evening. Release of growth hormone is also strongly associated with sleep. The circadian clock responsible for these diurnal rhythms appears to be located in the suprachiasmatic nuclei of the hypothalamus. With long-term stress, GnRH ac-

Fig. 7-7. Hypothalamic-pituitary axis. As an example of feedback, the long and short feedback loops from the somatomedins to the pituitary and hypothalamus have been drawn.

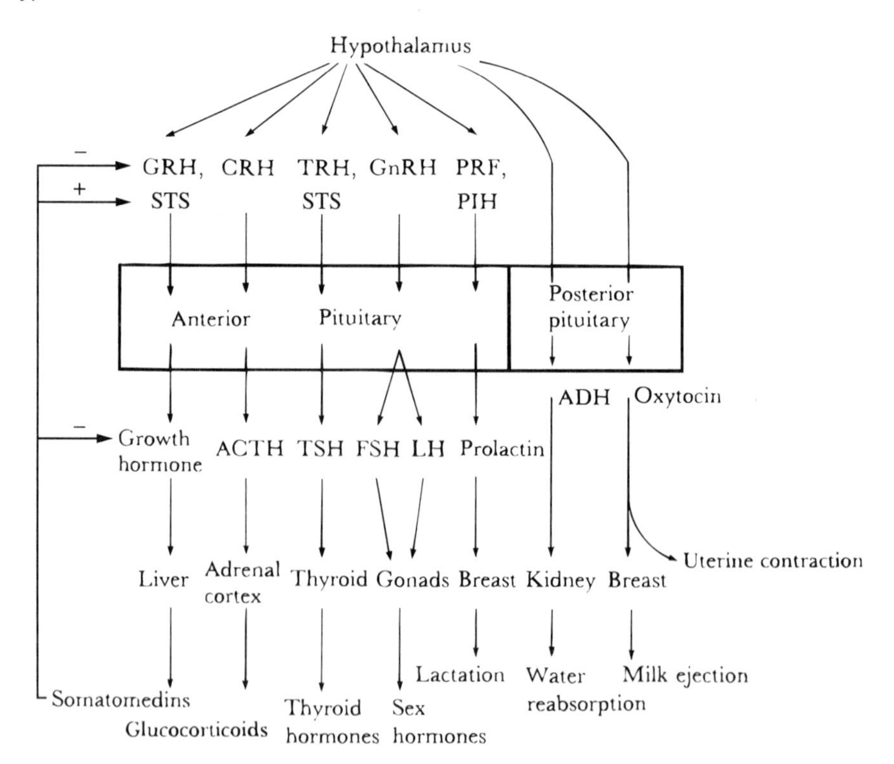

tivity is depressed (e.g., "boarding-school amenorrhea" in students who move away from home). Emotional stress, fear, or anxiety causes an increase in CRH and ACTH. These emotional signals are probably relayed to the median eminence of the hypothalamus via the limbic system.

Posterior Pituitary Hormones

Vasopressin (Antidiuretic Hormone)

Synthesis

A vasopressin polypeptide precursor, called prepropressophysin, is synthesized in the cell bodies of the hypothalamus. It is cleaved into vasopressin, neurophysin, and glycopeptide during transport down the axon to the posterior pituitary. The actions of the latter two peptide fragments are not known.

Actions

Antidiuretic hormone increases the water permeability of the collecting ducts in the kidneys and, thereby, increases reabsorption of solute-free water. This results in an increased urine osmolality, decreased urine volume, decreased plasma osmolality, and increased extracellular fluid volume. In pharmacologic doses, ADH raises arterial blood pressure by vasoconstriction, hence its other name vasopressin.

Regulation

Antidiuretic hormone is regulated by plasma osmolality. Osmoreceptor cells located in the anterior hypothalamus near the supraoptic and paraventricular nuclei are activated when plasma osmolality rises above 280 mOsm/kg. A signal is relayed to the adjacent nuclei to synthesize and release ADH. A hyperosmolar plasma stimulates other osmoreceptors in the hypothalamus, resulting in a sensation of thirst. Decreases in ECF volume stimulate ADH secretion via stretch receptors in the great veins, atria, and pulmonary vessels. Other stimulators of ADH secretion include angiotensin II, nicotine, nausea, pain, and some emotions. Secretion of ADH is depressed with decreased plasma osmolality, increased ECF volume, and alcohol.

Excess Hormone

In the syndrome of "inappropriate" hypersecretion of ADH **(SIADH),** ADH is autonomously released from the posterior pituitary (e.g., CNS disorders, drug induced) or from tumor tissue (e.g., oat-cell carcinoma of the lung). The excess ADH causes water retention, which results in hyponatremia. Urine is inappropriately concentrated and the plasma is hypoosmolar.

Insufficient Hormone

Diabetes insipidus is characterized by ADH deficiency. It can be caused by either a lesion anywhere along the hypothalamic-pituitary axis that is involved with ADH production, or an inability of the kidney to respond to ADH. Deficiency of ADH results in excretion of large amounts of hypoosmotic urine. The plasma becomes hyperosmotic and stimulates the thirst mechanism (polydipsia). It is the abnormally large water intake that prevents severe dehydration in these patients.

Oxytocin

Oxytocin causes contraction of the uterus during labor and contraction of the myoepithelial cells of the lactating breast. In late pregnancy, both uterine sensitivity to oxytocin and oxytocin secretion are increased. Complex neuroendocrine reflexes take place during labor, involving activation of afferent fibers to the oxytocin-containing neurons in the hypothalamus so that there is sufficient release of oxytocin to maintain uterine contractions until delivery is completed.

The milk ejection reflex includes the stimulation of touch receptors in the breast by infant suckling, activation of afferent fibers to the supraoptic (SO) and paraventricular (PV) nuclei, release of oxytocin, contraction of myoepithelial cells, and ejection of milk.

Growth Hormone

Synthesis and Growth Effects

Growth hormone (GH, or **somatotropin)** is a peptide hormone synthesized and secreted by the anterior pituitary. Its major action is to promote linear growth by its effect on the skeletal system. This action, however, is mediated by the induction of **somatomedins** (insulin-like growth factors), a group of small peptides produced in the liver that have insulin-like and mitogenic activity. Somatomedins stimulate the proliferation of chondrocytes by increasing the synthesis of DNA, RNA, protein, and hydroxyproline. As a result, the proliferating cartilage widens the epiphyseal plates, allowing new bone to be laid down, and thereby causes linear growth. The somatomedins are regulated by GH and transported in the blood by carrier proteins.

Metabolic Effects

Growth hormone also has many metabolic actions, the most important of which include

1. Increased protein synthesis
2. A positive nitrogen balance
3. Increased lipolysis
4. Insulin resistance (diabetogenic)
5. Increased intestinal absorption of calcium
6. Increased renal reabsorption of phosphate

These actions are aimed at acquiring building blocks for growth.

Growth and development require a net anabolic effect on protein metabolism. Lipolysis releases free fatty acids, which can be used by the body so that protein is spared from catabolism. The antiinsulin action of GH on carbohydrate metabolism includes increased hepatic glucose output and decreased glucose uptake in some tissues. The effects on calcium and phosphate increase their availability for skeletal growth. It is unclear which of these metabolic actions are exerted through GH directly and which through somatomedins.

Regulation

The secretion of GH is regulated by several factors. Secretion is increased by the hypothalamic GH-releasing hormone (GRH), sleep, physical or emotional stress, exercise, hypoglycemia, increase in blood amino acids following a protein meal, and dopaminergic agonists. Factors that decrease GH secretion include the hypothalamic GH-inhibiting hormone (GIH, somatostatin), somatomedins, obesity, hyperglycemia, GH, cortisol, and elevated free fatty acids.

Deficiency and Excess

Growth hormone deficiency in a child causes retardation of growth, of epiphyseal development, and of bone age. Excess growth hormone (e.g., in the case of anterior pituitary tumor) in children causes **gigantism** with massive skeletal growth. In adults, whose epiphyseal plates have fused, excess GH causes **acromegaly,** characterized by enlarged extremities, enlarged visceral organs, and coarsening of facial features.

The Adrenal Gland

Adrenal Medulla

Catecholamine Synthesis

The adrenal medulla is the inner core of the adrenal gland (Fig. 7-8) and secretes the catecholamines epinephrine, norepinephrine, and dopamine. Because they respond to stimulation by preganglionic sympathetic nerve fibers (splanchnic nerves), endocrine cells of the medulla can be considered as analogous to postganglionic sympathetic neurons with missing axons. Unlike other postganglionic neurons, the medullary cells can convert norepinephrine to epinephrine. The biosynthesis of catecholamines is shown in Fig. 7-9. The conversion of norepinephrine to epinephrine is catalyzed by phenylethanolamine-N-methyltransferase (PNMT). This enzyme is induced by glucocorticoids from the adrenal cortex, which are carried in the blood by a direct vascular connection between the cortex and medulla. The medulla secretes about 80 percent epinephrine and 20 percent norepinephrine.

Secretion

The catecholamines are stored in **chromaffin granules** bound to ATP and protein. Release of the granules is mediated by the release of acetylcholine from preganglionic

Fig. 7-8. Cross section through adrenal showing cortex and medulla. (Adapted from A. Despopoulos and S. Silbernagl. *Color Atlas of Physiology.* P. 269. © 1986 Georg Thieme Verlag, Stuttgart.)

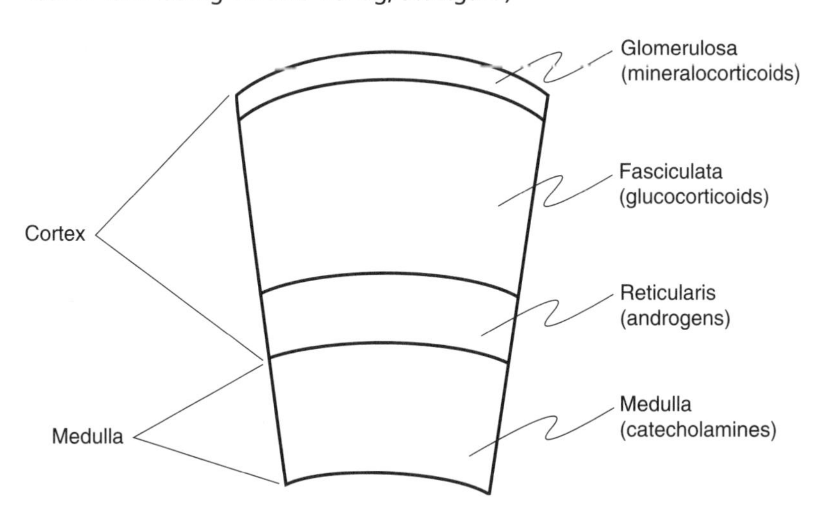

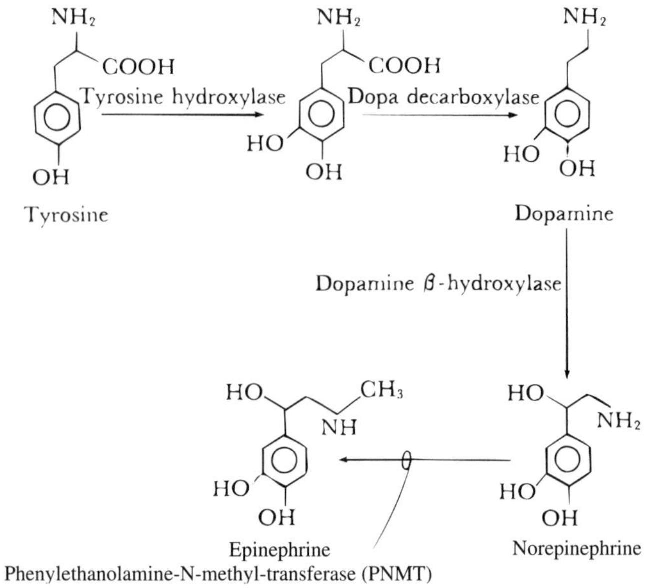

NH_2 COOH Tyrosine hydroxylase → NH_2 COOH HO OH Dopa decarboxylase → NH_2 HO OH

OH Tyrosine

Dopamine

Dopamine β-hydroxylase

HO NH CH_3 HO OH Epinephrine ← HO NH_2 HO OH Norepinephrine

Phenylethanolamine-N-methyl-transferase (PNMT)

Fig. 7-9. Catecholamine synthesis.

fibers. The acetylcholine increases the cell membrane's permeability to calcium, and the ensuing calcium influx triggers exocytosis of these storage vesicles. Also released are dopamine, ATP, dopamine beta-hydroxylase, and other proteins. The half-life of catecholamines in the circulation is about 2 minutes before they are metabolized and excreted.

Effects

The actions of catecholamines are mediated by two classes of receptors, alpha- and beta-adrenergic receptors, which are further divided into alpha-1, alpha-2, beta-1, and beta-2 receptors. The effects of circulating catecholamines on these receptors are identical to the effects of catecholamines released from noradrenergic nerve terminals of sympathetic postganglionic axons (see Chaps. 1 and 8) and are shown in Table 7-2. In summary, epinephrine decreases total peripheral resistance, increases heart rate, increases cardiac output, and does not change the mean blood pressure. (Note: At higher levels epinephrine causes an alpha-1–mediated increase in total peripheral resistance [TPR].) Norepinephrine increases TPR and increases the mean blood pressure, resulting in a reflex decrease in heart rate and cardiac output.

The metabolic effects of the catecholamines include increased glycogenolysis in liver and muscle producing a rise in blood glucose, increased lipolysis in adipose tissue and release of free fatty acids, and an increased metabolic rate (i.e., increased oxygen consumption and heat production). The adrenal medullary hormones also play a role in stimulating the CNS. They enhance alertness and arousal by lowering the threshold of neurons in the reticular formation in the brainstem. The physiologic function of circulating dopamine is not known.

Regulation

Adrenal medullary secretion is regulated by the sympathetic nervous system. Situations that evoke a **"fight-or-flight"** response include fear, anxiety, pain, trauma, hemorrhage, exercise, extreme cold or heat, hypoglycemia, and hypotension. The increase in medullary secretion is part of the diffuse sympathetic discharge that prepares the

Table 7-2. Adrenergic responses of selected tissues

Organ	Receptor	Effect
Heart	Beta-1	Increased inotropy
		Increased chromotropy
Blood vessels	Alpha	Vasoconstriction
	Beta-2	Vasodilation
Kidney	Beta	Increased renin release
Gut	Alpha, beta	Decreased motility
		Increased sphincter tone
Pancreas	Alpha	Decreased insulin release
		Increased glucagon release
	Beta	Increased insulin and glucagon release
Liver	Alpha, beta	Increased glycogenolysis
Adiopose	Beta	Increased lipolysis
Skin	Alpha	Increased sweating
Bronchioles	Beta-2	Bronchodilation
Uterus	Alpha, beta	Contraction, relaxation

Modified from N. Weiner and P. Taylor. Neurohumoral Transmission: The Autonomic and Somatic Motor Nervous Systems. In A. G. Gilman, L. S. Goodman, T. W. Rall, and F. Murad (eds.). *Goodman and Gilman's The Pharmacological Basis of Therapeutics* (7th ed.). New York: Macmillan, 1985. (Copyright © 1985 by Macmillan Publishing Company.)

individual for emergency situations. The body responds with an increase in heart rate and cardiac output, increased blood flow to skeletal muscles, decreased blood flow to the viscera, mobilization of glucose and free fatty acids to provide readily available energy sources, and enhanced CNS arousal. The role of circulating catecholamines in mediating vascular changes is less important than that of sympathetic innervation of the vessels. The metabolic actions of circulating catecholamines are more significant during such emergency situations, such as trauma. The adrenal medulla, however, is not essential for life.

Pheochromocytomas are rare tumors, arising from chromaffin cells in the sympathetic nervous system, which secrete catecholamines.

Adrenal Cortex

The Gland

The adrenal cortex is the outer part of the adrenal gland. It secretes glucocorticoids, which have widespread effects on carbohydrate and protein metabolism, and mineralocorticoids, which regulate the body's sodium, potassium, and fluid volume. It also secretes small amounts of androgens, which normally play a less important role than the androgens secreted by the gonads. The adrenal cortex is comprised of three zones: the **zona glomerulosa** (outermost), the **zona fasciculata,** and the **zona reticularis** (innermost). The zona glomerulosa produces aldosterone, the zona fasciculata produces glucocorticoids, and the zona reticularis produces androgens.

Synthesis

The synthesis of adrenocortical steroids (Fig. 7-10) begins with cholesterol, which is converted to pregnenolone. This rate-limiting step is mediated by a cholesterol hydroxylase and desmolase. ACTH via the adenylate cyclase–cyclic adenosine

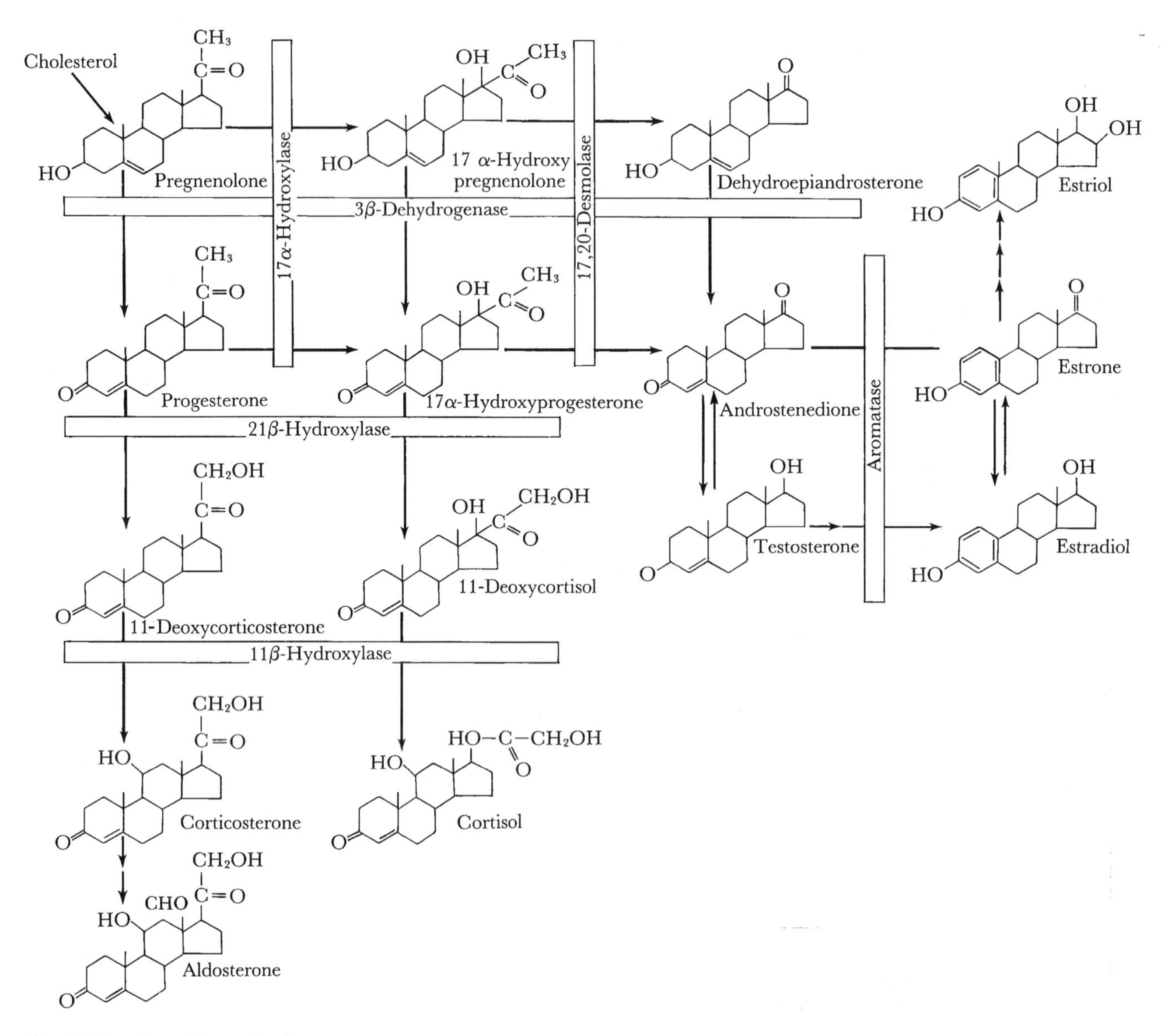

Fig. 7-10. Steroid synthesis.

monophosphate (cAMP) system stimulates this conversion. ACTH also increases the uptake of lipoprotein (the major source of adrenal cholesterol) by the adrenal cortex and the hydrolysis of stored cholesterol esters to free cholesterol. Many of the intermediates of steroid synthesis are secreted to some extent, but the steroids that are secreted in physiologically significant amounts include aldosterone, cortisol, corticosterone, dehydroepiandrosterone (DHEA), and androstenedione.

Glucocorticoids

Glucocorticoids in the Body

About 10 percent of circulating cortisol is free, 75 percent is bound to **corticosteroid-binding globulin (CBG,** or **transcortin),** and 15 percent is bound to albumin. Only the free form is biologically active. The CBG has high-affinity and low-capacity binding. The opposite is true of albumin. Estrogen increases CBG synthesis in the liver while protein deficiency states such as cirrhosis or nephrosis decrease CBG production. The hypothalamic-pituitary-adrenal axis is regulated by the free cortisol level, not the total cortisol level. **Cortisol** is largely metabolized and conjugated in the liver before excretion in the urine.

Effects

The physiologic effects of glucocorticoids are as follows:

1. Intermediary metabolism
   Increased hepatic glycogenesis
   Increased hepatic gluconeogenesis
   Increased blood glucose
   Increased protein catabolism
   Increased plasma amino acids
   Increased lipolysis
   Increased plasma lipids and ketone bodies
2. Feedback inhibition of ACTH secretion
3. Cardiovascular
   Maintenance of vascular sensitivity to catecholamine's vasoconstricting effects
4. Growth and development
   Increased pulmonary surfactant production
   Development of hepatic and gastrointestinal enzymes
5. Hematologic
   Increased circulating neutrophils
   Decreased circulating lymphocytes, eosinophils, and monocytes
   Inhibition of migration of white blood cells
   Lympholysis
   Decreased size of lymph nodes and thymus
6. Renal
   Increased ability to excrete free water
7. CNS
   Personality changes
8. Resistance to stress
   Adrenalectomized animals die when exposed to noxious stimuli

The antiinflammatory and antiallergic effects of pharmacologic doses of steroids make them therapeutically useful. Glucocorticoids inhibit the conversion of membrane phospholipids to **arachidonic acid.** This reduces the formation of mediators of inflammation such as **leukotrienes, prostaglandins,** and **prostacyclin.** Steroids also prevent the release of histamine from mast cells.

Secretion

Adrenocorticotropic hormone stimulates both the basal and the stress-induced secretion of glucocorticoids. As mentioned earlier, ACTH comes from a larger precursor protein, proopiomelanocortin (POMC), which is cleaved after secretion into ACTH, beta-lipotropin, and gamma-MSH. The adrenals respond to an increase in ACTH with a rapid synthesis and secretion of steroids. Chronic stimulation with ACTH leads to adrenocortical hyperplasia and hypertrophy, while chronic depletion of ACTH results in adrenal atrophy and decreased adrenal responsiveness. Other actions of ACTH include melanocyte stimulation and stimulation of acute aldosterone secretion, although the renin-angiotensin axis is far more important in the regulation of aldosterone.

Regulation

Adrenocorticotropic hormone is released from the pituitary in response to hypothalamic corticotropin-releasing hormone (CRH). Three factors control ACTH secretion: (1) severe stress, (2) circadian rhythm, and (3) negative feedback from cortisol.

In dealing with severe stress, afferent signals relayed to the hypothalamus include emotions via the limbic system or traumatic injury via the spinothalamic pathways. Secondly, ACTH is secreted episodically as well as with a diurnal rhythm. It is released in irregular bursts throughout the day, but most frequently in the early morning and least frequently in the evening. The increased morning secretion occurs before waking up. (Perhaps the thought of waking up is traumatic.) The biologic clock driving the circadian rhythm is located in the suprachiasmatic nuclei of the hypothalamus. Lastly, ACTH secretion is inhibited through negative feedback by cortisol. The inhibition takes place mostly at the pituitary level.

Excess Hormone

The effects of high (nonphysiologic) levels of glucocorticoids can be seen with chronic administration of exogenous steroids, glucocorticoid-producing adrenocortical tumors, and hypersecretion of ACTH. The result is **Cushing's syndrome,** which consists of the following clinical manifestations:

1. Connective tissue—thinning of skin, easy bruising, **stria** formation, and poor wound healing due to decreased collagen and inhibition of fibroblasts
2. Bone—inhibition of intestinal absorption of calcium, decreased serum calcium, increased PTH secretion, decreased bone formation, and increased bone resorption leading to osteoporosis
3. Muscle—muscle wasting, weakness, and fatigability
4. Fat—redistributed to trunk **(truncal obesity),** face **(moon facies),** and upper back **(buffalo hump)**
5. Skin—**hirsutism** and **acne** secondary to increased adrenal androgen secretion
6. Endocrine—impaired glucose tolerance, **amenorrhea**
7. Renal—salt and water retention and hypokalemia due to the mineralocorticoid activity of the excess glucocorticoids
8. Cardiovascular—hypertension
9. CNS—**euphoria,** irritability, emotional lability, and depression.

The most common cause of Cushing's syndrome is **iatrogenic.** The most common noniatrogenic cause is Cushing's disease, which is the hypersecretion of ACTH from the pituitary. Nonpituitary (ectopic) ACTH hypersecretion and glucocorticoid-producing adrenal tumors are less common causes (Table 7-3).

Table 7-3. Cushing's syndrome

Type	Hormone profile	Causes	Source
ACTH dependent	Cortisol ↑ ACTH ↑	Cushing's disease, ectopic ACTH	Problem outside adrenals
ACTH independent	Cortisol ↑ ACTH ↓	Adrenal tumors	Problem within adrenals

Insufficient Hormone

Adrenal insufficiency is called **Addison's disease.** Primary Addison's disease refers to adrenal disorders and secondary Addison's disease refers to hypothalamic or pituitary disorders (Fig. 7-11). Clinical manifestations include

1. Weakness
2. Fatigue
3. Anorexia
4. Weight loss
5. Hyperpigmentation (only in primary Addison's caused by increased ACTH and MSH)
6. Hypotension
7. Fasting hypoglycemia

In patients receiving long-term exogenous steroid therapy, not only does the adrenal gland become atrophied and less responsive to ACTH, but the pituitary also becomes suppressed and is unable to secrete normal amounts of ACTH. Thus, the termination of steroid therapy should involve a careful tapering schedule to prevent a fatal Addisonian crisis. In an Addisonian crisis, the adrenals cannot secrete sufficient glucocorticoids or mineralocorticoids to survive a stressful situation such as trauma or disease (Table 7-4).

Mineralocorticoids

Aldosterone in the Body and its Effects

Aldosterone is the major **mineralocorticoid hormone.** In the plasma, it is bound to protein only to a slight extent. Aldosterone is metabolized before excretion. The main action of aldosterone is increasing the renal reabsorption of sodium in the distal tubules and collecting ducts in exchange for potassium and hydrogen ions. This leads to sodium retention, increased ECF volume, increased blood pressure, and even mild hypokalemic alkalosis. As with other steroids, aldosterone acts at the DNA level, stimulating synthesis of specific mRNAs, but the exact function of the aldosterone-induced protein is not known.

Fig. 7-11. Comparison of primary and secondary Addison's disease.

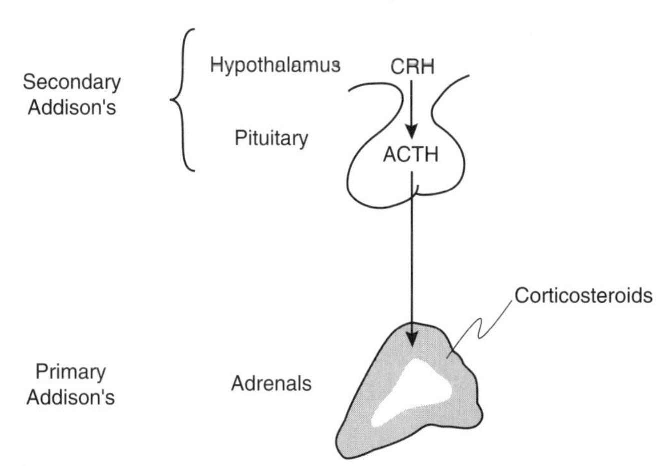

Endocrine

Table 7-4. Addison's disease*

Type	Hormone profile	Causes	Source
Primary	Corticoids ↓ ACTH ↑	Idiopathic, infection, surgery, cancer	Problem in adrenals
Secondary	Corticoids ↓ ACTH ↓	Hypothalamic-pituitary disease, hypothalamic-pituitary inhibition (iatrogenic or ectopic steroid)	Problem in hypothalamic/pituitary axis

*Note: Whereas Cushing's disease is an excess of glucocorticoids, Addison's disease is a lack of *all* adrenocorticoids.

Regulation

The **renin-angiotensin** system is the most important regulator of aldosterone secretion. (Figure 7-12 illustrates the renin-angiotensin system.) **Renin** is a peptide hormone secreted from specialized cells in the kidney, the **juxtaglomerular (JG)** cells, which are located in the renal afferent arterioles as they enter the glomerulus (see

Fig. 7-12. Renin-angiotensin axis. (Redrawn from R. M. Berne and M. N. Levy. *Physiology* [3rd ed.]. P. 967. © 1993 Mosby–Year Book, St. Louis.)

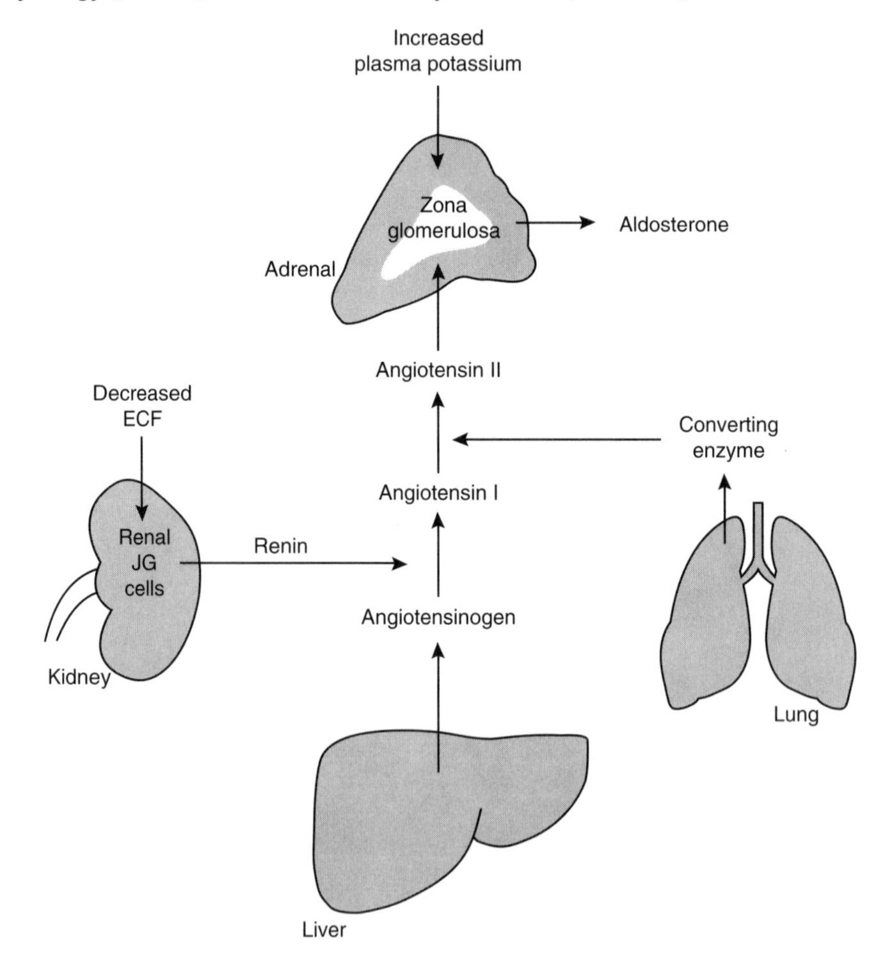

Chap. 3). It is an acid-protease whose action is to form angiotensin I by cleaving angiotensinogen. The secretion of renin is stimulated by the following factors:

1. Decreased ECF volume (e.g., dehydration, diuretics, sodium depletion)
2. Decreased systemic blood pressure (hemorrhage) or decreased renal arterial pressure (constriction of renal artery)
3. Increased renal sympathetic nerve discharge or circulating catecholamines
4. Decreased delivery of chloride and sodium to the distal tubule sensed by the macula densa

Angiotensinogen is a glycoprotein produced by the liver. Renin causes the liver to release angiotensin I, a decapeptide. Angiotensin I is then converted to the octapeptide angiotensin II (AII). Most of this conversion takes place as blood passes through the lungs and is mediated by angiotensin-converting enzyme (ACE) located in endothelial cells. The AII acts directly on the adrenal cortex to stimulate aldosterone secretion.

Angiotensin II is also the most potent vasoconstrictor known and causes arteriolar constriction in all vascular beds. Angiotensin II acts on the CNS to increase thirst and vasopressin secretion. It is rapidly metabolized by a group of enzymes called angiotensinases. One such enzyme, aminopeptidase, cleaves AII to form the heptapeptide **angiotensin III** (AIII), which is equally potent as AII in stimulating aldosterone secretion but only partially as potent in pressor activity.

Other stimulators of aldosterone secretion include ACTH, increased plasma potassium, and decreased plasma sodium. The increase in aldosterone release from ACTH lasts only 1 to 2 days. The mechanism by which aldosterone secretion is regulated is at the level of biosynthesis. ACTH, AII, and potassium stimulate the conversion of cholesterol to pregnenolone, which, as mentioned earlier, is a precursor to the synthesis of steroids including aldosterone. Potassium and AII also stimulate the conversion of corticosterone to aldosterone.

Excess Hormone

Primary hyperaldosteronism (Conn's syndrome) is caused by aldosterone-secreting adrenal tumors or hyperplasias. The effects of chronic mineralocorticoid excess include increased ECF volume, hypertension, potassium depletion, and metabolic alkalosis. When the ECF expansion reaches a certain point, sodium retention is decreased in spite of continued aldosterone stimulation. This **"aldosterone-escape" phenomenon** can be seen after 2 or 3 days of continued mineralocorticoid stimulation. Urinary excretion of sodium may return to normal levels, so that patients with Conn's are not volume expanded to the point of edema. No "escape" is seen with potassium excretion.

Secondary hyperaldosteronism may be caused by congestive heart failure (CHF), cirrhosis with ascites, and nephrosis. In these edematous states, ECF volume is distributed to extravascular spaces. The kidneys interpret this as a decrease in "effective" intravascular volume and respond by releasing renin. This increases aldosterone, which causes even further retention of sodium and fluid. This only makes matters worse because the extra fluid accumulates as edema fluid (Table 7-5).

Deficient Hormone

In hypoaldosteronism (e.g., Addison's disease), sodium is lost in the urine and potassium is retained. Plasma volume decreases and the resulting hypotension may develop to the degree of circulatory collapse. In an Addisonian crisis, the lack of mineralocorticoids is more significant to the patient's illness than is the lack of glucocorticoids.

Table 7-5. Hyperaldosteronism

Type	Cause	Source	Effects
Primary	Adrenal tumor or adrenal hyperplasia	Problem within adrenals	↑ ECF, alkalosis, hypertension, K^+ depletion
Secondary	Edematous states, CHF, ascites, nephrosis	Adrenals responding to low ECF	Edema, ↑ ECF, alkalosis, hypertension, K^+ depletion

Sex Hormones

Normally, the sex hormones produced in the adrenal cortex have minor physiologic significance. The gonadal production of sex steroids plays the major role; the physiologic function of these hormones is discussed in the next section. As mentioned earlier, **dehydroepiandrosterone (DHEA)** and **androstenedione** are the only sex steroids that are normally secreted in physiologically significant amounts from the adrenals. Both are weakly androgenic and exert their effects by their peripheral conversion to more potent androgens. Estrogen is secreted in physiologically insignificant amounts.

Enzyme Deficiencies

Adrenal sex steroids become important in patients with congenital enzyme deficiencies. For example, congenital deficiencies in either 21-beta-hydroxylase or 11-beta-hydroxylase lead to a deficiency in cortisol secretion (see Fig. 7-10). Because feedback inhibition is removed, ACTH secretion is increased. The excess ACTH causes adrenal hyperplasia as well as continued steroid synthesis. Due to the block in a biosynthetic pathway, the steroid intermediates "pile up" behind the block and are diverted to the remaining open pathways—in this case androgen synthesis. In genetic females, the excess of androgens may cause varying degrees of virilization. **Virilization,** or masculinization, consists of hirsutism, deepened voice, acne, clitoral enlargement, and male-pattern baldness. These females may also have ambiguous external genitalia. From Fig. 7-10, one can predict that a deficiency in 17,20-desmolase or 17-alpha-hydroxylase produces a deficiency in androgen synthesis. As a consequence, genetic males develop female or ambiguous genitalia and genetic females develop female genitalia. In addition, these enzyme deficiencies result in **sexual infantilism,** in which prepubertal sexual characteristics persist into adult life. The glucocorticoid deficiency in some of the above enzyme deficiencies may be severe enough to cause a fatal Addisonian crisis. Consequences of either mineralocorticoid excess or deficiency may also occur, depending on where the biosynthetic block is and if the intermediates that build up have mineralocorticoid activity.

Regulation of the Reproductive System

Female Reproductive System

Gametogenesis

The female reproductive system functions to accommodate both fertilization and pregnancy. It undergoes periodic changes known as the **menstrual cycle** (Fig. 7-13). **Menarche,** the first menstrual cycle, occurs at a mean age of 13 years. The cessation of menstrual cycles is called **menopause,** and occurs between the ages of 45 and 55

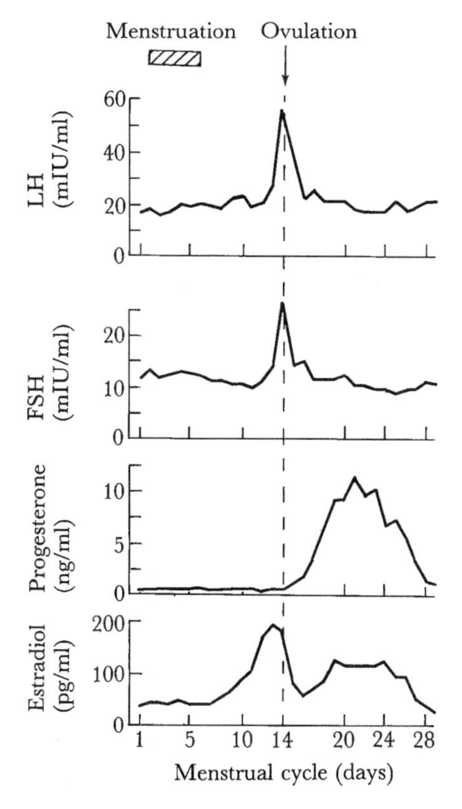

Fig. 7-13. Hormone concentrations during the menstrual cycle. (From I. H. Thorneycroft et al. The relation of serum 17-hydroxyprogesterone and estradiol-7β levels during the human menstrual cycle. *Am. J. Obstet. Gynecol.* 111:950, 1921.)

years. The length of each cycle is 28 days but can be quite variable even in the same woman. The first half of the cycle is much more variable than the second half. By convention, the first day of vaginal bleeding, or menstruation, is counted as the first day of the cycle. Each cycle results from the interaction of the hypothalamus, pituitary, and ovaries. No new ova are formed in the female gonads (ovaries) after birth. At birth, the ova are arrested at prophase of the first meiotic division and are called **primary oocytes.** The primary oocyte plus a single layer of surrounding cells comprise a **primordial follicle.** At puberty, the primordial follicles mature into **primary follicles.** In contrast, the primary oocyte does not change until called on during a menstrual cycle. During each menstrual cycle, approximately 20 ova begin to mature, but only one **ovum** is stimulated to mature fully. Just before ovulation, the first meiotic division is completed to form a **secondary oocyte** and the first **polar body.** The secondary oocyte then begins the second meiotic division up to the metaphase stage. The division is completed at fertilization and the second polar body extruded.

The Menstrual Cycle

Figure 7-14 shows the major hormone changes in the cycle. The menstrual cycle is divided into two phases, characterized by changes that take place in the ovaries. The **follicular phase** begins with the first day of menses and ends with ovulation. The rise in levels of FSH late in the preceding menstrual cycle promotes the growth and recruitment of a group of primary follicles (each consisting of a primary oocyte and surrounding granulosa cells) in the ovary. By about day seven, one follicle becomes

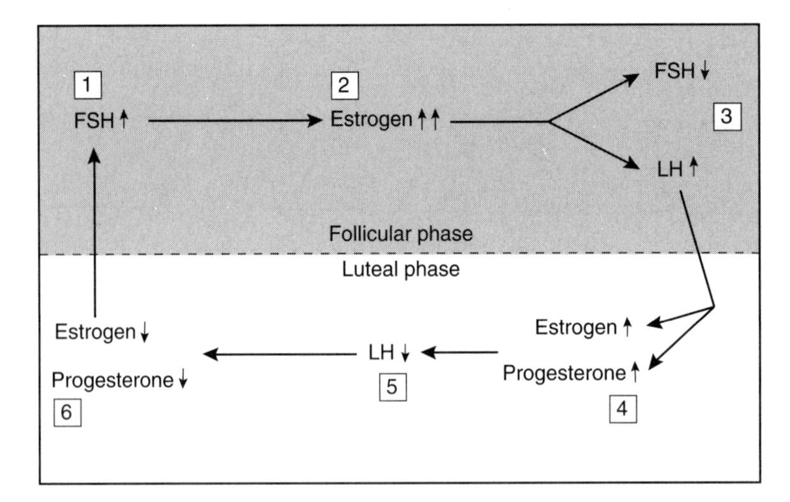

Fig. 7-14. Hormonal changes in the menstrual cycle. 1. Beginning of cycle—loss of inhibition from estrogen. 2. Follicle is releasing very high levels of estrogen because it is being stimulated by FSH. 3. High levels of estrogen stimulate LH release and inhibit FSH release; leads to ovulation. 4. LH stimulates corpus luteum to release another (lower) wave of estrogen and progesterone. 5. Lower levels of estrogen inhibit LH. 6. Corpus luteum regresses. Note: Just before ovulation FSH approximately doubles. The reason is unknown.

dominant, while the other recruited follicles begin to degenerate **(atresia).** In the dominant follicle there is an accelerated proliferation of granulosa cells and the development of a fluid-filled antrum. Follicle-stimulating hormone also stimulates the activity of aromatase enzymes in the granulosa cells converting androstenedione to estrogen. (The androstenedione came from the **theca cells** around the follicle under the influence of LH. In other words, LH causes the theca cells to produce androstenedione, which travels to the granulosa cells and is converted to estrogen under the influence of FSH. This is called the two-cell, two-gonadotropin theory of estrogen synthesis.) With further proliferation of granulosa cells, estrogen production increases, and FSH secretion is inhibited by the rising estrogen levels. Estrogen levels peak just before ovulation. It is believed that although low estrogen levels inhibit LH secretion (negative feedback), high estrogen levels stimulate LH secretion from the anterior pituitary (positive feedback). Thus, an LH peak results (see Fig. 7-13) that mediates the final maturation of the follicle and ovulation itself. The follicle is ruptured and the ovum extruded. The ovum is picked up by the fimbriated ends of the fallopian tubes (oviducts) and transported to the uterus. There is also a midcycle FSH peak, the significance of which is not clear.

The ruptured follicle becomes the **corpus luteum.** The **luteal phase** spans the interval from ovulation to the beginning of menses. The corpus luteum secretes both estrogen and progesterone, giving rise to another estrogen peak and a progesterone peak during the luteal phase. The LH stimulates progesterone and estrogen secretion from the corpus luteum. The FSH and LH levels then begin to decline, because estrogen suppresses the release of these gonadotropins. It is important to realize that the reason we see suppression is that estrogen levels are too low to give the positive feedback we saw in the follicular phase just prior to ovulation. Because LH is required to maintain the corpus luteum, the decline in LH results in the regression of the corpus luteum unless conception occurs. The corpus luteum regresses into the **corpus albicans.** Estrogen and progesterone secretion fall and the loss of inhibition permits the pituitary to secrete FSH again to begin a new cycle.

The Reproductive Tract

During the menstrual cycle, changes take place in the reproductive tract as directed by estrogen and progesterone. After menses, estrogen stimulates glandular growth in the endometrium; therefore, this part of the cycle, from the menses to ovulation, is also called the **proliferative phase,** corresponding to the follicular phase in the ovaries. Estrogen also stimulates the endocervical glands to secrete a thin and alkaline mucus that promotes survival and transport of sperm. After ovulation, progesterone and estrogen secreted by the corpus luteum act on the endometrium to stimulate glandular secretion and curling of the spiral arterioles (blood supply to the endometrium). This **secretory phase,** corresponding to the luteal phase in the ovaries, begins at ovulation and ends at the onset of menses. The endometrium is prepared for implantation during this phase. Progesterone causes the cervical mucus to be thick, tenacious, and cellular to prevent sperm transport. Progesterone also mediates a rise in basal body temperature of 0.3 to 0.5°C during the luteal phase. The fall in estrogen and progesterone levels late in the luteal phase causes vasospasm of the spiral arterioles. This results in tissue ischemia, desquamation and sloughing of the more superficial endometrial layer, and hemorrhage. Menses occurs on days one through five of the cycle.

The Female Sexual Response

The sexual response in women can be divided into four events: **arousal, erection and lubrication, orgasm,** and **resolution.** Sexual arousal, which causes secretion of a lubricating fluid from the vestibular glands through the vaginal mucosa, is initiated through a combination of psychic and physical stimuli. The stimuli travel afferently through the central limbic structures or through a reflex arc that begins with the pudendals. The efferent signals pass through the sacral parasympathetic plexus. From here they cause engorgement of erectile tissues including the clitoris and vestibular bulbs. Another effect is to stimulate Bartholin's glands to secrete mucus for lubrication. With continued stimulation, a reflex is begun in which the perineal muscles contract rhythmically and the cervical canal dilates. At the same time, many of the muscles throughout the body become tense. The orgasm refers to this physiologic reflex and the accompanying pleasurable psychic experience. Over the course of the next few minutes, resolution occurs, which includes relaxation of the muscles and disengorgement of the erectile tissues.

Fertilization, Implantation, and the Placenta

By their own motility, as well as by uterine and oviduct contractions, sperm reach the extruded ovum at about the midpoint of the oviduct. The **fertilized egg (blastocyst)** is transported to the uterus, where it becomes implanted about 6 days after fertilization (approximately day 20 of the cycle). A placenta eventually develops, which secretes **human chorionic gonadotropin (HCG).** This hormone structurally and functionally resembles LH. The HCG is detectable 8 days after ovulation and reaches a maximum in 7 weeks. Its detection constitutes a positive pregnancy test. HCG takes over for LH and maintains the corpus luteum, which in turn secretes progesterone and estrogen to support the uterus and prevent endometrial sloughing. Thus, a "missed period" is another indicator of pregnancy. The HCG supports the corpus luteum during the first trimester. During the subsequent two trimesters, the placenta is capable of producing its own estrogen and progesterone and the corpus luteum is no longer needed. While the ovaries produce mainly estradiol, the placental estrogen is largely estriol. Both fetus and placenta are required for the production of estriol. The fetal adrenal glands secrete DHEA, which is ultimately converted to estriol in the placenta.

Estrogen

Estrogen is secreted by the ovary, placenta, adrenal cortex, and testis. Under nonpregnant conditions, the ovary is the major source of estrogens. As described above, there are two peaks of estrogen secretion, during the late-follicular phase and during the midluteal phase. Estrogen secretion dramatically declines during menopause and it remains low thereafter. The biosynthesis of estrogen is shown in Fig. 7-10 and involves the aromatization of androgens. Estradiol is the major estrogen produced by the ovary. It is in equilibrium with estrone, which can be converted to estriol. Of the three, estradiol is the most potent, and estriol is the least potent. In the circulation, 2 percent of estradiol is free, 37 percent is bound to steroid hormone–binding globulin (SHBG), and 61 percent is bound to albumin.

Estrogen's major actions include

1. Growth of the ovarian follicle and promotion of the LH surge prior to ovulation
2. Endometrial proliferation and (with progesterone) endometrial glandular secretion
3. Enhanced excitability of uterine myometrium
4. Development of vagina, uterus, and oviducts, and enlargement of external genitalia (mons pubis, labia majora, and labia minora)
5. Stromal development and ductal growth of breasts
6. Fat deposition in breasts, buttocks, and thighs
7. Libido
8. Hepatic synthesis of several hormone transport proteins
9. Feedback inhibition of FSH secretion

Progesterone

Progesterone is secreted by the corpus luteum, placenta, adrenal cortex, and testis. Its biosynthesis is shown in Fig. 7-10. Its actions include

1. Stimulation of endometrial glandular secretions
2. Maintenance of pregnancy
3. Decreased excitability of the myometrium
4. Production of thick endocervical mucus
5. Development of breast lobules
6. Increased basal body temperature
7. Inhibition of gonadotropin secretion

Regulation of Estrogen and Progesterone

The regulation of estrogen and progesterone secretion has already been described to some extent. To summarize, **FSH** and **LH** are required for ovarian follicle maturation, during which estrogen is secreted. In the luteal phase, LH maintains the corpus luteum, which secretes estrogen as well as progesterone. The secretion of FSH and LH by the pituitary is stimulated by gonadotropin-releasing hormone (GnRH), also called luteinizing hormone–releasing hormone (LHRH), which is released in a pulsatile fashion from the hypothalamus. A negative feedback control of the gonadotropin secretion occurs at lower estrogen levels, while a positive feedback control occurs at higher estrogen levels. The secretion of both gonadotropins is inhibited by estrogen and progesterone during the luteal phase. It is not known whether the steroid hormones exert their feedback control at the hypothalamic or pituitary level.

Prolactin

Prolactin is a polypeptide hormone produced in and released from the anterior pituitary. Its primary action is to stimulate lactation (milk secretion) in the postpartum period. An excess of prolactin disturbs the hypothalamic pituitary control of gonadotropin secretion and ultimately leads to **hypogonadism** in both sexes. The secretion of prolactin is tonically inhibited by dopamine (the prolactin-inhibiting hormone) from the hypothalamus. Thyrotropin-releasing hormone (TRH) stimulates prolactin secretion. A yet unidentified prolactin-releasing factor (PRF) is also thought to be released by the hypothalamus. Physiologic factors that stimulate prolactin secretion include pregnancy, nursing, nipple stimulation, stress, and sleep. During pregnancy, the high levels of estrogen and progesterone inhibit the secretion of milk. It is only after birth, with the loss of estrogen and progesterone from the placenta, that the prolactin can have its effects.

Male Reproductive System

Testicular Structure and Function

The testis is made up of two components, **seminiferous tubules** for spermatogenesis and **Leydig cells** for androgen production. The seminiferous tubules are composed of **Sertoli cells** and **germinal cells.** The Sertoli cells line the epithelial basement membrane and form tight junctions with other Sertoli cells to create a blood-testis barrier through which testosterone, but not protein, penetrates easily. The Sertoli cells surround developing **germ cells** with their extensive cytoplasmic processes and provide nourishment essential for germ cell differentiation. Sertoli cells also phagocytose damaged germ cells and unused portions of germ cell cytoplasm during differentiation. Finally, Sertoli cells secrete two substances: **androgen-binding protein (ABP),** which enters the tubular lumen and ensures a high concentration of testosterone to the germ cells, and **inhibin,** which inhibits FSH secretion from the anterior pituitary. The secretion of both substances is stimulated by FSH.

Gametogenesis

In males, **gametogenesis** (the production of haploid spermatozoa) begins at puberty and persists until death. Under the influence of FSH and testosterone, the **spermatogonia,** or primitive germ cells, mature into **primary spermatocytes.** Through meiotic division, primary spermatocytes divide into **secondary spermatocytes,** and then into **spermatids,** which contain the haploid number of 23 chromosomes. The spermatids mature into **spermatozoa (sperm),** which are transported into the head of the epididymis; however, they are still infertile and immobile. The spermatozoa undergo further morphologic and functional changes during their transit through the epididymis so that they are both fertile and mobile on reaching the tail of the epididymis. They next enter the vas deferens, a muscular duct that peristaltically moves the spermatozoa into the ejaculatory duct. The ejaculatory duct also receives fluid from the seminal vesicles, which contain fructose for nourishment to the spermatozoa and other substances such as ascorbic acid and prostaglandins. The sperm then enters the prostatic urethra, into which prostatic secretions are also emptied. The prostate secretes citric acid, acid phosphatase, spermine, and other substances. The major components of fluid that is ejaculated at orgasm, the **semen,** include sperm, seminal vesical secretions, and prostatic secretions. An average ejaculate volume is 2.5 to 3.5 ml and an average sperm concentration is 100 million sperm per milliliter of semen. A concentration of less than 20 million sperm per milliliter usually results in sterility.

The Male Sexual Response

The male sexual response can be divided into five events: **arousal, erection, ejaculation, orgasm,** and **resolution.** Arousal, or sexual desire, is produced by psychic factors and testosterone. Penile erection occurs as a result of dilatation of arteriolar vessels that supply the corpora cavernosa and spongiosa, engorgement of these erectile tissues with blood, and passive venous congestion. This vascular phenomenon is an involuntary reflex that can be mediated by either of the following mechanisms. Psychogenic stimuli are transmitted to the limbic system, which, in turn, stimulates the parasympathetic innervation to the penile vessels. The second mechanism involves a **spinal reflex arc** where direct genital stimuli are transmitted by the pudendal nerves (afferent limb) and activate sacral parasympathetic nerves (efferent limb). Ejaculation is mediated by the sympathetic nervous system and consists of two processes: the emission of semen into the urethra and the true ejaculation of semen out of the urethra. Orgasm is a rhythmic contraction of the bulbocavernosus and ischiocavernosus muscles and an accompanying pleasurable feeling. During the resolution phase, blood rapidly leaves the erectile tissue and the penis becomes flaccid.

Testosterone

Testosterone is the major hormone secreted by the testis, and it is synthesized from cholesterol in the **Leydig cells** (Fig. 7-10). In plasma, only 2 percent of testosterone exists in the free form and the remainder is protein bound, mainly to SHBG and albumin. The actions of testosterone include

1. Differentiation of internal and external genitalia in boys during fetal development
2. Enlargement of the penis, the scrotum, and the testis during puberty **(primary sex characteristics)**
3. Development and maintenance of male **secondary sex characteristics** (hair growth on face, chest, and pubis; male-pattern baldness; deeper voice; increased sebaceous gland secretion; increased musculature; broad shoulders; narrow pelvic outlet)
4. Anabolic effects on protein metabolism, that is, increased synthesis and decreased breakdown of protein, causing an increase in the rate of growth (growth spurt at puberty)
5. Fusion of epiphyses to long bones to stop growth
6. Maintenance of gametogenesis (along with FSH)
7. Inhibitory feedback effect on pituitary LH secretion
8. Libido

In most target cells, testosterone is converted to the more potent dihydrotesterone, which binds to a cytoplasmic protein, and the resulting complex is translocated into the nucleus.

Regulation

Testosterone secretion is stimulated by LH, which is also trophic to the Leydig cells. Testosterone feedback inhibits the secretion of LH both at the pituitary level and at the hypothalamic level. The FSH stimulates Sertoli cells to secrete ABP and inhibin. The latter feedback inhibits FSH secretion at the pituitary level. Figure 7-15 summarizes the regulation.

Thyroid Hormones

Synthesis

The **thyroid hormones** are produced in the thyroid gland, which is made up of **follicles.** Each follicle consists of a single layer of thyroid cells surrounding a viscous gel called **colloid.** The synthesis of thyroid hormones requires **iodide,** which is obtained

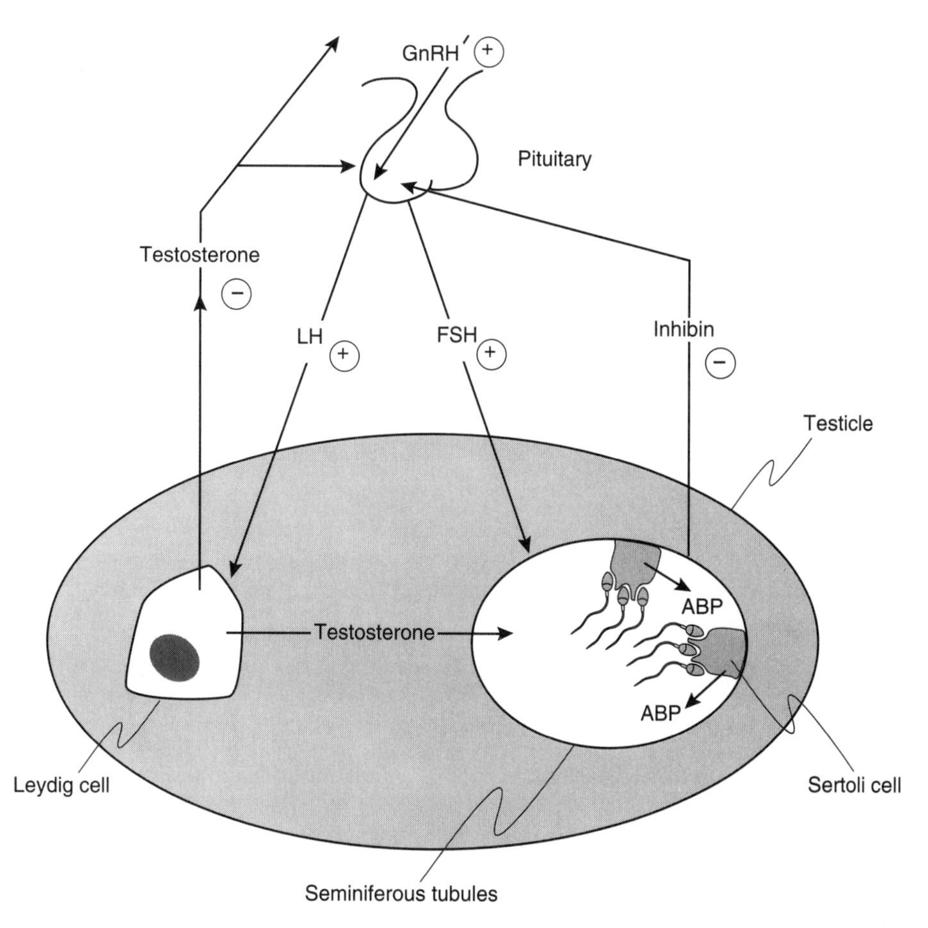

Fig. 7-15. Summary of regulation of male reproductive system. LH stimulates the Leydig cells to produce testosterone. FSH stimulates Sertoli cells to produce androgen-binding protein (ABP) to bind the testosterone. The Sertoli cells also feedback inhibit FSH release, whereas testosterone feedback inhibits LH release. (Adapted from F. S. Greenspan and J. D. Baxter. *Basic and Clinical Endocrinology.* P. 396. © 1994 Appleton & Lange, E. Norwalk, CT.)

from the diet and absorbed by the gastrointestinal tract. Free iodide in the extracellular space is removed either through uptake by the thyroid gland or by renal excretion. Iodide is actively taken up (trapped) by thyroid cells using a Na^+-K^+ adenosine triphosphatase (ATPase) pump. **Thyroglobulin** is a glycoprotein synthesized in the thyroid cells and secreted into the colloid by exocytosis of granules. Intracellular iodide is oxidized by thyroid peroxidase to uncharged iodine. It is then bound to tyrosine residues of the **thyroglobulin molecule,** within the colloid. The result is monoiodotyrosine (MIT). A second iodination forms diiodotyrosine (DIT). An oxidative condensation reaction (coupling) takes place between two DIT residues to form **thyroxine (T_4). Triiodothyronine (T_3)** comes from coupling of an MIT residue with a DIT residue. Both T_4 and T_3 are still linked to thyroglobulin by peptide bonds. By endocytosis, thyroid cells take up the thyroglobulin stored in the colloid and the resulting vesicle is joined by lysosomes. Proteases from the lysosomes hydrolyze the peptide bonds of thyroglobulin and thereby release T_4, T_3, MIT, and DIT into the intracellular space. Iodide is removed from MIT and DIT by deiodinase and returned to the intracellular iodide pool. Both T_4 and T_3 are released into the bloodstream. The thyroid gland secretes about 15 T_4 molecules for every T_3 molecule. Figure 7-16 summarizes the production of thyroid.

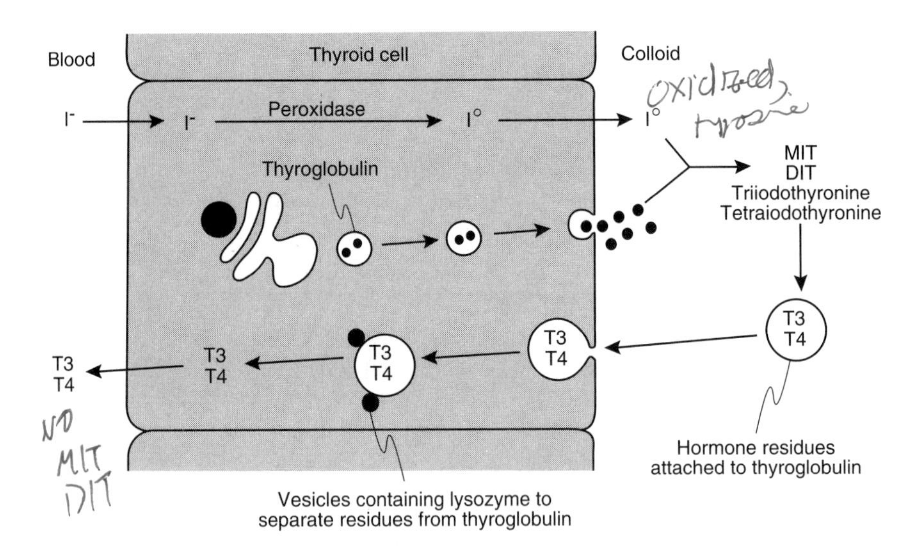

Fig. 7-16. Production of thyroid hormones. (Adapted from A. Despopoulos and S. Silbernagl. *Color Atlas of Physiology.* P. 251. © 1986 Georg Thieme Verlag, Stuttgart.)

Thyroid Hormones in the Body

Only free (unbound) thyroid hormones in the circulation are biologically active. Only about 0.04 percent of the T_4 and 0.4 percent of the T_3 are free in the bloodstream. Most of T_4 is bound to **thyroxine-binding globulin (TBG).** The remainder is bound to **thyroxine-binding prealbumin (TBPA)** and albumin. The majority of T_3 is bound to TBG and the rest to albumin. Although the serum level of free T_4 is about seven times that of free T_3, T_3 is about four times as potent as T_4. The half-life of T_4 is eight times that of T_3. Thyroxine is metabolized by deiodination, and it is deiodinated to form either T_3 or reverse T_3 (rT_3). While only 20 percent of the T_3 in circulation comes from thyroid secretion, 80 percent is derived from T_4 deiodination. Reverse T_3 does not have any known biologic activity. Both T_3 and rT_3 are further deiodinated to diiodothyronines. T_3 is commonly thought of as the "effective hormone" while T_4 is considered a prohormone.

Effects

Thyroid hormone has the following actions:

1. Stimulates calorigenesis
2. Increases oxygen consumption
3. Stimulates protein synthesis
4. Stimulates glycogenolysis, gluconeogenesis, and intestinal absorption of glucose
5. Stimulates synthesis, mobilization, and degradation of fat
6. Increases heart rate and force of contraction
7. Enhances the sensitivity of tissues to catecholamines
8. Required for fetal development, erythropoiesis, and vitamin A synthesis

Targets

Thyroid hormone has its primary action at several sites within its target cell, including RNA transcription, protein translation, membrane-bound Na^+-K^+ ATPase, and mitochondria.

Regulation

The secretion of thyroid hormone is stimulated by **thyroid-stimulating hormone (TSH)** from the anterior pituitary. The TSH stimulates the organification of iodide, exocytosis of thyroglobulin, endocytosis of colloid, proteolysis of thyroglobulin, and release of thyroid hormone. Thyroid hormone exerts a negative feedback control of TSH secretion. In the pituitary, T_4 is converted to T_3, which inhibits TSH secretion. The TSH-releasing hormone (TRH), a tripeptide secreted from the hypothalamus, stimulates TSH secretion. Thyroid hormone may also exert feedback inhibition at the level of the hypothalamus.

The thyroid gland itself exhibits autoregulation. In a state of intrathyroid organic iodine deficiency, the iodide transport mechanism becomes more active. An excess of intrathyroid organic iodine inhibits iodide transport. When the intracellular iodide concentration rises above a critical level, the iodide organification step is blocked **(Wolff-Chaikoff block).** Pharmacologic doses of iodide result in a rapid and transient inhibition of thyroid hormone secretion by an unknown mechanism (different from the Wolff-Chaikoff block).

Insufficiency of Hormone

Hypothyroidism may be **primary, secondary,** or **tertiary,** depending on whether it is due to thyroid disease, pituitary disease, or hypothalamic disease, respectively. In adults, **hypothyroidism** results in cold intolerance, slow mentation, a husky voice, weight gain, decreased reflexes, and skin changes referred to as myxedema. Neonatal hypothyroidism results in **cretinism,** characterized by growth retardation as well as mental retardation.

Excess of Hormone

Symptoms of hyperthyroidism **(thyrotoxicosis)** include heat intolerance, fatigue, palpitations, increased appetite, sweating, weight loss, nervousness, and hand tremors. The most common cause of hyperthyroidism is **Graves' disease,** in which thyroid-stimulating immunoglobulins stimulate the TSH receptors on thyroid cells and produce the same hormone action as TSH.

Goiter is any enlargement of the thyroid gland and results from chronic stimulation of the gland by excessive amounts of TSH (or thyroid stimulating immunoglobulin). In iodine-deficiency goiter, the dietary iodine is inadequate to maintain synthesis of thyroid hormone. Thyroid hormone levels decline and feedback inhibition of TSH is decreased; thus, increased TSH secretion leads to hypertrophy of the thyroid gland.

Endocrine Function Testing

Numerous tests are used to examine endocrine function. Evaluation for Cushing's disease, diabetes mellitus, and disorders of the thyroid is discussed below.

Dexamethasone Suppression Test

Dexamethasone is a potent glucocorticoid that feedback inhibits pituitary release of ACTH even at low levels. If these low levels of dexamethasone fail to suppress the glucocorticoid plasma levels, Cushing's syndrome is suspected and a higher level of dexamethasone is tried. The higher level will suppress glucocorticoid levels if the

Table 7-6. Summary of hormones discussed in chapter 7

Hormone	Source	Type	Stimulates release	Inhibits release	Effects	Excess	Deficiency
Insulin	Pancreas beta cells	Protein	↑Blood glucose Amino acids Intestinal hormones	Catecholamines Sympathetic tone Somatostatin	Liver: Glycogen synthesis, ↓gluconeogenesis, lipogenesis Muscle: Glycogen synthesis, amino acid uptake, glucose transport Adipose: Lipogenesis, glucose and triglyceride uptake	Hypoglycemia	Diabetes mellitus
Glucagon	Pancreas alpha cells	Protein	↓Blood glucose Exercise Cortisol Beta-adrenergic tone	↑Blood glucose Somatostatin	Gluconeogenesis, proteolysis, lipolysis, glycogenolysis, ketogenesis	Hyperglycemia	Hypoglycemia
PTH	Parathyroid chief cells	Protein	↓Serum Ca^{2+}	↑Serum Ca^{2+}	↑Bone resorption, ↑renal Ca^{2+} absorption, ↓phosphate reabsorption, ↑vitamin D	Hypercalcemia, hypercalciuria, ↓phosphate bone resorption	Abnormal conduction, hypocalcemia, tetany, CNS calcification
Calcitonin	Thyroid parafollicular cells	Protein	↑Serum Ca^{2+}	↓Serum Ca^{2+}	↓Serum Ca^{2+}, ↓bone resorption, ↑calciuria, ↑phosphaturia	No known effects	No known effects
Vitamin D	Skin-liver-kidney, diet	Steroid	PTH ↓Phosphate ↓Vitamin D	↓PTH ↑Phosphate ↑Vitamin D	↑Gut Ca^{2+} absorption, ↑gut phosphate absorption	Hypercalcemia	Rickets—children Osteomalacia—adults
ADH	Posterior pituitary	Protein	Osmolality > 280 ↓ECF volume Angiotensin II Nicotine Nausea Pain	↓Osmolality ↑ECF volume Ethanol	↑Permeability of collecting ducts, reabsorption of water, vasoconstriction, ACTH release	SIADH, hyponatremia, hypertension	Diabetes insipidus
Oxytocin	Posterior pituitary	Protein	Suckling Dilation of uterus		Milk letdown, uterine contraction		
GH	Anterior pituitary acidophils	Protein	GRH Sleep Stress Exercise Hypoglycemia Amino acids	Somatostatin Cortisol Somatomedins Free fatty acids Obesity Hyperglycemia	↑Somatomedins, ↑protein synthesis, ↑lipolysis, diabetes, ↑Ca^{2+} absorption	Gigantism—children Acromegaly—adults	Dwarfism
ACTH	Anterior pituitary basophils	Protein	CRH Stress Circadian rhythm	Cortisol	↑Cortisol, ↑androgens, maintains adrenal cortex	Cushing's syndrome	Secondary Addison's disease
Cortisol	Adrenal zona fasciculata	Steroid	ACTH	Cortisol	↑Glycogenesis, ↑gluconeogenesis, ↑protein and fat catabolism, antiinflammatory, ↑stress tolerance, ↓WBC migration	Cushing's syndrome	Stress intolerance Addison's disease (in part)

(continued)

Hormone	Source	Type	Stimulates release	Inhibits release	Effects	Excess	Deficiency
Aldosterone	Adrenal zona glomerulosa	Steroid	Angiotensin II ACTH ↑K^+ ↓Na^+	Dopamine	Na^+ retention in kidneys	Conn's syndrome, ↑ECF volume, hypertension, ↓K^+, alkalosis	Hypotension, ↑K^+
Adrenal androgens	Adrenal zona reticularis	Steroid	ACTH		Contribute to secondary sex characteristics	Virilism	Poor secondary sex characteristic development, hypogonadism
LH	Anterior pituitary basophils	Protein	GnRH Very high estrogen	Estrogen Testosterone	↑Testosterone from Leydig cells, ovulation	Amenorrhea	Amenorrhea, infertility
FSH	Anterior pituitary basophils	Protein	GnRH	Estrogen Inhibin	Synthesis of estrogen, spermatogenesis, follicular recruitment	Amenorrhea	Amenorrhea, infertility
Progesterone	Ovaries, placenta, adrenals	Steroid	LH, HCG, high estrogen	Low estrogen	Follicular growth, libido, endometrial proliferation, breast growth, primary sex characteristics, ↑uterine excitability	Abnormal bleeding, infertility	Abnormal bleeding, infertility
Estrogen	Ovaries, placenta, adrenals	Steroid	LH, FSH, high estrogen, HCG	Low estrogen	Endometrial gland secretion, maintenance of pregnancy, ↓uterine excitability, breast development, increased temperature, secretion of endocervical mucus	Abnormal bleeding, precocious puberty	Amenorrhea, infertility
Testosterone	Testes, Leydig cells	Steroid	LH	Testosterone (via pituitary)	Sexual differentiation, development of primary and secondary sex characteristics, protein synthesis, epiphyseal fusion, gametogenesis, libido	Hypergonadism	Hypogonadism
Prolactin	Anterior pituitary acidophils	Protein	Suckling TRH Pain Stress Sleep Pregnancy	Dopamine	Lactation	Amenorrhea, infertility, menstrual irregularity	Failure to lactate
TSH	Anterior pituitary basophils	Protein	TRH Cold	T_3, T_4 Cortisol Somatostatin	↑T_4/T_3 release, iodide organification, thyroglobulin exocytosis, colloid endocytosis	Hyperthyroidism	Hypothyroidism
T_3/T_4	Thyroid	Amino acid	TSH	Iodine excess	↑Growth, calorigenesis, ↑O_2 consumption, protein synthesis, ↑glycogenolysis, gluconeogenesis, ↑fat mobilization, ↑heart rate, ↑contractility, vitamin A synthesis	Heat intolerance, fatigue, palpitations, ↑appetite, weight loss, nervousness, hand tremors	Cold intolerance, depression, memory defect, slow mentation, myxedema, cretinism, husky voice

problem is in the pituitary (i.e., Cushing's disease). However, if the patient's glucocorticoid levels remain high with the higher dexamethasone test, the problem is either an adrenal tumor that is producing too many glucocorticoids or a source of ACTH outside the pituitary. Neither the tumor nor the ectopic source would respond normally to negative feedback.

Oral Glucose Tolerance Test

The patient is instructed to fast overnight. A solution containing 75 mg glucose is then administered orally. Blood samples are taken every half hour for 2 hours. If the patient's blood sugar exceeds 200 mg/dl, the diagnosis of diabetes mellitus is made.

Thyroid-Stimulating Hormone and Thyroxine Levels

In order to begin to evaluate a patient with a thyroid abnormality, it is important to know whether the problem is arising in the pituitary or the thyroid. A radioimmunoassay will test TSH levels. Evaluating thyroxine levels is trickier because the amount of free T_4 is a function of the total T_4 and the serum-binding capacity. The serum-binding capacity increases and decreases with different situations. For example, in pregnancy, hepatitis, and cirrhosis, thyroxine-binding globulin is increased. With large protein losses, steroid administration, and phenytoin or aspirin ingestion, the binding capacity decreases. Therefore, after the total T_4 level is determined, it is important to determine the binding capacity. This is done with a resin T_3 uptake test. Using the total T_4 and the resin T_3 uptake the free T_4 index is produced. The T_4 index is proportional to the free T_4 concentration.

For a summary of hormones discussed in this chapter, see Table 7-6.

Questions

Directions: Each of the numbered items or incomplete statements in this section is followed by answers or by completions of the statement. Select the one lettered answer or completion that is best in each case.

1. The term neuroendocrine refers to
   A. Hormones that act across a synapse
   B. Hormones that act on the cell releasing them
   C. Hormones that are released from a nerve directly into the blood
   D. Hormones of the peripheral nervous system
2. Which of the following stimulates release of a hormone from the anterior pituitary?
   A. Hyperosmolality
   B. PIH
   C. Somatomedins
   D. Somatostatin
   E. Suckling during lactation
3. Which of the following **cannot** be used to make glucose?
   A. Amino acids
   B. Free fatty acids
   C. Glycerol
   D. Lactate

4. Which of the following **does not** stimulate insulin release?
   A. Amino acids
   B. Increased blood glucose
   C. Fats
   D. Gastric inhibitory protein
5. Which of the following **does not** increase in the presence of glucose?
   A. Amino acid uptake in skeletal muscle
   B. Glucose entry into the brain
   C. Glucose entry into the liver
   D. Glycerol synthesis in adipose tissue
   E. Glycogenesis
6. Secretion of growth hormone has which of the following effects?

	Lipolysis	Calcium Absorption	Glucose Uptake
A.	↑	↑	↓
B.	↑	↑	↑
C.	↓	↓	↓
D.	↓	↑	↑
E.	↑	↓	↓

7. A 45-year-old female patient comes to you complaining of easy bruising, weakness, and increasing coarse facial hair. She also shows moon facies and a buffalo humpback. Suspecting Cushing's syndrome, you do a dexamethasone suppression test. The low-dose test is abnormal but the high-dose test shows ACTH at normal levels. You suspect that she
   A. Has an adrenal tumor
   B. Has Cushing's disease
   C. Has iatrogenic Cushing's syndrome
   D. Has an oat-cell carcinoma of the lung
8. Which of the following **does not** contribute to calcium absorption in the intestines?
   A. High parathyroid hormone (PTH)
   B. Low serum calcium
   C. High serum phosphate
   D. A vacation in the sun in the Virgin Islands

9–10. A 56-year-old male smoker develops squamous cell carcinoma of the larynx. On diagnosis, the cancer has already directly spread to the thyroid.

9. If the serum calcitonin level doubles by hyperactivity of the parafollicular cells, which of the following is most likely to happen?
   A. Hypocalcemic tetany will develop
   B. Massive calciuria will develop
   C. Rickets will develop
   D. There will be no defect in bone or calcium metabolism
10. The patient undergoes surgery and in the process all the parathyroid glands are removed. The patient is **unlikely** to develop which of the following?
   A. Cardiac conduction abnormalities
   B. Cataracts
   C. Hypercalcemia
   D. Tetany
11. In which of the following would ADH secretion be decreased?
   A. A continuous intravenous infusion of hypertonic NaCl solution
   B. A patient who has lost 4 units of blood from a duodenal ulcer

C. A sedentary person with a psychiatric disorder that causes her to drink 12 liters of water per day
D. A person with a plasma osmolality of 300 mOsm/kg

12. The figure below shows the change in plasma aldosterone levels over time. Which best describes what is happening over the interval from A to B?
    A. The person is becoming dehydrated
    B. The person is being administered an angiotensin-converting enzyme inhibitor
    C. The person is receiving an intravenous solution of isotonic solution
    D. The person is eating a bag of potato chips

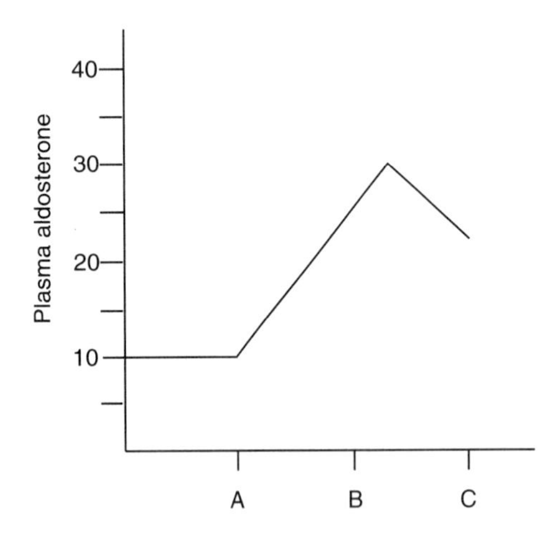

13. Metyrapone is a drug used to control cortisol secretion in Cushing's syndrome. It blocks 11-beta-hydroxylase. One of the side effects is hirsutism in women. This occurs because
    A. The drop in cortisol stimulates ACTH release, which creates more DHEA
    B. Low levels of cortisol are still released, with heightened ability to cause hirsutism due to receptor up-regulation
    C. Metyrapone causes increased quantities of aldosterone, which has androgenic activity
    D. Metyrapone is an adrenergic agonist
14. Which of the following is **false** about Sertoli cells?
    A. They are part of the seminiferous tubules
    B. They are phagocytic
    C. They form a blood-testes barrier that testosterone easily penetrates
    D. They secrete androgen-binding protein, which stimulates testosterone release from Leydig cells
    E. They secrete inhibin, which inhibits FSH secretion
15. Which of the following is **not** found in semen?
    A. Acid phosphatase
    B. Ascorbic acid
    C. Citric acid
    D. Fructose
    E. Testosterone
16. Which of the following is **false?**
    A. At birth, all ova are arrested at the prophase of meiosis I
    B. First meiotic division is completed just before ovulation
    C. Second meiotic division is completed at fertilization
    D. Primordial follicles contain a primary oocyte and a single layer of follicular cells
    E. Several fully mature ova are produced during each menstrual cycle

17. A patient with a *hypo*active thyroid shows all the following **except**
    A. Cretinism in neonates
    B. Coarse dry hair and skin
    C. Dull expression
    D. Heat intolerance
    E. Myxedema
18. Which of the following is true?
    A. Iodide (I^-) is bound to a thyroglobulin molecule intracellularly
    B. Iodine (neutral) is bound to a thyroglobulin molecule intracellularly
    C. Iodide (I^-) is bound to a thyroglobulin molecule in the colloid
    D. Iodine (neutral) is bound to a thyroglobulin molecule in the colloid

Directions: Each group of items in this section consists of lettered options followed by a set of numbered items. For each item, select the one lettered option that is most closely associated with it. Each lettered option may be selected once, more than once, or not at all.

19–21. Match each of the following hormones to the appropriate statement concerning its signal transduction.

A. Antidiuretic hormone (ADH)
B. Insulin
C. T_4 (thyroxine)
D. Vitamin D

19. Causes effects via a second-messenger system
20. Binds its receptor in the cytoplasm
21. Receptor is a tyrosine kinase

22–24. The effects of starvation on major food stores is shown below. Match each of the statements about the food stores to the appropriate line in the graph.

22. The body's reliance on this nutrient in ketosis
23. Carbohydrate
24. The use of free fatty acids allows us to minimize use of this energy store

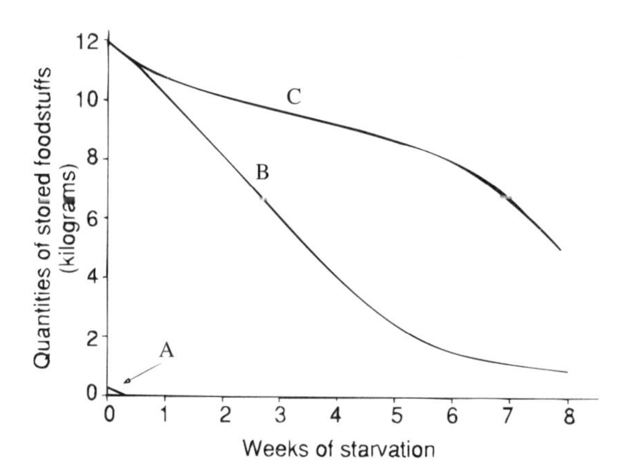

25–27. All of the following hormones are involved in maintaining blood glucose levels. Match the appropriate hormone to the following statements.

A. Catecholamines
B. Glucagon

C. Growth hormone
D. Somatostatin

25. Dropping insulin levels is the most important way to increase plasma glucose concentration; this hormone functions in the second most important method
26. This acts acutely on fat, muscle, and liver to maintain blood glucose
27. This mobilizes fat stores, but does not act acutely in maintaining blood glucose

28–31. The figure below shows a wedge from an adrenal gland. Match the section labeled on the wedge that is responsible for each of the following:

28. Responsible for producing androgens
29. Acts like postganglionic sympathetic neurons
30. Secretes high quantities of hormone with renal artery obstruction
31. Can cause mild alkalosis even under physiologic conditions

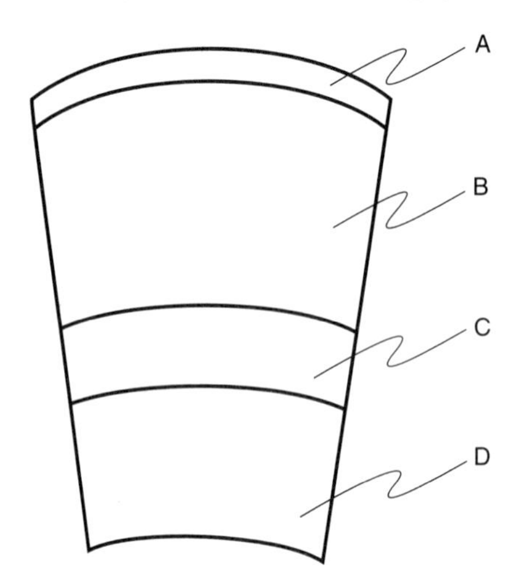

32–36. Match each numbered event with the lettered hormone that is most responsible for that event.

A. Estrogen
B. FSH
C. LH
D. Progesterone

32. Ovulation
33. Basal body temperature increase
34. Secretion of thin endocervical mucus
35. Maintenance of corpus luteum during menstrual cycle
36. Recruitment of a group of primary follicles

Answers

1. C Neuroendocrine hormones are released from a nerve into the bloodstream and act at a distance. Answer A refers to neurotransmitters, not to neuroendocrine hormones. Answer B refers to paracrine secretion. Answer D again refers to neurotransmitters, not neuroendocrine hormones.

2. E Suckling causes oxytocin release from the posterior pituitary *and* prolactin release from the anterior pituitary. Hyperosmolality causes ADH release from the *posterior* pituitary. PIH inhibits release of prolactin. The somatomedins feedback negatively and therefore suppress the release of prolactin. Somatostatin inhibits release of TSH and GH.

3. B Free fatty acids cannot be converted to glucose because the oxidation of pyruvate is irreversible. Amino acids, glycerol, and lactate can all be used as gluconeogenic substrates.

4. C Fats do not stimulate insulin release. Amino acids, increased blood glucose, and GIP all stimulate insulin release.

5. B Answer B is the correct answer because glucose uptake into the brain is not facilitated by insulin. However, insulin does facilitate the uptake of glucose by muscles, its entry into the liver, and the formation of glycerol, and it also increases glycogenesis.

6. A Growth hormone increases lipolysis and calcium absorption while decreasing glucose uptake.

7. B The higher dexamethasone levels will suppress even an overactive pituitary. An adrenal tumor would not be suppressed even with the low levels of ACTH caused by the high levels of dexamethasone. Iatrogenic Cushing's syndrome can only be reversed by removing the external source of glucocorticoids. An oat-cell carcinoma is considered an ectopic source for ACTH and would not be suppressed by high dexamethasone levels.

8. C A high serum phosphate stimulates the formation of 24, 25-dihydroxycholecalciferol, which is an inactive vitamin D metabolite. High PTH causes 1,25-dihydroxycholecalciferol production, which does contribute to calcium absorption. A low serum calcium contributes to PTH secretion, which contributes to 1,25-dihydroxycholecalciferol production and calcium absorption. The exposure to adequate amounts of ultraviolet sunlight allows for the formation of vitamin D_3, a precursor for 1,25-dihydroxycholecalciferol.

9. D It is unclear exactly what calcitonin does because neither an excess nor a deficiency causes bone or calcium defects. Although calcitonin increases urinary calcium excretion, an excess of the hormone does not cause any defects in bone or calcium metabolism. Hypocalcemic tetany is a result of hypoparathyroidism. Rickets is from vitamin D deficiency and is a childhood disease.

10. C PTH increases serum calcium levels. A complete absence of this would produce a hypocalcemia, not a hypercalcemia. Cardiac conduction abnormalities are a result of lack of parathormone due to altered calcium levels. Cataracts and other intracranial calcifications occur during a lack of parathormone as a result of the altered plasma calcium levels. Tetany also develops as a result of the hypocalcemia.

11. C A person who drinks that much water would have an increased ECF and a decreased osmolality, both of which inhibit ADH release. A continuous intravenous infusion of hypertonic NaCl solution would cause an increase in osmolality, resulting in increased ADH secretion. A loss of blood will decrease ECF volume and stimulate ADH secretion via stretch receptors in the great veins, atria, and pulmonary vessels. Any plasma osmolality above 280 mOsm/kg stimulates ADH release.

12. A As the person is becoming dehydrated, the decreased ECF volume is detected by the kidneys stimulating the renin-angiotensin axis. This stimulates the release of aldosterone. An ACE inhibitor would only decrease the amount of aldosterone circulating. It would not triple it. An isotonic solution would not cause an increase in the renin-angiotensin axis. Eating a bag of potato chips would not cause an increase in the renin-angiotensin axis. Increased renin secretion is caused by things such as decreased ECF or decreased delivery of chloride and sodium to the distal tubule.

13. A The hypothalamic-pituitary axis is primed by the lack of inhibition from cortisol. As a result, high quantities of ACTH are secreted, resulting in high production of androgens. Cortisol, assuming it were released in low levels, would not cause hirsutism. Aldosterone production would be inhibited by the 11-beta-hydroxylase block as well. In addition, aldosterone has no significant androgenic activity. Metyrapone is not an adrenergic agonist.

14. D Although Sertoli cells secrete androgen-binding protein (ABP), ABP does not stimulate testosterone release from Leydig cells. Sertoli cells are part of the seminiferous tubules, are phagocytic, form a blood-testis barrier penetrable by testosterone, and secrete inhibin, which inhibits FSH secretion.

15. E Testosterone is secreted by the Leydig cells and acts in the Sertoli cells, but is not a component of the semen. Acid phosphatase is secreted by the prostate, ascorbic acid by the seminal vesicles, citric acid by the prostate, and fructose by the seminal vesicles.

16. E Only one mature ova is produced each menstrual cycle, although several follicles are stimulated to mature.

17. D Heat intolerance is a result of *hyper*thyroidism. Cold intolerance is a result of hypothyroidism. Neonatal hypothyroidism can result in cretinism, and coarse dry hair and skin, dull expression, and myxedema are all also associated with hypothyroidism.

18. D Iodide is oxidized to iodine intracellularly and then bound to thyroglobulin in the colloid. The binding of iodine (neutral) occurs extracellularly.

19. A Antidiuretic hormone, being a peptide, binds a surface receptor and works through a second-messenger system.

20. D Vitamin D, behaving like a steroid, binds a receptor in the cytoplasm.

21. B The insulin receptor, as well as the receptors for growth factors, are tyrosine kinases themselves.

22. B This line represents fat stores. It is the free fatty acids from the lipolysis that are converted to ketone bodies. Notice that the fat supplies are used steadily throughout the period of starvation.

23. A This line represents carbohydrate stores. These stores are usually depleted within a day.

24. C This line represents protein stores. Initially, these are used but eventually the body becomes accustomed to using ketone bodies and proteolysis is markedly slowed. With exhaustion of fat stores, rapid proteolysis is resumed.

25. B An increase in glucagon is the next most important counterregulatory mechanism for maintaining blood glucose.

26. A Only glucagon and catecholamines act acutely. Glucagon works only in the liver whereas catecholamines work in liver, fat, and muscle.

27. C Growth hormone and glucocorticoids act in the long term to maintain blood glucose. Glucocorticoids increase protein metabolism whereas growth hormone mobilizes fat.

28. C The zona reticularis is responsible for making androgens.

29. D The adrenal medulla receives preganglionic sympathetic nerve fibers and secretes catecholamines. This makes them like postganglionic sympathetic neurons.

30. A A renal artery obstruction would cause decreased renal blood flow. This would cause an increase in the renin-angiotensin axis, which would cause an increase in aldosterone. Aldosterone is made in the zona glomerulosa.

31. A Aldosterone primarily causes potassium secretion in exchange for sodium uptake. It also causes some hydrogen ion exchange for sodium uptake. This results in a mild alkalosis.

32. C The midcycle LH peak is responsible for ovulation.

33. D Progesterone is responsible for the increase in basal body temperature.

34. A Estrogen is responsible for the thin endocervical mucus.

35. C LH is responsible for the maintenance of the corpus luteum during the menstrual cycle.
36. B FSH is responsible for the recruitment of primary follicles.

Bibliography

Ganong, W. F. *Review of Medical Physiology* (16th ed.). E. Norwalk, CT: Appleton & Lange, 1993.

Greenspan, F. S., and Baxter, J. D. (eds.). *Basic and Clinical Endocrinology.* E. Norwalk, CT: Appleton & Lange, 1994.

Guyton, A. C. *Textbook of Medical Physiology* (8th ed.). Philadelphia: Saunders, 1991.

Isselbacher, K. J., et al. (eds.). *Harrison's Principles of Internal Medicine* (13th ed.). New York: McGraw-Hill, 1994.

School of Medicine Faculty. *Endocrine Physiology Syllabus.* Lecture handout, Stanford University, 1993.

Wilson, J. D., and Foster, D. W. (eds.). *Williams Textbook of Endocrinology* (8th ed.). Philadelphia: Saunders, 1992.

8

Neurophysiology

Leslie C. Andes

The neurology questions on the national boards cover neuroanatomy, neurophysiology, neuropathology, and biochemical aspects of neurochemical function. This chapter reviews the essential neurophysiology with minimal references to neuroanatomy; therefore, a concurrent review of neuroanatomy is strongly recommended. In particular, this text does not emphasize histology, specific neurologic pathways, or the localization of lesions.

The Components of the Nervous System

The nervous system senses and interprets the environment in an attempt to produce behavior appropriate to that environment. The nervous system is divided into the **peripheral nervous system** and the **central nervous system** (CNS). The peripheral nervous system is further subdivided into the **somatic** and **autonomic nervous systems;** the peripheral nervous system acts as an interface between the CNS and the environment, and serves to detect and process sensory information. The CNS has two components, the spinal cord and the brain. The spinal cord has five sections: cervical, thoracic, lumbar, sacral, and coccygeal. The brain also is separated into five parts: **myelencephalon** (medulla), **metencephalon** (pons and cerebellum), **mesencephalon** (midbrain), **diencephalon** (thalamus and hypothalamus), and **telencephalon** (basal ganglia and cerebrum). The CNS performs higher-level informational processing, including learning and memory, and produces the behaviors of an organism (Table 8-1).

The Neuronal Environment

Cerebrospinal Fluid

Approximately 600 to 700 ml of **cerebrospinal fluid** (CSF) is formed each day, with about 150 ml circulating in the **subarachnoid space.** The CSF is formed in the **choroid plexuses** and in the ependymal lining of the ventricles, as well as in the capillary endothelium of the brain. Cerebrospinal fluid is not an ultrafiltrate of plasma, as it is actively secreted by the choroid plexuses. The concentrations of Na^+, Cl^-, and Mg^{2+} are higher than plasma, whereas K^+, glucose, and Ca^{2+} are lower. It is almost protein free, and is normally clear and colorless. The CSF in the ventricles flows through the foramina of Magendie and Luschka into the cerebellomedullary cistern and then to the subarachnoid space around the brain and spinal cord, and is absorbed into the venous sinuses via the **arachnoid granulations,** or **villi.** If an obstruction occurs in the ventricular system or foramina, the result is called **noncommunicating**

Table 8-1. Parts and functions of the CNS

Region	Subdivision	Function
Spinal cord		Sensory input, reflex organization, somatic and autonomic motor output
Myelencephalon	Medulla	Cardiovascular control, respiratory control, brainstem reflexes
Metencephalon	Pons	Respiratory and urinary bladder control, vestibular control of eye movements
	Cerebellum	Motor control, motor learning
Mesencephalon	Midbrain	Acoustic relay, control of eye movements, motor control
Diencephalon	Thalamus	Sensory and motor relay to cerebral cortex
	Hypothalamus	Autonomic and endocrine control
Telencephalon	Basal ganglia	Motor control
	Cerebral cortex	Sensory perception, cognition, learning and memory, motor planning and voluntary movement

From R. M. Berne and M. N. Levy (eds.). *Physiology* (3rd ed.). St. Louis: Mosby–Year Book, 1993.

hydrocephalus; if the obstruction is at the arachnoid villi, it is called **communicating hydrocephalus.** One of the functions of the CSF is to support the brain in the cranium. Removal of CSF (for example, during a lumbar puncture) can cause the brain to exert traction on the blood vessels and nerves, resulting in a severe headache. The normal pressure of CSF as measured in the lumbar region is 70 to 180 mm Hg. Various drugs inhibit CSF production, most notably diuretics (furosemide, acetazolamide) and corticosteroids.

The Blood-Brain Barrier

The **blood-brain barrier** (BBB) is composed of the capillary endothelium and basement membrane of the vasculature supplying the brain. The capillaries of the brain have tight junctions but no gap junctions or channels, and thus exclude most substances from entry into brain cells. Exceptions to this are water, carbon dioxide, and oxygen, which can diffuse into the brain with ease. No substance is completely excluded; however, substances that are lipid soluble cross more easily. Transport systems exist in endothelial cells for glucose, K^+, and organic and amino acids. Areas of the brain that do not have a BBB are the pituitary gland, the pineal gland, the choroid plexuses, and some parts of the hypothalamus. The absence of a BBB in these areas facilitates systemic uptake and release of hormones. The BBB can be disrupted temporarily by sudden severe increases in blood pressure or by intravenous injection of hypertonic fluids (used to extract cerebral edema fluid).

Cerebral Blood Flow

Cerebral blood flow (CBF) is about 80 ml/100 g/min in gray matter and about 20 ml/100 g/min in white matter. As in the coronary circulation, CBF is **autoregulated,** meaning that it remains constant over a wide range of (normal) blood pressures (Fig. 8-1). Autoregulation is active between a mean blood pressure of 50 and 150 mm Hg. When the mean pressure is greater than 150 mm Hg, the BBB may be disrupted. The curve is shifted to the right in people with chronic hypertension. Regional metabolic activity helps determine regional CBF, as do arterial concentrations of oxygen and carbon dioxide. Unlike the coronary circulation, cerebral resistance vessels are more sensitive to PCO_2 than to PO_2. Carbon dioxide is an important regulator of CBF: Even a slight increase in PCO_2 will cause a large increase in CBF. If CBF decreases, activity on the electroencephalogram (see later section on electrical activity of the brain) slows, and when CBF is less than 10 ml/100 g/min, irreversible tissue damage can occur at normal body temperatures. Many drugs have effects on CBF: Barbiturates

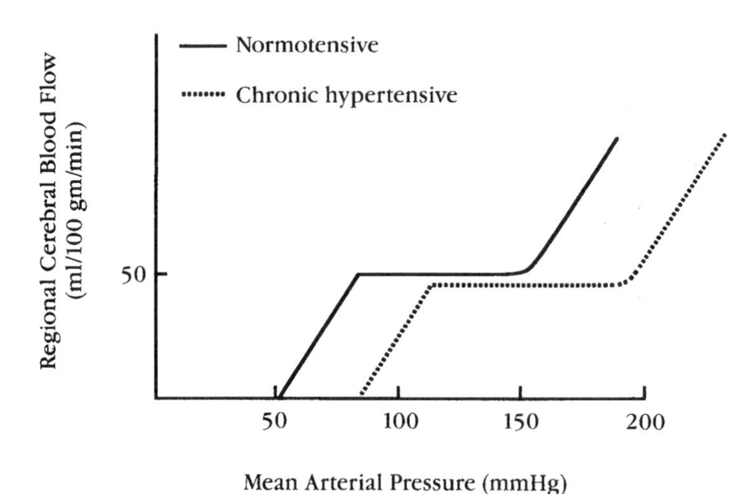

Fig. 8-1. Cerebral autoregulation in both normotensive and chronic hypertensive patients whose actual limits of mean arterial pressure depend on the degree of hypertension. Autoregulation occurs normally from a mean arterial pressure of about 50 to 150 mm Hg and keeps average cerebral blood flow relatively constant at about 50 ml/100g/min over this range. (From M. F. Newman, J. G. Reves, and R. D. McKay. Cardiovascular Therapy. In P. Newfield and J. E. Cottrell [eds]. *Neuroanesthesia: Handbook of Clinical and Physiologic Essentials* [2nd ed.]. Boston: Little, Brown, 1991.)

constrict cerebral blood vessels, for example, while volatile anesthetic agents dilate them. Constriction of the cerebral vasculature can help decrease intracranial pressure (ICP), and dilation can increase ICP.

Intracranial Pressure

The brain has three compartments: the CSF, brain tissue, and intravascular space. The contents of the cranium are essentially noncompressible, although an increase in pressure in one area can sometimes be compensated for by a decrease in another. Blood and CSF are readily **displaceable** and can thus partially offset the effect of an increase in **intracranial pressure** (ICP) as seen in tumors or hemorrhage into brain tissue. The counterbalancing measures of decreased CSF volume and cerebral blood flow may themselves be dangerous. ICP is measured supratentorially as surface CSF pressure or as pressure in the anterior horn of a lateral ventricle. **Intracranial compliance** is described by Fig. 8-2; note that as intracranial volume increases, the ICP remains low (high compliance) until the compensatory mechanisms are overcome, at which time ICP increases fairly rapidly. ICP can be acutely increased by coughing or by Valsalva maneuvers, which are well tolerated in the normal brain but may cause herniation of the vault contents through the foramen magnum if compliance is on the steep upward part of the curve.

The Somatic Sensory System

Nerve Fiber Types

There are two systems of nomenclature for nerve fiber types. Sensory neurons are sometimes classified with a numerical system as follows: Ia fibers are from the annulospiral endings of muscle spindles. Ib travel from the Golgi tendon organ (both

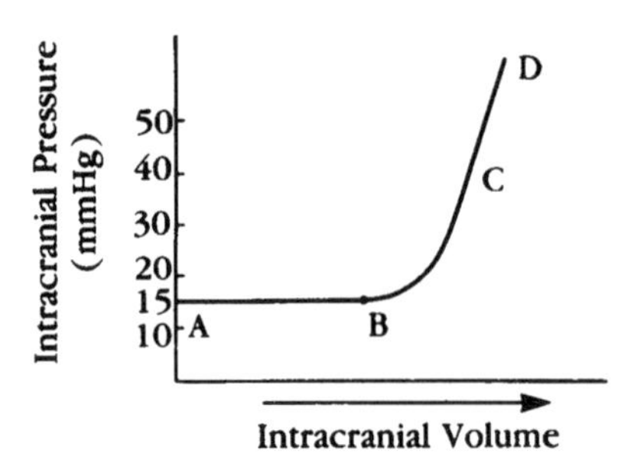

Fig. 8-2. The relationship between changes in intracranial volume and intracranial pressure. The line between A and B on the curve shows that the increase in volume of one of the three intracranial compartments (blood, brain, cerebrospinal fluid) was compensated successfully by a decrease in volume of one or both of the other two compartments. The line between B and C is the knee of the curve; intracranial compliance is decreased. Between C and D, a relatively small increase in intracranial volume leads to a large increase in intracranial pressure. The slope of the curve from B to C is also influenced by the rate at which the intracranial volume is increased. (From M. F. Newman, J. G. Reves, and R. D. McKay. Cardiovascular Therapy. In P. Newfield and J. E. Cottrell [eds]. *Neuroanesthesia: Handbook of Clinical and Physiologic Essentials* [2nd ed.]. Boston: Little, Brown, 1991.)

are A-alpha in the system delineated below), II fibers are from the flower spray endings of the muscle spindle as well as from touch and pressure receptors (corresponding to A-beta fibers), III carry pain and temperature and some touch sensations (A-delta), and IV fibers carry pain and temperature and correspond to C fibers. In a more comprehensive system, fibers are divided up on the basis of size, velocity of conduction, myelination, and differences in sensitivity to pressure, hypoxia, and anesthesia. Type A fibers are large myelinated fibers that are fast conducting. They are subdivided into four groups: A-alpha (proprioception, somatic motor information), A-beta (touch and pressure), A-gamma (motor to muscle spindles), and A-delta (pain, temperature, and touch). B fibers are medium-sized myelinated fibers with moderate conduction speed, and are seen in preganglionic autonomic fibers. C fibers carry pain, temperature, and postganglionic sympathetic information. The somatosensory and somatomotor systems are discussed in greater detail below.

Sensory Receptors

Sensory receptors are transducers that convert energy in the environment into **action potentials** in neurons. Anatomically, a sensory receptor may be either part of a modified sensory nerve ending, such as a **pacinian corpuscle,** or a specialized cell, such as a **cochlear hair cell.** Receptors respond to a given stimulus by increasing permeability to Na^+ and K^+ and producing a **receptor potential,** a nonpropagating local potential. The high sensitivity of a receptor to a particular form of energy accounts for the extraordinary specificity of receptors. As is true throughout the nervous system, when the local receptor reaches threshold, an action potential is produced. This action potential travels along a neuronal pathway specific for both sensory modality and, to varying degrees, location. Stimulation of a nerve fiber anywhere along this pathway, electrically or chemically, centrally or peripherally, produces a conscious sensation for that modality and that location.

All sensory receptors adapt either partially or completely to their stimuli over time. Thus a constant stimulus produces a progressively decreasing response. Two processes contribute to **adaptation:** adjustments in the structure of the receptor itself and modifications in the ion channels of the nerve fibers, a process termed **accommodation.** Depending on their rate of adaptation, receptors can be classified as either **phasic** or **tonic.** Phasic receptors adapt rapidly and react strongly while a change is taking place, thus transmitting valuable information about a change in stimulus strength. Impulse frequency is proportional to the rate of change. When combined with information about the body's current status, information about rate change allows the nervous system to predict the future status of the body. Phasic receptors include **pacinian corpuscles** and **Meissner's corpuscles.** Tonic receptors adapt slowly and incompletely and thus fire continuously to apprise the brain of a given stimulus. Examples of tonic receptors are **Ruffini's corpuscles** and **Merkel's disks** (Fig. 8-3). All of the cutaneous sensory receptors discussed below have group II afferent nerve fibers. Proprioceptive pathways are discussed in the sensory pathway section, while proprioceptors are discussed in the somatomotor section.

Cutaneous Sensory Receptors

The **pacinian corpuscle** consists of the unmyelinated tip of a sensory nerve surrounded by concentric lamellations of connective tissue that resemble an onion. When a vibratory stimulus is applied to the corpuscle, an electrical potential is produced in the nerve. The magnitude of the receptor potential increases proportionately as acceleration of the vibration occurs, and if threshold is attained an action potential is generated. Further increase in vibration results in a higher frequency of action potentials and may even spread to activate surrounding receptors.

Fig. 8-3. The encapsulated organs of the skin are all innervated by a single, large afferent fiber. A. The pacinian corpuscle is a rapidly adapting receptor that encodes vibration. B. Ruffini's corpuscle is a slowly adapting receptor that encodes pressure. C. Meissner's corpuscle is a rapidly adapting receptor responsible for detecting the speed of stimulus application. D. Merkel's disk is a slowly adapting receptor capable of encoding the location of a stimulus. Its receptor cells communicate with the primary afferent neurons by synaptic transmission. (From M. B. Wang. Neurophysiology. In J. Bullock, J. Boyle, and M. B. Wang. *Physiology* [2nd ed.]. Baltimore: Williams & Wilkins, 1991.)

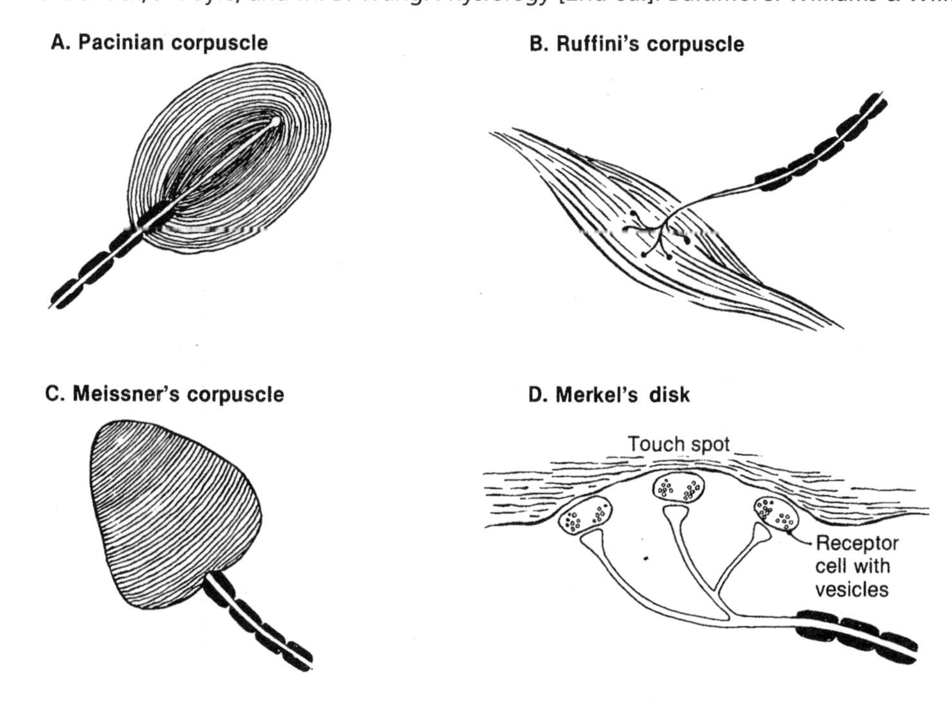

Pressure is detected by **Ruffini's corpuscles,** which are deep skin receptors with a collagen-filled capsule. Strands of collagen connect skin with nerve fibers; deformation of the skin stretches the collagen, causing depolarization and an action potential. The intensity of the pressure applied is encoded by the firing frequency.

Touch is mediated by three receptor types: **Meissner's corpuscles** (touch/flutter), **Merkel's disks** (touch/pressure), and peritrichial arborizations **(hair follicle receptors).** Meissner's corpuscles are dermal receptors and encode velocity of stimulus application, which gives information on object identification (e.g., discriminatory touch). Merkel's disks are disk-like nerve ending expansions that form synaptic connections with small receptive fields and are used for localization of a stimulus.

Temperature sensation is encoded by tonic **thermoreceptors** located on free endings of small myelinated (A-delta) and unmyelinated (C) fibers. Warm fibers begin to respond at a skin temperature of 30°C, with a maximum response at 45°C and cessation of activity at 47°C. Cold fibers begin to discharge at 10°C, with a maximal response at 25°C; they cease firing at 40°C. They will again start firing at 45°C. At skin temperatures in the comfort zone (31–36°C), awareness of temperature disappears (also known as a sensation of thermal neutrality). The skin can detect changes of less than 0.5°C when the skin temperature is in the comfort zone. Temperatures below 15°C and above 45°C are painful due to stimulation of pain receptors.

Pain Receptors

Nociceptors, the receptors for pain, are free nerve endings found in almost every tissue in the body (the brain being a notable exception). They report **skin pain, visceral pain,** and **deep somatic pain** (e.g., headache). These receptors may be activated by mechanical, chemical, or thermal stimuli, and adapt extremely slowly, thus assuring that a painful stimulus will not be ignored. Chemicals such as bradykinin and histamine are released from damaged cells and activate nociceptors. The intensity of a pain response is related to the extent of tissue damage. **Fast pain** is acute, discretely localized pain (e.g., pinprick) transmitted by small myelinated A-delta fibers. It is followed approximately 0.5 to 1 second later by **slow pain,** a more diffuse, chronic pain (e.g., burning) transmitted by small unmyelinated C fibers. Fast pain elicits the withdrawal reflex (described in the somatomotor section) and sympathetic responses to pain, while slow pain induces nausea, sweating, and changes in blood pressure and muscle tone. Viscera are much more sensitive to distention, traction, and ischemia than to incision or temperature. Pain receptors in the viscera are sparsely distributed and visceral pain is thus poorly localized. Because visceral pain fibers travel with fibers of the autonomic nervous system, autonomic symptoms such as nausea and vomiting are associated with visceral pain. Visceral pain, like deep somatic pain, can initiate reflex contraction of nearby skeletal muscles, as in the rigid abdominal wall seen with peritoneal pain. Deep pain is often poorly localized, and may be associated with autonomic symptoms as well.

Sensory Pathways

The two major pathways for transmitting sensation from cutaneous sensory receptors are the **dorsal columns (lemniscal system)** and the **anterolateral spinothalamic tracts.** These systems are contrasted in Table 8-2. A key feature of the dorsal columns is the faithfulness of transmission, in terms of both stimulus intensity and geographic location. Large myelinated fibers transmit this information rapidly. The information from the spinothalamic tract is more crude. It generally requires a less rapid response from the organism, and is transmitted by smaller, more slowly conducting fibers.

Table 8-2. Comparison of the dorsal column/medial lemniscal and anterolateral/spinothalamic systems

	Dorsal column	Anterolateral
Submodalities	Touch-pressure Position sense Kinesthesia Vibration	Pain Temperature Crude touch
Location in spinal cord	Dorsal columns	Anterolateral columns
Type of fibers	Large myelinated	Small myelinated Small unmyelinated
Somatotopic organization	Strictly maintained throughout pathway	Present, but not strictly maintained
Level of decussation	Medulla	Spinal cord
Brainstem terminations	Ventral posterior lateral nucleus and posterior nuclear group of thalamus	Brainstem reticular formation Ventral posterior lateral nucleus and posterior nuclear group of thalamus
Cerebral terminations	Primary and secondary somatic sensory cortices and somatic sensory association area	Primary and secondary sensory cortices and somatic sensory association area

First-order neurons carrying fine touch information from Meissner's corpuscles, Merkel's disks, pacinian corpuscles, and peritrichial arborizations (hair follicles) enter the **dorsal columns** (fasciculus cuneatus and gracilis) and continue on to synapse in the **nuclei cuneatus** and **gracilis** in the medulla. Fibers of the **second-order neurons** decussate (cross the midline) and ascend in the **medial lemniscus** to the **ventral posterolateral** (VPL) **nucleus** in the thalamus. **Third-order fibers** pass through the **internal capsule** and end in the somatosensory area of the parietal lobe cortex (postcentral gyrus). Crude touch information travels in similar pathways, although some fibers may cross the midline on entering the spinal cord and ascend in the **spinothalamic tracts** with pain and temperature fibers.

Pain and temperature afferent fibers synapse in the spinal cord **dorsal horns** within the **substantia gelatinosa;** axons of second-order neurons cross the midline and ascend in the **lateral spinothalamic** and **spinotectal tracts** to the VPL nucleus of the thalamus and the brainstem **reticular formation** (RF). Third-order neurons continue from the VPL nucleus to the cortex via the internal capsule.

Fibers in these two sensory systems are joined in the brainstem by fibers that mediate sensation from the head. Pain and temperature information from the head is relayed via the **spinal nucleus of cranial nerve V** to the **ventroposteromedial** (VPM) **nucleus** in the thalamus, and touch and proprioception via the **sensory and mesencephalic nuclei of cranial nerve V.**

Higher levels of the nervous system also participate in sensory processing. For example, two-point discrimination depends on cells in the dorsal column nuclei that have small excitatory cortical receptive fields with inhibitory surrounds. Such receptive fields can be explained by **lateral inhibition.** The effect is to sharpen contrast between adjacent stimuli and is depicted graphically in Fig. 8-4. **Stereognosis,** the ability to identify objects by handling them, depends on information of touch and pressure but also requires a large cortical component. Sensory input to the cortex can be inhibited by descending cortical pathways at each relay station (thalamus, medulla, etc.).

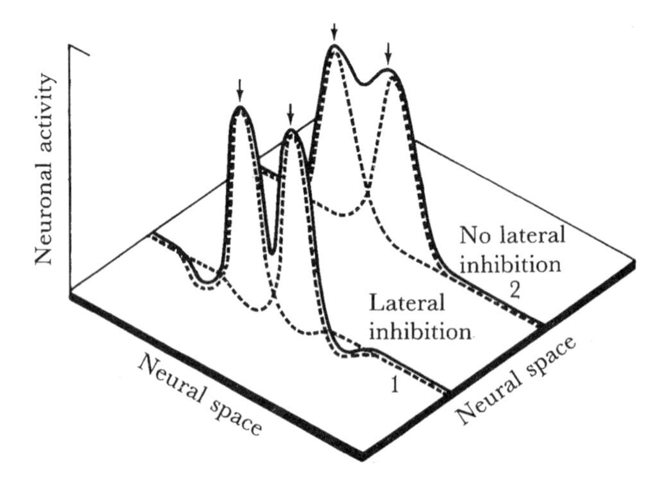

Fig. 8-4. Lateral inhibition. In the absence of lateral inhibition, two closely applied stimuli (*dotted lines*) are perceived as the sum of their pressures, and it is difficult to determine if one large blunt stimulus is applied or two sharp stimuli. With lateral inhibition, the border between the two stimuli is sharpened. (Reprinted by permission of the publisher from *Principles of Neural Science* [2nd ed.], by E. K. Kandel and G. H. Schwartz. P. 325, 1985, by Elsevier Science Publishing Company, Inc.)

Pain Sensations

The neurotransmitter at the synapses in the dorsal horn from primary pain afferents is **substance P.** Glutamate (an excitatory amino acid), calcitonin gene-related peptide, and vasoactive intestinal polypeptide may also be active.

Opiate receptors are widely distributed in the brain and spinal cord. Endogenous opioid peptides **(enkephalins and endorphins)** found in the brain, spinal cord, pituitary, and adrenal medulla can react with opiate receptors to suppress pain. Endorphin receptors are found in many sensory relay nuclei in the brainstem and medulla, and in the periaqueductal gray matter. Opioid receptors seen in the substantia gelatinosa may be on the substance P–containing terminals of nociceptive afferents and may inhibit substance P release. Placebos, acupuncture, and transcutaneous electrical nerve stimulation (TENS) units can cause release of endogenous opioids and may provide long-term analgesia.

Although pain perception may be a function of subcortical centers (i.e., the cortex is not necessary to perceive pain), it is believed that the cortex plays an important role in interpreting the quality of pain. "Stress analgesia" (when pain disappears during a period of stress or concentration on the matter at hand) and the ability of tactile stimulation to reduce pain (e.g., rubbing an injured area) can be explained by higher-order processing of pain signals. Tactile stimuli and descending signals from the brain are believed to activate enkephalinergic interneurons in the substantia gelatinosa of the dorsal horn, which act presynaptically to inhibit transmission of impulses from the dorsal root pain fibers to spinothalamic neurons. This is the basis of the **gate control theory,** the dorsal horn being the "gate" through which pain impulses reach the lateral spinothalamic tract.

Referred Pain

Pain referred from a viscus to a somatic structure (usually skin) that shares the same embryologic dermatome results from visceral and somatic afferents converging on the same spinothalamic neurons in the dorsal horns. Because dorsal horn cells are usually activated by cutaneous input, and skin is topographically mapped in the cortex whereas viscera are not, the cortex misinterprets the activity of these dorsal horn neurons as somatic rather than visceral pain. Angina, chest pain due to cardiac muscle ischemia, is often referred to the anterior chest wall. **Projected pain** is not the same as **referred pain.** Projected pain results from actual stimulation of a pain pathway anywhere along the path, which causes that pain to manifest as if it were occurring at the periphery of the pain pathway. Common examples are phantom limb pain, or chronic back pain after a herniated disk.

Chronic Pain and Pain Modulation

Chronic pain is not well understood. There may be parallel pain pathways outside the spinothalamic tracts, or chronic pain may result from spontaneous activity of pain centers in the CNS. Pain receptors may also develop a background discharge and produce spontaneous pain and signs of inflammation in the affected area, as seen in **reflex sympathetic dystrophy.** Also, irritation of nerves can cause sympathetic discharge, which makes pain more intense and unpleasant. Reverberating circuits that develop because of continual pain input may fail to stop firing when pain input is removed. Sensitization of nociceptors after destruction by a noxious stimulus can cause fibers to become more responsive and discharge more vigorously after a given stimulus with a lower activation threshold; **denervation hypersensitivity,** increased sensitivity to circulating neurotransmitter, can result when normal synaptic input is removed from a neuron. Chronic irritation of nerves increases the number of substance P receptors in the spinal cord on the side of the injury and causes A-beta fibers from mechanoreceptors to grow into areas where nociceptors end.

Somatic Motor Function

Skeletal muscle functions include performance of basic involuntary reflex movements in response to environmental stimuli, maintenance of posture and position in the face of gravity, and performance of voluntary movements initiated by higher centers of the nervous system.

A **motor unit** is defined as a motor neuron and all the muscle fibers it innervates. **Lower motor neurons** are the cells and fibers of motor cranial and spinal nerves, while **upper motor neurons** are the motor cells and fibers from the cortex to the cranial nerve nuclei and anterior horn cells. Lesions of motor neurons are described in Table 8-3.

Reflex Arcs

The **spinal reflex arc** is the basic unit of integrated neural activity. Reflex arcs consist of afferent sensory input, central integration and processing, and efferent output.

Table 8-3. Motor neuron lesions

Feature	Upper motor neuron	Lower motor neuron
Tone	Increased	Decreased
Reflexes	Hyperreflexic	Areflexic
Muscle mass	No or slight decrease	Atrophy
Spontaneous activity	Fasciculation	None

The **monosynaptic stretch reflex** (e.g., the knee jerk) is the simplest reflex arc (Fig. 8-5). Stretch is detected by a sensory receptor in muscle, the **muscle spindle.** It lies parallel to the **extrafusal** muscle fibers and consists of small **intrafusal** muscle fibers with a noncontractile central portion that receives sensor nerve endings (Fig. 8-6). The ends of the intrafusal fibers are innervated by small-diameter A-gamma motor neurons, while the extrafusal fibers of the main muscle are innervated by large-diameter A-alpha motor neurons. When the muscle is stretched (i.e., by tapping on the tendon), the spindle organ is also stretched because it is aligned parallel to the extrafusal fibers. Stretching the spindle excites its Ia sensory afferents, which end in the spinal cord on the A-alpha motor neurons of the same muscle, and causes reflex contraction of the extrafusal fibers, all in about 20 msec. The spindle Ia afferents are ex-

Fig. 8-5. The monosynaptic stretch reflex. Also illustrated is inhibition of the antagonist muscles through an inhibitory *(shaded)* interneuron. This is an example of a polysynaptic reflex. (Reprinted by permission of the publisher from *Principles of Neural Science* [2nd ed.], by E. K. Kandel and G. H. Schwartz. P. 459, 1985, by Elsevier Science Publishing Company, Inc.)

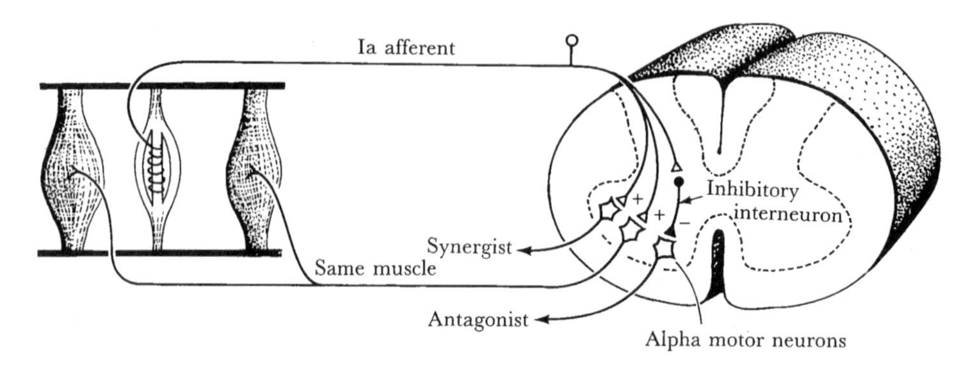

Fig. 8-6. The muscle spindle. The muscle spindle is seen to lie in parallel to the extrafusal fibers. Sensory fibers arise from the spindle to provide information about stretch. The group Ia (annulospiral) primary nerve endings are sensitive to both absolute stretch and to changes in the rate of stretch of the spindle. The group II secondary nerve endings are sensitive to absolute stretch only. Gamma motor neurons innervate the intrafusal fibers located at the end of the spindles. (From A. C. Guyton. *Textbook of Medical Physiology* [7th ed.]. Philadelphia: Saunders, 1985. P. 608.)

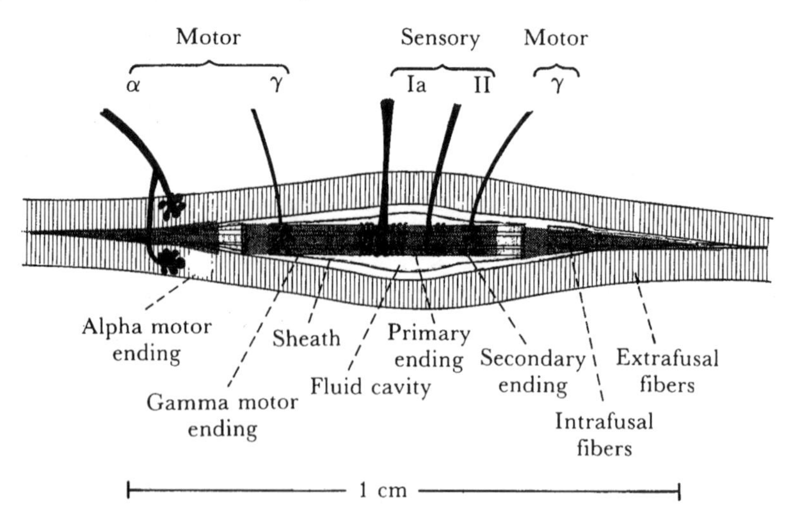

cited at a rate proportional to the amount of stretch. If the muscle contracts without contraction of the intrafusal fibers, however, the spindle becomes slack because the muscle has shortened while the spindle itself has not. In this case the frequency of afferent impulses decreases and the extrafusal fibers relax, also known as **unloading.** Thus, the spindle and its reflex connections (Fig. 8-7) constitute a feedback device that operates to maintain muscle length. Muscles involved in more precise movements contain a proportionately higher number of muscle spindles.

In most motor behaviors, alpha and gamma motor neurons are activated together, a process termed **alpha-gamma coactivation.** The gamma motor neurons directly

Fig. 8-7. Effect of various conditions on muscle spindle discharge. (From W. F. Ganong. *Review of Medical Physiology* [16th ed.]. East Norwalk, CT: Appleton & Lange, 1993.)

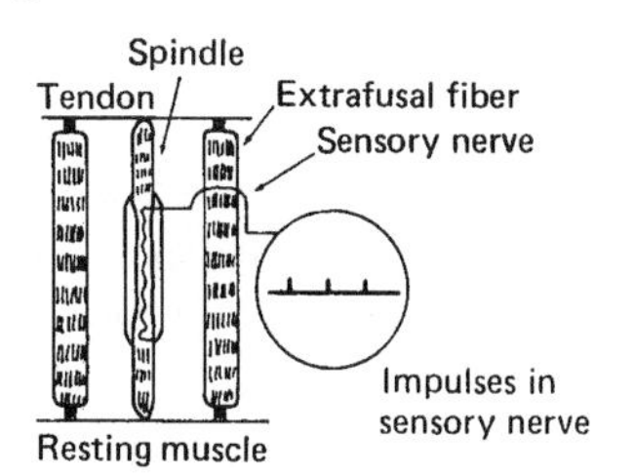

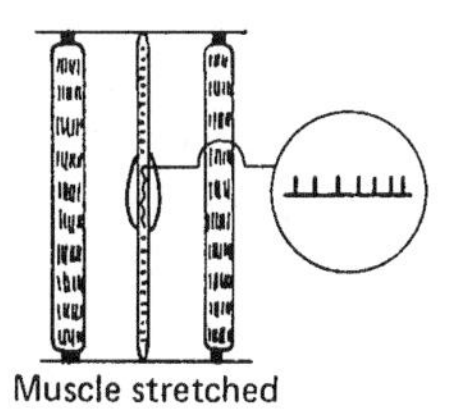

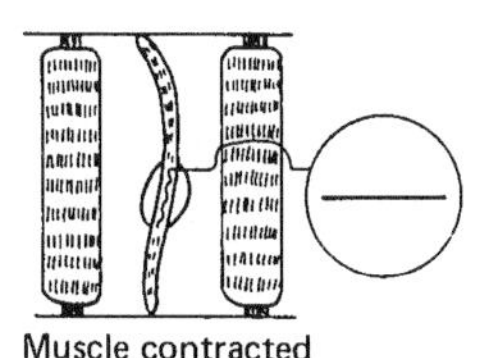

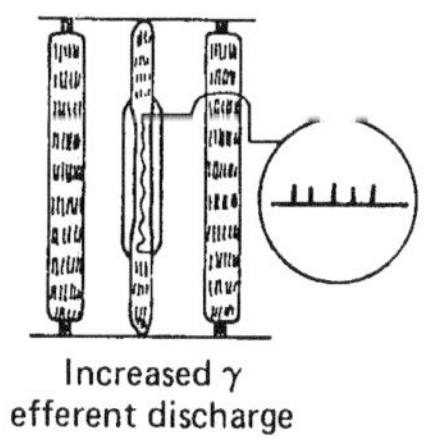

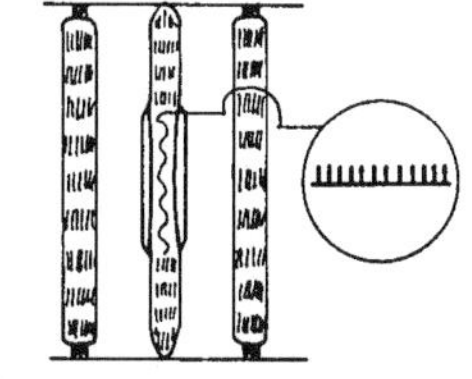

cause contraction (i.e., shortening) of the intrafusal fibers as the extrafusal fibers shorten, but the central portion of the spindle remains stretched. The subsequent activation of Ia fibers leads to a reflex contraction of the extrafusal fibers. Therefore, no unloading occurs; the spindle shortens with the muscle throughout the contraction so that it is able to respond to stretch continuously, and to help reflexively adjust motor neuron discharge throughout the contraction. The gamma efferent system is important in the maintenance of muscle tone; motor neurons of the gamma loop are regulated by cortical descending pathways to adjust the sensitivity of the muscle spindles as needed for postural control.

The **Golgi tendon organ** (GTO) is another stretch receptor found in muscle, which is in series (as opposed to muscle spindles, which are in parallel) with the muscle in its connection with the tendon. It is stimulated by either passive stretch or active contraction of a muscle. When the GTO is stretched strongly, its Ib fibers to the spinal cord are activated and cause inhibitory input to the alpha motor neurons to the same muscle, causing relaxation and preventing a potential rupture of the muscle.

Polysynaptic reflexes make use of centrally located **interneurons,** excitatory or inhibitory, converging or diverging, to provide for more complex reflex behavior. Examples include **reciprocal innervation** of antagonist muscles in the stretch reflex and the **crossed extensor response** of the withdrawal reflex. In reciprocal innervation, when the stretch reflex occurs, the muscles that antagonize the action of the muscle involved relax; impulses in the Ia fibers from the muscle spindles of the muscle stretched cause inhibition of alpha motor neurons to its antagonists. Afferents to the spinal cord may follow one of several paths, depending on the strength of the initial stimulus, which will help to reinforce the initial reflex arc. Suppose one steps on a nociceptive stimulus, a sharp object. The basic response is via a flexor reflex of the ipsilateral leg, which withdraws the leg from the stimulus. Other potential paths, all of which serve to make the basic reflex action more efficient by assisting in withdrawing the leg, include (1) to inhibitory interneurons to the *ipsilateral extensors* (relaxation), so they will not oppose the flexors; (2) to excitatory interneurons to the *contralateral extensors* (contract), extending the opposite limb to bear weight; and (3) to inhibitory interneurons to the *contralateral flexors* (relaxation).

Normally these reflexes are subordinated to supraspinal centers. Descending tracts from the brainstem (red nucleus, vestibular nuclei, and reticular formation) exert an inhibitory influence on alpha and gamma motor neurons of extensor muscles and stimulate the flexors, or vice versa, depending on which nuclei are involved. Transection of the spinal cord causes temporary loss of peripheral reflexes **(spinal shock),** then recovery of reflexes in a hyperactive fashion **(spasticity)** due to loss of inhibitory input by higher centers.

Proprioception Pathways

First-order neurons carrying proprioception information from Golgi tendon organs, muscle spindles, and joint receptors (pacinian corpuscles and Ruffini's endings) in the lower extremities travel to the dorsal gray horns in levels C8 to L3 of the spinal cord, and synapse in the nucleus dorsalis in the posterior horn. Fibers from below L3 travel in the fasciculus gracilis and synapse in the nucleus dorsalis. The second-order neurons from this nucleus travel with neurons from the dorsal nucleus in the anterior and posterior spinocerebellar tracts to the cerebellum. Fibers from the upper extremity travel in the fasciculus cuneatus to synapse in the accessory cuneate nucleus in the medulla, continuing onward via the cuneocerebellar tract, which can be thought of as the upper extremity equivalent of the posterior spinocerebellar tract. Fibers from the face travel to the VPM nucleus of the thalamus via the trigeminothalamic tract and to the reticular formation via the mesencephalic tract of cranial nerve V, then to the post-central gyrus in the cortex.

Control of Posture and Movement

Somatic motor activity ultimately depends on the discharge of the spinal motor neurons and the corresponding motor nuclei of the cranial nerves. These motor neurons receive input from a variety of higher centers to control the initiation, coordination, and maintenance of motor activity. These pathways are depicted schematically in Fig. 8-8. The physiology of these systems is less well understood than that of the sensory systems, and much of the available information comes from lesion studies and disease states. The three main sources of higher input are the pyramidal system, the extrapyramidal system (EPS), and the cerebellum.

The **pyramidal system** consists of the **corticospinal** (CST) and **corticobulbar tracts.** Additionally, the **rubrospinal tract** (with fibers originating in the red nucleus) is closely related in function to the corticospinal tract. Fibers arise primarily from the precentral motor cortex, descend through the internal capsule, decussate in the pyramids, and descend to the bulbar region or spinal cord to synapse with either interneurons or lower motor neurons directly. Most CST fibers end on interneurons, which allows for the signal to spread to a variety of motor neurons. This helps provide integrated muscular activity and permits fine muscular control, as is necessary for delicate hand movements. The human motor **homunculus** has greatly enlarged hands and mouth, representing the motor cortex necessary for the fine control of manual manipulations and speech. The rubrospinal tract provides for relatively discrete signals (i.e., fine control) from the cortex to the spinal cord via the red nucleus. In the cord it lies just anterior to the CST. Together with the CST it forms the lateral motor system of the cord, which is primarily concerned with movement of the distal extremities. The pyramidal tract also includes descending fibers from the postcentral sensory cortex. These fibers allow the cortex to modify ascending sensory input and serve as a feed-forward system for motor control.

Extrapyramidal is a clinical term that includes those systems regulating movement aside from the pyramids and cerebellum, namely the **basal ganglia** and the **brainstem.** Within the brainstem (pons, midbrain, and medulla), the ventromedial motor system controls the proximal and axial musculature to stabilize the pelvic and shoulder girdles, primarily through the rubrospinal, reticulospinal, tectospinal, and

Fig. 8-8. Control of voluntary movement. (From W. F. Ganong. *Review of Medical Physiology* [16th ed.]. East Norwalk, CT: Appleton & Lange, 1993.)

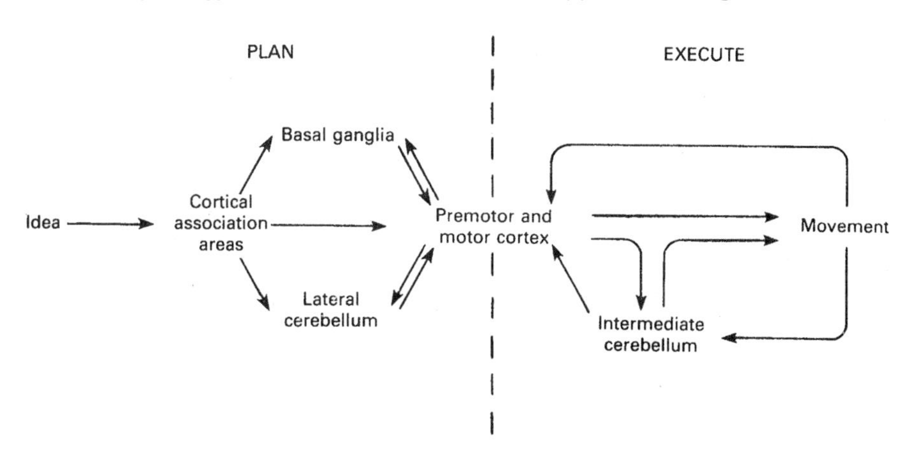

vestibulospinal tracts. A variety of reflexes and inputs are essential to this control. Lesion studies have been useful in demonstrating the integration and hierarchy of postural control from brainstem centers. Results of these studies are displayed in Table 8-4.

The basal ganglia (the caudate, putamen, globus pallidus, substantia nigra, and subthalamic nucleus) are believed to play a major role in the initiation and direction of voluntary movement, particularly of large muscle masses. Their precise function remains unknown, but prior to the initiation of muscle activity, action potentials appear in the basal ganglia before appearing in the motor cortex. As demonstrated in Fig. 8-8, the basal ganglia do not project to the spinal cord. They form a closed feedback loop with the motor cortex, receiving input from the motor cortex and discharging their output to the motor cortex via the thalamus. Disorders of the basal ganglia result in increased tone and **dyskinesias.** The increased tone, as in brainstem animals (Table 8-4, medullary pontine lesion) is a result of decreased gamma motor neuron inhibition. Various dyskinesias and their associated lesion sites are shown in Table 8-5.

The **cerebellum** functions to modulate and coordinate motor activity. The medial portions of the cerebellum help the medial motor system maintain equilibrium and tone. The lateral cerebellum is more concerned with the lateral motor system and the distal extremities. The cerebellum has extensive connections with the spinal cord, brainstem reticular formation, vestibular nuclei, and cerebral cortex, which enable it to perform its role. The circuitry of the cerebellum (Fig. 8-9) helps explain its func-

Table 8-4. Motor deficits from lesion studies

Level of lesion	Motor deficits or capabilities
Spinal cord	Paralysis Initially reflexes are depressed (spinal shock) but then become hyperactive No righting reflexes
Junction of medulla and pons	Decerebrate rigidity (extension in all extremities) due to decreased inhibition from higher centers, causing increased gamma efferent discharge and increased excitability of the motor neuron pool No righting reflexes
Midbrain	Can rise to standing position, walk, right self
Neocortex	Decorticate rigidity (extensor in legs, moderate flexion in arms) from decreased inhibition of gamma efferents Great deal of movement is possible, but animal cannot learn

Table 8-5. Disorders of the basal ganglia

Disease	Abnormal movements	Muscle tone	Primary anatomic locus
Parkinsonism	Tremor at rest	Rigidity	Substantia nigra
Huntington's disease	Chorea	Hypotonicity	?
Athetosis	Athetosis	Spasticity Paresis	Lenticular nucleus
Hemiballismus	Ballismus	Marked hypotonia	Subthalamic nucleus

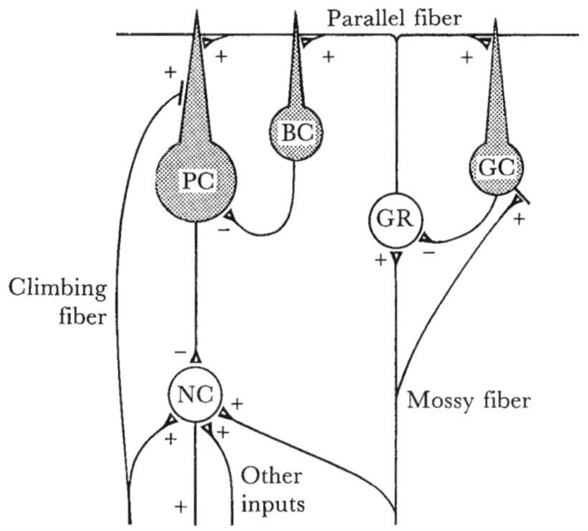

Fig. 8-9. Cerebellar circuitry. See text for details. BC = basket cell; GC = Golgi cell; GR = granule cell; NC = nuclear cell; PC = Purkinje cell. (Modified from J. C. Eccles, M. Itah, and J. Szentágothai. *The Cerebellum as a Neuronal Machine.* New York: Springer, 1967.)

tion in modulating motor activity. Input to the cerebellum is excitatory from **mossy** and **climbing fibers.** Climbing fibers arise from the inferior olivary nuclei (in the medulla) and synapse directly with **Purkinje cells,** providing a strong excitatory stimulus. All other afferent projections to the cerebellum (spinocerebellar, cuneocerebellar, trigeminocerebellar, reticulocerebellar) end as mossy fibers that synapse on **granule cells.** These granule cells then stimulate Purkinje cells and other inhibitory interneurons (e.g., basket cells and Golgi cells). Both climbing and mossy fibers also send out collateral fibers that excite the deep cerebellar nuclei. The output of the deep cerebellar nuclei to either the cord or thalamus is always excitatory. The final output of the cerebellum is thus a balance of the excitatory effects of the afferent collaterals of the mossy and climbing fibers and the inhibitory effects of the Purkinje cells. Note that there is a built-in time delay between the initial excitement mediated by the collaterals and the subsequent inhibition from the Purkinje cells. This balanced circuitry and extensive input enable the cerebellum to modulate and coordinate motor activity, providing smooth and properly timed motions. Lesions of the cerebellum result in hypotonia (from decreased gamma efferent discharge), marked ataxia, the breakdown of movements into their component parts, dysdiadochokinesis, dysmetria, and intention tremors.

The Autonomic Nervous System

The **autonomic nervous system** (ANS) is the part of the nervous system that innervates smooth muscle, cardiac muscle, arrector pili muscle (hair follicles), and myoepithelial cells (glands) (Fig. 8-10). Together with the endocrine system it regulates the visceral functions of the body. The ANS operates via reflex arcs that are modified by higher centers in the brainstem, hypothalamus, limbic system, and cortex. Although it is considered an involuntary system, conscious input can also influence autonomic output and contribute to such conditions as ulcers, palpitations, angina, myocardial infarction, hypertension, and constipation or diarrhea. Unlike the somatomotor system, the ANS can function normally in the absence of CNS input.

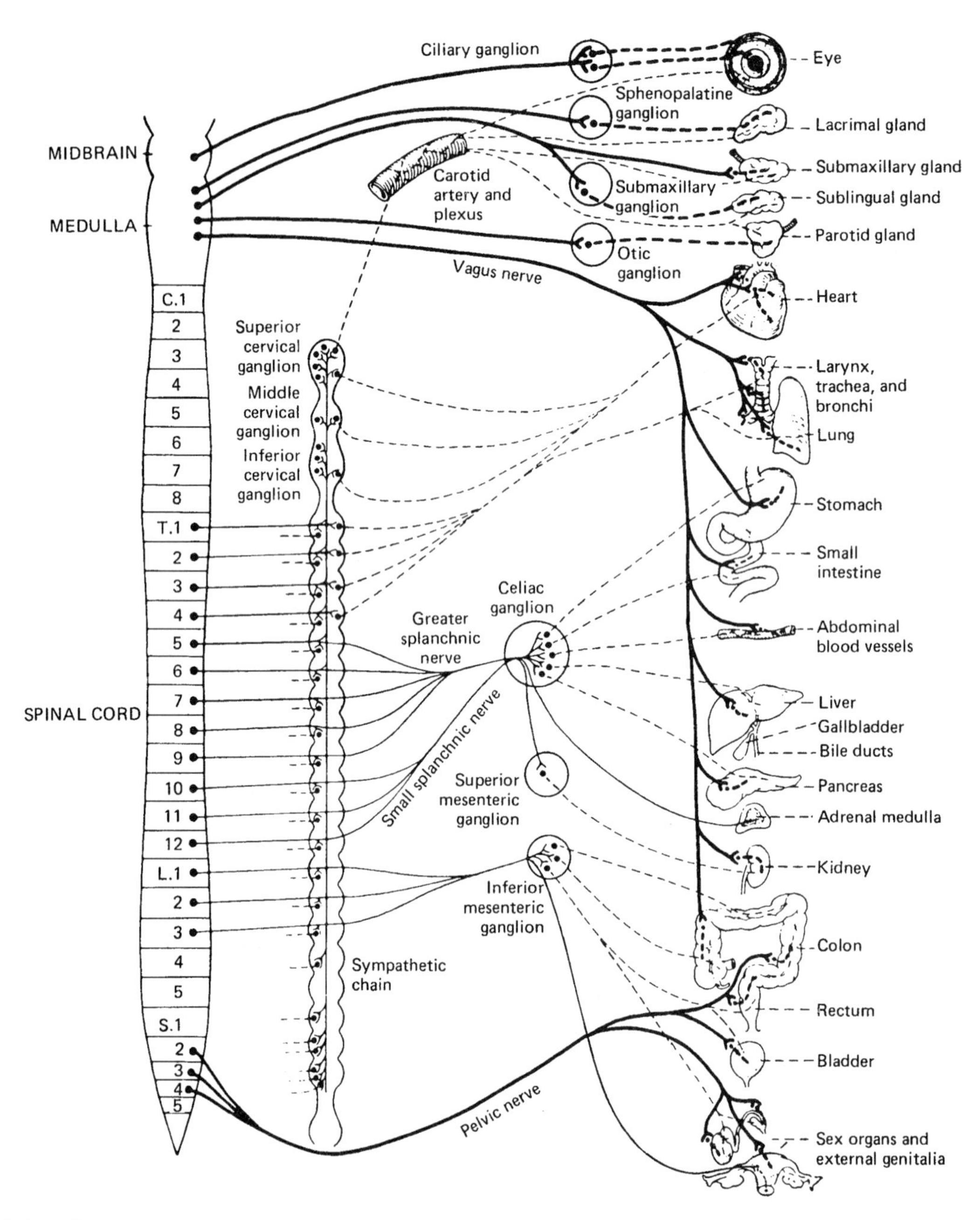

Fig. 8-10. Diagram of the efferent autonomic pathways. Preganglionic neurons are shown as solid lines and postganglionic neurons as dashed lines. The heavy lines are parasympathetic fibers; the light lines are sympathetic. (From W. F. Ganong. *Review of Medical Physiology* [16th ed.]. East Norwalk, CT: Appleton & Lange, 1993.)

Information from visceral receptors travels in visceral afferent neurons running with the visceral efferents (motor fibers) to the dorsal gray matter of the spinal cord. Unlike the somatic motor system, in which skeletal muscle is innervated directly by a centrally located motor neuron, the ANS is a two motor neuron system. Centrally located **preganglionic neurons** send out myelinated axons that synapse with peripherally located **postganglionic neurons.** The postganglionic neurons send out small unmyelinated axons to the effector organs (Fig. 8-11). Anatomically, the ANS is traditionally partitioned into **parasympathetic** (craniosacral) and **sympathetic** (thora-

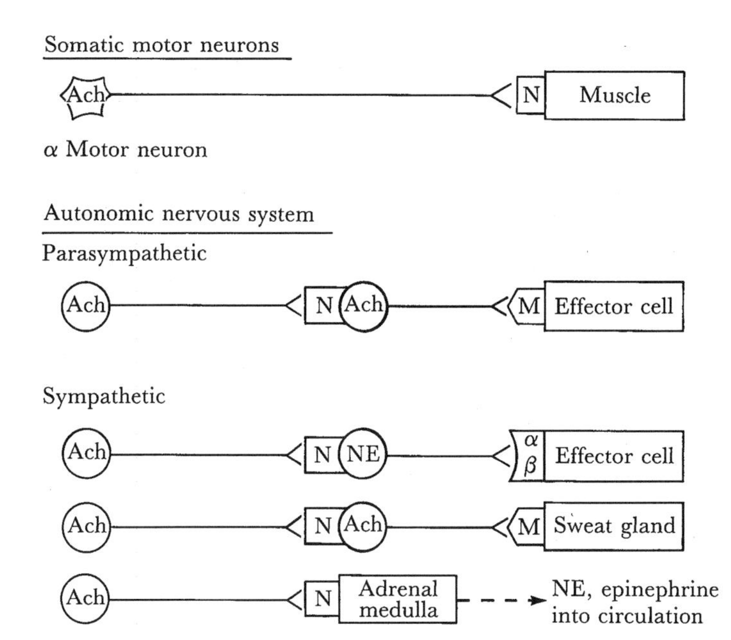

Fig. 8-11. Transmitters and receptors in somatic and autonomic motor neuron transmission. Note that the autonomic nervous system is a two motor neuron system. Ach = cholinergic neuron; NE = noradrenergic neuron; N = nicotinic receptor; M = muscarinic receptor.

columbar) divisions; most physiologists now include a third division, the **enteric nervous system.**

The Parasympathetic Nervous System

In the **parasympathetic nervous system** (PNS), preganglionic fibers travel with the third, seventh, ninth, and tenth cranial nerves to supply the head, the thorax, and the bulk of the abdominal viscera, as well as from the second through fourth sacral spinal roots to supply the pelvic viscera and the remaining abdominal viscera (see Fig. 8-10). The vagus nerve (cranial nerve X) alone carries approximately 75 percent of preganglionic parasympathetic fibers. These preganglionic axons take a long but direct path to synapse onto short postganglionic neurons in or near the target organ. Preganglionic parasympathetic fibers release **acetylcholine** onto **nicotinic receptors** of parasympathetic postganglionic neurons. The postganglionic neurons in turn release acetylcholine onto **muscarinic receptors** of the effector organ (see Fig. 8-11). Cholinergic stimulation is terminated with the hydrolysis of acetylcholine into acetate and choline by **acetylcholinesterases** in the synaptic cleft. The small amount of acetylcholine that diffuses away is quickly and completely degraded by serum and erythrocyte cholinesterases. This assures that cholinergic stimulation is short and discrete. Drugs that inhibit cholinesterases **(anticholinesterases)** act as **indirect parasympathomimetics,** and can cause nicotinic and muscarinic side effects due to acetylcholine accumulation in the heart, lungs, smooth muscle, and CNS. Neostigmine, edrophonium, and parathion (an insecticide) are all examples. Carbachol and pilocarpine are direct cholinergic agonists, whose breakdown by cholinesterases is slower than that of acetylcholine.

The actions of the PNS are discussed in more detail throughout various chapters of this text. As shown in Table 8-6, parasympathetic stimulation may be either excitatory or inhibitory depending on the specific organ system. In general, the PNS can be

Table 8-6. Responses of effector organs to autonomic nerve impulses and circulating catecholamines

Effector organs	Cholinergic impulses response	Noradrenergic impulses	
		Reactor type[a]	Response
Eye			
Radial muscle of iris	—	α_1	Contraction (mydriasis)
Sphincter muscle of iris	Contraction (miosis)		—
Ciliary muscle	Contraction for near vision	β_2	Relaxation for far vision
Heart			
SA node	Decrease in heart rate, vagal arrest	β_1	Increase in heart rate
Atria	Decrease in contractility and (usually) increase in conduction velocity	β_1	Increase in contactility and conduction velocity
AV node	Decrease in conduction velocity	β_1	Increase in conduction velocity
His-Purkinje system	Decrease in conduction velocity	β_1	Increase in conduction velocity
Ventricles	Decrease in conduction velocity	β_1	Increase in contractility
Arterioles			
Coronary	Constriction	α_1, α_2	Constriction
		β_2	Dilation
Skin and mucosa	Dilation	α_1, α_2	Constriction
Skeletal muscle	Dilation	α_1	Constriction
		β_2	Dilation
Cerebral	Dilation	α_1	Constriction
Pulmonary	Dilation	α_1	Constriction
		β_2	Dilation
Abdominal viscera	—	α_1	Constriction
		β_2	Dilation
Salivary glands	Dilation	α_1, α_2	Constriction
Renal	—	α_1, α_2	Constriction
		β_1, β_2	Dilation
Systemic veins	—	α_1	Constriction
		β_2	Dilation
Lung			
Bronchial muscle	Contraction	β_2	Relaxation
Bronchial glands	Stimulation	α_1	Inhibition
		β_2	Stimulation
Stomach			
Motility and tone	Increase	$\alpha_1, \alpha_2, \beta_2$	Decrease (usually)
Sphincters	Relaxation (usually)	α_1	Contraction (usually)
Secretion	Stimulation		Inhibition
Intestine			
Motility and tone	Increase	$\alpha_1, \alpha_2, \beta_2$	Decrease (usually)
Sphincters	Relaxation (usually)	α_1	Contraction (usually)
Secretion	Stimulation	α_2	Inhibition
Gallbladder and ducts	Contraction	β_2	Relaxation
Urinary bladder			
Detrusor	Contraction	β_2	Relaxation (usually)
Trigone and sphincter	Relaxation	α_1	Contraction
Ureter			
Motility and tone	Increase (?)	α_1	Increase (usually)
Uterus	Variable[b]	α_1, β_2	Variable[b]
Male sex organs	Erection	α_1	Ejaculation

(continued)

Effector organs	Cholinergic impulses response	Noradrenergic impulses	
		Reactor type[a]	Response
Skin			
Pilomotor muscles	—	α_1	Contraction
Sweat glands	Generalized secretion	α_1	Slight, localized secretion[c]
Spleen capsule	—	α_1	Contraction
		β_2	Relaxation
Adrenal medulla	Secretion of epinephrine and norepinephrine		—
Liver	—	α_1, β_2	Glycogenolysis
Pancreas			
Acini	Increased secretion	α	Decreased secretion
Islets	Increased insulin and glucagon secretion	α_2	Decreased insulin and glucagon secretion
		β_2	Increased insulin and glucagon secretion
Salivary glands	Profuse, watery secretion	α_1	Thick, viscous secretion
		β_2	Amylase secretion
Lacrimal glands	Secretion	α	Secretion
Nasopharyngeal glands	Secretion		—
Adipose tissue	—	β_1, β_3	Lipolysis
Juxtaglomerular cells	—	β_1	Increased renin secretion
Pineal gland	—	β	Increased melatonin synthesis and secretion

[a]Where a receptor subtype is not specified, data are as yet inadequate for characterization.
[b]Depends on stage of menstrual cycle, amount of circulating estrogen and progesterone, pregnancy, and other factors.
[c]On palms of hands and in some other locations ("adrenergic sweating").
From W. F. Ganong. *Review of Medical Physiology* (16th ed.). East Norwalk, CT: Appleton & Lange, 1993. Pp. 204–205.

considered a "vegetative" system, conserving energy (e.g., decreasing heart rate) and mediating various "housekeeping" functions (e.g., digestion). Also in contrast to the sympathetic nervous system (see next section), the various components of the PNS act relatively independently of each other, and there is no mass PNS discharge.

The Sympathetic Nervous System

In the **sympathetic nervous system** (SNS), the cell bodies of the preganglionic sympathetic neurons are located in the **interomediolateral gray horns** of spinal cord segments T1 to L3–4 (see Fig. 8-10). Preganglionic axons reach the **pre-** and **paravertebral sympathetic ganglia** via the spinal nerves and **white rami communicantes.** Prevertebral ganglia are located near the origins of large arteries in perivascular plexi (e.g., celiac, superior mesenteric, aortic). The paravertebral chain extends from cranium to coccyx. Unlike parasympathetic preganglionic axons, sympathetic ganglion connections diverge and converge extensively. A sympathetic preganglionic axon may (1) synapse with postganglionic neurons in the paravertebral ganglion it first enters, (2) travel up or down the sympathetic trunk to synapse in other ganglia, and/or (3) pass through the paravertebral chain to synapse in one of the outlying (prevertebral) sympathetic ganglia. Postganglionic axons exiting the paravertebral ganglia pass via the **gray rami communicantes** to rejoin the spinal nerve and travel with peripheral nerves to all segments of the body; postganglionic fibers from prevertebral ganglia travel with their associated blood vessels toward the effector organs.

Preganglionic sympathetic fibers release **acetylcholine** onto **nicotinic receptors** of postganglionic sympathetic nerve fibers. Postganglionic sympathetic effector compounds include **acetylcholine** (sweat glands and skeletal muscle blood vessels) and the catecholamines **epinephrine** and **norepinephrine** (see Fig. 8-11). Preganglionic sympathetic neurons also innervate chromaffin cells of the adrenal medulla; these cells can be considered modified postganglionic neurons that secrete epinephrine (80%) and norepinephrine (20%) into the bloodstream. These circulating catecholamines can also activate adrenergic receptors.

Adrenergic receptors are pharmacologically classified into alpha-1, alpha-2, beta-1, and beta-2 receptors on the basis of the receptor's affinity for various adrenergic agonists and antagonists (Tables 8-6 and 8-7). Epinephrine and norepinephrine both act strongly at alpha receptors. Epinephrine significantly stimulates both beta-1 and

Table 8-7. Some drugs and toxins that affect autonomic activity[a]

Site of action	Compounds that augment autonomic activity	Compounds that depress autonomic activity
Sympathetic and parasympathetic ganglia	**Stimulate postganglionic neurons** Nicotine Dimethylphenylpiperazinium **Inhibit acetylcholinesterase** DFP (disopropyl fluorophosphate) Physostigmine (Eserine) Neostigmine (Prostigmin) Parathion	**Block conduction** Hexamethonium (C-6) Mecamylamine (Inversine) Pentolinium Trimethaphan (Arfonad) High concentrations of acetylcholine Anticholinerase drugs
Endings of postganglionic noradrenergic neurons	**Release norepinephrine** Tyramine Ephedrine Amphetamine	**Block norepinephrine synthesis** Metyrosine (Demser) **Interfere with norepinephrine storage** Reserpine Guanethidine (Ismelin) **Prevent norepinephrine release** Bretylium (Bretylol) Guanethidine (Ismelin) **Form false transmitters** Methyldopa (Aldomet)
α receptors	**Stimulate α_1 receptors** Methoxamine (Vasoxyl) Phenylephrine (Neo-Synephrine) **Stimulate α_2 receptors** Clonidine (Catapres)[b]	**Block α receptors** Phenoxybenzamine (Dibenzyline) Phentolamine (Regitine) Prazosin (Minipress) (blocks α_1) Yohimbine (blocks α_2)
β receptors	**Stimulate β receptors** Isoproterenol (Isuprel)	**Block β receptors** Propanolol (Inderal) and others (block β_1 and β_2) Atenolol (Tenormin) and others (block β_1) Butoxamine (blocks β_2)

[a]Only the principal actions are listed. Note that guanethidine is believed to have two principal actions.

[b]Clonidine stimulates α receptors in the periphery, but along with other α_2 agonists that cross the blood-brain barrier, it also stimulates α_2 receptors in the brain that decrease sympathetic output. Therefore, the overall effect is decreased sympathetic discharge.

From W. F. Ganong. *Review of Medical Physiology* (16th ed.). East Norwalk, CT: Appleton & Lange, 1993. P. 207.

beta-2 receptors while norepinephrine has weak beta-1 and essentially no beta-2 activity.

After norepinephrine is released into the synaptic cleft, it is removed by (1) active reuptake into the noradrenergic nerve ending, (2) diffusion, and (3) less importantly, degradation by the enzymes **monoamine oxidase** (MAO) and **catechol O-methyl transferase** (COMT), both of which are found principally in the liver. Catecholamines released into the circulation from the adrenal medulla are inactivated primarily by MAO and COMT metabolism; their half-lives are approximately 10 times greater than that of norepinephrine released by the noradrenergic nerve terminal. Despite the long half-lives of the adrenal catecholamines, norepinephrine released from the nerve terminals is the prime effector of sympathetic tone, providing 75 percent of the circulating catecholamines in the basal state. Adrenal epinephrine (20%) and norepinephrine (5%) make up the remaining circulating catecholamines.

As was true for the parasympathetic actions, sympathetic end-organ effects are a function of the effector cell and not of the receptor type (see Table 8-6). For any sympathomimetic agent, the response depends on the agent's affinity for the various receptors, the amount released or dose administered, and reflex homeostatic responses. For example, norepinephrine-mediated vasoconstriction (alpha-1) may increase total peripheral resistance to the point that the baroreceptor reflex induces a vagally mediated bradycardia that overcomes the noradrenergic-induced tachycardia (beta-1).

Most sympathetic discharge (for example, that concerned with the maintenance of blood pressure) is discrete; however, the SNS is also capable of a mass discharge, the so-called **fight-or-flight response.** This mass activation of the SNS is facilitated by both (1) the extensive divergence, convergence, and overlapping of preganglionic sympathetic neurons and (2) the adrenal release of large amounts of catecholamines. The net effect is to prepare the organism for emergency activity: Heart rate, contractility, blood pressure, and stroke volume are increased (beta-1); minute ventilation increases; circulation to skeletal muscle is increased (muscarinic and beta-2 vasodilation) while that to the splanchnic beds and skin is decreased (alpha vasoconstriction), thereby preserving the blood flow for the most vital organs and limiting surface bleeding. Glycogenolysis in muscle and in liver is accelerated (beta-2); pupillary dilation occurs; the basal metabolic rate is increased (mediated primarily by epinephrine); and the overall state of alertness is increased (by noradrenergic lowering of the threshold of the reticular activating system).

Special Features of Autonomic Neural Transmission

Within the autonomic ganglia extensive neuronal interaction modifies the simple action potential and thereby modulates future responses to neuronal stimulation. The best-studied example of this occurs in the sympathetic ganglia. Here, the usual fast **excitatory postsynaptic potential** (fast EPSP), the action potential, combines with both a prolonged **inhibitory** (slow IPSP) and excitatory (slow EPSP) **postsynaptic potential** to produce a compound or **complex potential** (Fig. 8-12).

The ANS generally functions with less precise temporal control than does the somatic motor system. Skeletal muscle movements may be completed in fractions of a second while smooth muscle contraction or glandular secretion may occur over minutes. This longer autonomic response time is due in part to the contrasting properties of the end effector organs, skeletal and smooth muscle differences, and the process of glandular secretion. Additionally, the structure of the postganglionic neuron contributes to the slower response time. First, postganglionic neurons are unmyelinated. Second, unlike the neuromuscular junction in which there is tight apposition of the

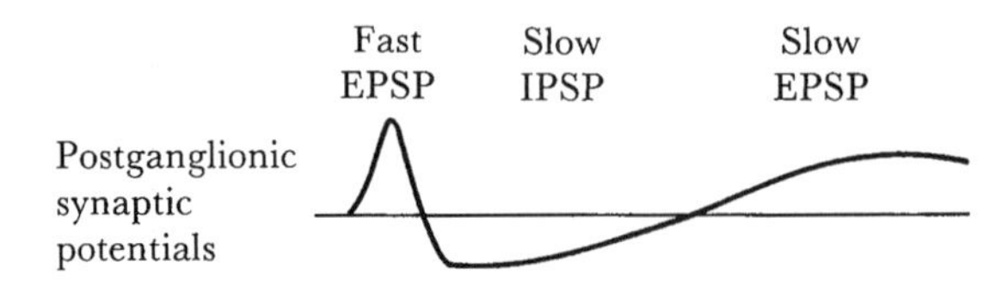

Fig. 8-12. Synaptic potentials in postganglionic sympathetic neurons. The fast EPSP, slow IPSP, and slow EPSP combine to produce a compound or complex potential. (From W. F. Ganong. *Review of Medical Physiology* [11th ed.]. P. 177. © 1985 by Lange Medical Publications, Los Altos, CA.)

presynaptic cell to the postsynaptic cell, postganglionic autonomic nerve terminals end as varicosities or swellings that are relatively distant from the postsynaptic effector cell. Thus, more time is required for the neurotransmitters to diffuse this greater distance, slowing transmission. Such varicosities are more characteristic of adrenergic than cholinergic terminals. This greater distance between cells in the adrenergic junction also helps explain the increased reliance on diffusion for the termination of noradrenergic stimulation.

Interactions of the Parasympathetic and Sympathetic Nervous Systems

The parasympathetic and sympathetic nervous systems are tonically and simultaneously active and generally have reciprocal effects on organs innervated by both divisions of the ANS. Where present, dual innervation allows a mutual antagonism that provides for fine control of body function. There are three basic mechanisms by which SNS and PNS actions oppose one another. First the PNS and SNS may operate at the same site to produce opposite results. This is illustrated at the sinoatrial (SA) node, where beta-1 stimulation (SNS) increases the heart rate while vagal stimulation (PNS) slows the heart rate. These effects are mediated by increasing or decreasing the slope of phase 4 depolarization in pacemaker cells (see Chap. 2). Second, the PNS and SNS may act at different sites to produce physiologically opposing actions. For example, alpha stimulation (SNS) of the radial muscle of the iris causes pupillary dilation while muscarinic (PNS) stimulation of the sphincter or circular muscle of the iris causes pupillary constriction. Here, each muscle is innervated by only one division of the ANS, but fine control of pupil diameter is achieved through the two reciprocal systems. Third, as a general principle for dually innervated organs, synapses of parasympathetic and sympathetic neurons act reciprocally both centrally and peripherally to decrease (or increase) parasympathetic release when sympathetic neurons are stimulated (or inhibited). It should be noted that autonomic regulation of organ function is modulatory, not initiatory; thus, organs deprived of their autonomic innervation (e.g., the heart after being transplanted) will still function adequately, though differently.

Basal rates (i.e., **tone**) of parasympathetic and sympathetic activity allow the body to reach a given state through activation or inhibition of one system without altering the other system. Returning to the example of the SA node, heart rate may be decreased by reducing sympathetic tone or increasing vagal activity, or both. If there were no basal sympathetic discharge, changes in sympathetic release could only result in tachycardia, never bradycardia. This resting tone is crucial to the regulation of organs innervated by only one division of the ANS such as the thermoregulatory sweat glands.

The features of the PNS and the SNS are compared in Table 8-8.

Table 8-8. Comparison of the parasympathetic and sympathetic nervous systems

Feature	Parasympathetic	Sympathetic
Location of central nuclei	Cranial, sacral	Thoracolumbar
Preganaglionic fiber	Long, travels directly to innervated organ	Short, makes multiple synapses along the sympathetic trunk
Neurotransmitter of preganglionic neuron	Acetylcholine (nicotinic receptor)	Acetylcholine (nicotinic receptor)
Postganglionic neuron	Short, often found in target organ	Long, cell body in sympathetic trunk (some cell bodies are in outlying ganglia)
Postganglionic neurotransmitter(s)	Acetylcholine (muscarinic receptor)	Norepinephrine Acetylcholine (muscarinic receptor)
Termination of action	Acetylcholinesterase	Acetylcholine: acetylcholinesterase Catecholamines: re-uptake, diffusion, MAO and COMT
Circulating agonists	None	Adrenal catecholamines and diffused norepinephrine
Mass discharge	None	Fight-or-flight response

MAO = monoamine oxidase; COMT = catechol O-methyl transferase.

The Enteric Nervous System

The enteric nervous system consists of neural plexi within the walls of the esophagus, stomach, and large and small intestines. It has an intrinsic pattern of activity that is important in the control of most functions of the gastrointestinal tract, including motility, secretion, and transport. Neurotransmitters known to be active within these plexi include enkephalins and endorphins, serotonin, vasoactive intestinal peptide, somatostatin, adenosine triphosphate (ATP), and adenosine.

The Special Senses

Vision

The eye contains the same three elements as a camera for recording visual images: a lens system, a means for controlling aperture (the **pupil**), and film (the **retina**). The **cornea** and **lens,** which are avascular and therefore transparent, focus the visual image on the retina by bending or refracting incoming light rays. The change in refractive index at the air-cornea interface contributes most of this focusing power, which amounts to +60 diopters (D). Tears maintain the surface of the cornea to ensure an undistorted image; they contain lysozyme and immunoglobulin A, which may be an indication of an immunologic function as well. **Glaucoma,** an increase in intraocular pressure above the normal level of 15 to 20 mm Hg as measured at the cornea, is due to an impediment in outflow of aqueous humor. It can compromise the integrity of the optic nerve and retina, leading to blindness if untreated.

The Lens and Image Formation

The image that is formed on the retina is inverted and reversed due to the path of the light rays on entering the eye. If the axial length of the eyeball is too long in relation to the refractive power, the image of faraway objects will be focused in front of the retina (**myopia,** or nearsightedness); this can be corrected with a diverging lens placed in front of the eye. If the eyeball is too short, near objects will be focused behind the retina (**hyperopia,** or farsightedness); correction is with a converging lens. The lens contributes only 25 percent of the eye's refractive power in the unaccommodated state; its importance lies in its ability to double its refractive power to +12 D by increasing its curvature and thus providing for near focus. The lens has an inherent tendency to assume a spherical shape because of the elasticity of the lens capsule. **Accommodation** is a reflex mediated by the cortex that allows one to voluntarily focus on a nearby object. In the unaccommodated state, tension of the lens (zonular) ligaments keeps the lens relatively flattened. In accommodation, (1) contraction of the ciliary muscle reduces the tension of the lens ligaments and the lens thickens, increasing refraction; (2) the pupil constricts, which helps to increase depth of focus (the degree to which the image can form anterior or posterior to the retina and still be in focus) and decrease stray light rays; and (3) gaze of each eye shifts medially **(convergence)** in order to keep the object focused on both **foveas.** Because accommodation is an active process that requires muscular effort, it can be tiring. **Presbyopia** is a loss in refractive power of the lens due to its decreased elasticity with age, until the lens is almost nonaccommodating. **Astigmatism** is a refractive problem due to an unevenly shaped cornea or lens in which the refractive power varies in different axes, causing distorted images. It is correctable with cylindrical lenses, which add refractive power in only one axis and allow the lens system to behave as if it were spherical.

Photoreceptor Cells and the Retina

The **rod** and **cone** photoreceptor cells of the retina contain photosensitive compounds stored in the membrane disks in the outer segments. These pigments decompose on exposure to light to produce a change in ion permeability and a corresponding change in potential in the receptor cell. Rods contain the pigment **rhodopsin,** composed of an **opsin** (a large protein molecule) and a **chromatophore** molecule, retinene$_1$ (11-cis-retinal, an aldehyde of vitamin A_1). The three cone pigments also contain retinene$_1$, but each has a different opsin, which confers the specific light sensitivities. Incoming light is absorbed by the photoreceptors and isomerizes the retinene$_1$, which dissociates from the opsin. Metarhodopsin II is formed and reacts with a G-protein (transducin in rods, G_{t2} in cones) to activate phosphodiesterase to decrease cyclic guanosine monophosphate (cGMP) levels. The cell becomes hyperpolarized by the closure of sodium channels due to the decreased cGMP levels. This information is transmitted as a local potential to the synaptic end of the receptor cell and decreases the rate of transmitter release, which hyperpolarizes horizontal cells and either depolarizes or hyperpolarizes bipolar cells. These responses are also local graded potentials. A sufficiently high receptor potential will elicit action potentials in the ganglion cells, the axons of which make up the optic nerve. The neural arrangement of the retina is shown in Fig. 8-13. Metarhodopsin II breaks down to the aldehyde and opsin, and photopigment is regenerated. A chronic deficiency of vitamin A_1 can lead to **night blindness** through insufficient rhodopsin regeneration; eventually anatomic changes in the rods and cones and subsequent retinal degeneration can occur. **Retinitis pigmentosa** is a congenital defect in which defective pigment is not removed normally by phagocytosis and instead accumulates in the retina, leading to blindness.

Rods absorb light from the entire visual spectrum and are responsible for black and white vision. The three types of cone pigments absorb light over most of the visible spectrum; however, each type of cone has a maximum absorption at a particular wavelength. Red cones absorb maximally at 570 nm, green ones at 535 nm, and blue cones at 445 nm. A given color is encoded by varying combinations of these three component colors. The eye is able to distinguish wavelength changes of just 1 to 2 nm in the visible spectrum. **Color blindness** is due to an inherited inability to produce one or more of the cone photopigments.

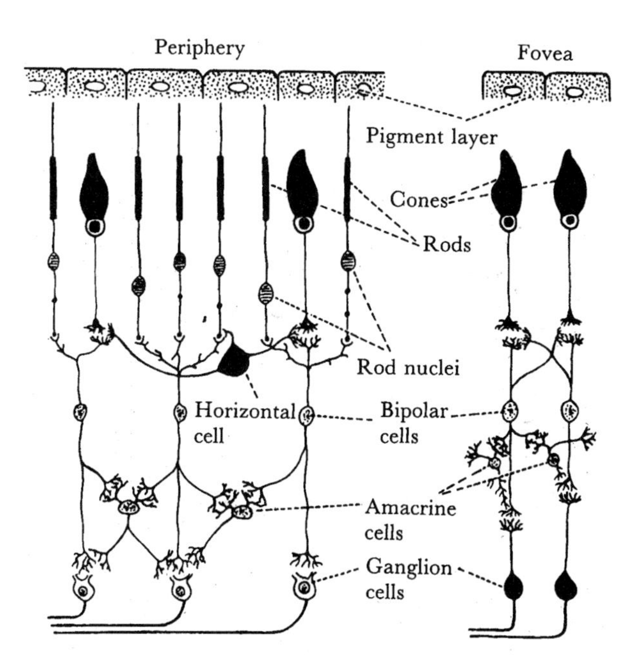

Fig. 8-13. Neural arrangement of the retina. (From A. C. Guyton. *Textbook of Medical Physiology* [7th ed.]. Philadelphia: Saunders, 1985. P. 720.)

Rods adapt more completely to dark than do cones, albeit more slowly. Adaptation to variations in light intensity involves (1) changes in pupillary size, (2) a reestablishment of the equilibrium between breakdown and synthesis of photopigment available, (3) a retinal feedback mechanism that can change the number of receptors available to be stimulated (i.e., increases in "gain" in dim light), and (4) prolongation of stimulus duration in order to summate available light.

There is a high degree of **convergence** of rod receptor cells onto **ganglion cells:** A photon hitting any one of many rods will activate a ganglion cell, which accounts for the low acuity (exactness of representation) and high sensitivity (likelihood of activation) of the rod system. A photon must strike one of only a few cones to activate a specific ganglion cell because of the low degree of convergence in the cone system; thus, sensitivity is low and acuity high. **Visual acuity** is greatest at the **fovea,** which has no rods, because convergence and the number of structures overlying the photoreceptors are both minimal. At the **optic disk,** there are no receptor cells; thus, there is a corresponding blind spot in the visual field.

Receptive Fields and Visual Pathways

The receptive field of a ganglion cell consists of all the photoreceptor cells that connect (via **bipolar cells**) to that ganglion cell. The ganglion cells have circular receptive fields functionally divided into a **center** and a **surround.** The center is termed *on* if light shining on the center increases the rate of action potentials and *off* if light decreases the frequency of action potentials. The surround responds oppositely to light than the center, so that "on" centers have "off" surrounds and vice versa. The inhibition of the center response by the surround is probably due to inhibitory feedback from one receptor to another mediated by the **horizontal cells** (which connect receptor cells). This is an example of **lateral inhibition** enhancing the organism's ability to detect contrast. Ganglion and bipolar cells respond best to a difference in illumination between the center and the surround. Higher-order color cells also have center-surround receptive fields, but instead of being on-off, they are blue-yellow or red-green. The basic principle is the same.

Axons from the ganglion cells project a detailed spatial representation of the retina via the optic nerves. At the **optic chiasm,** the fibers from each nasal hemiretina cross to join the opposite optic tract, which travels to the **lateral geniculate body** (LGB) in the hypothalamus. The LGB in turn projects a similar point-for-point representation via the geniculocalcarine tract **(optic radiation)** to the primary visual cortex (Brodmann's area 17) in the occipital lobe. This is where most processing and visual perception occur. Areas 18 and 19 of the cortex are visual association areas, which respond to more complicated patterns and designs.

Branches of ganglion cell axons pass from the optic tract to the **pretectal region** of the midbrain and to the **superior colliculus,** where they form connections that mediate visual reflexes. The **pupillary reflex** is initiated by a sudden change in light intensity. Impulses from the pretectal region pass to the oculomotor (Edinger-Westphal) nucleus, from which parasympathetic fibers travel in the usual fashion (see Fig. 8-10) to the constrictor muscle of the iris and cause miosis. Due to connections across the midline in the pretectum, it is a consensual reflex, in that both pupils react even if the stimulus reaches only one retina. If the pupillary response to light is absent but accommodation is intact (Argyll-Robertson pupil), there is a lesion in the tectal region. The superior colliculus serves in reflex control and regulation of many head and eye movements (e.g., conjugate eye movements, with the horizontal and vertical gaze centers in the brainstem RF; saccades) and certain aspects of vision (e.g., attention). Other axons pass directly from the optic chiasm to the suprachiasmatic nuclei in the hypothalamus, forming connections that synchronize a variety of endocrine functions and circadian rhythms.

Visual signals also reach the cerebellum, which helps integrate vertical and horizontal motions of visual targets and their surroundings with smooth eye and head movements. In searching a given field (e.g., reading), the eye performs small rapid jerky saccadic movements. To follow an object, the eye makes slow smooth pursuit movements that are followed by rapid movements in the opposite direction (known as **nystagmus**). Pathologic nystagmus can result from damage to the cerebellum or vestibular system.

The receptive fields of both LGB neurons and the cortical pyramidal cells resemble the center-surround receptive fields of ganglion cells. In the cortex, higher-order cells respond to line segments rather than single points, are orientation specific, and receive binocular input. These higher-order cells consist of a hierarchy of **simple, complex,** and **hypercomplex** cells. The receptive field of simple cells is an orientation- and position-specific line segment or bar. Presumably, each simple cell receives its input from a group of LGB-like cells whose receptive fields are arranged in a row as illustrated in Fig. 8-14. Complex cells are orientation specific but are less dependent on position and respond best to a moving linear stimulus. Their receptive fields are believed to be produced by input from simple cells that share the same orientation specificity but differ in position specificity. Hypercomplex cells resemble complex cells but are specific for certain lengths of stimuli and their edges; they integrate input from complex cells and are more prominent in areas 18 and 19 than in the primary visual cortex. The columnar organization of the cortex facilitates this structure. The cortex is arranged in orientation-specific columns such that cells in any given column respond maximally to stimuli of a specific orientation.

These columns can be further subdivided into ocular dominance columns. **Binocular input** is first present at the level of complex cells and is important for **stereoscopic** (3-dimensional) vision. The image must fall on corresponding points of both retinas if it is to be seen as a single object with both eyes. **Depth perception** is based

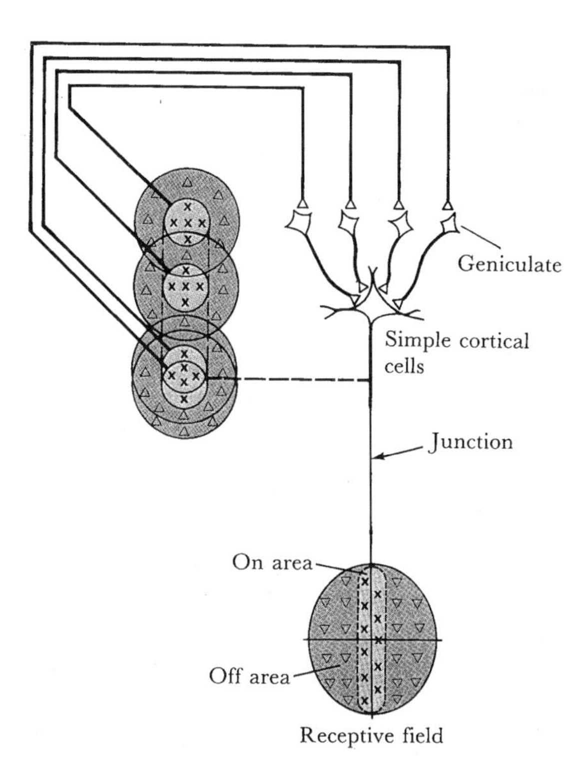

Fig. 8-14. Receptive fields of simple cells. By combining the input from a set of "aligned" LGB-like cells, each with circular–center surround patterns, the receptive field of the simple cell is produced. The simple cell is most responsive to orientation-specific line segments. (Reprinted by permission of the publisher from *Principles of Neural Science* [2nd ed.], by E. K. Kandel and G. H. Schwartz. P. 374, 1985, by Elsevier Science Publishing Company, Inc.)

in part on a horizontal disparity between two hemiretinas (e.g., left retinal and right temporal) viewing the same visual field and is thus somewhat limited to binocular vision. **Diplopia** results when an image on the retina of one eye no longer falls on the corresponding point of the other eye.

Higher-order processing of color vision is less well understood, but appears to involve cell clusters (called **blobs**) in layers 2 and 3 of the visual cortex. Ganglion cells in the retina subtract input from one type of cone from the input of another type of cone, and this information is relayed to layer 4 of the visual cortex and on to the blobs in layers 2 and 3. From here color vision information travels to the association areas for further processing.

The Auditory System

The human ear is able to detect sound frequencies ranging from 20 to 20,000 Hz over a 140-dB range of sound intensity, though **impedance matching** by the middle ear is most efficient for sounds in the 1000-to-5000-Hz range. Sounds above 140 dB are painful and potentially damaging to the inner ear. Sound waves are transmitted from the air to the **tympanic membrane** and **auditory ossicles** in the middle ear, an air-filled cavity in the petrous portion of the temporal bone. The lever action of the ossicles and the large size of the tympanic membrane compared to the **oval window** provide the 22-fold amplification of sound pressure at the oval window necessary to transfer the vibrations across the air-fluid interface to the fluid-filled **cochlea** in the inner ear (**labyrinth**).

The Cochlea

The cochlea is the sense organ for hearing. The motion of the end plate of the **stapes** on the oval window establishes a wave-like motion in the perilymph of the **scala vestibuli** in the cochlea, which sets up a traveling wave along the **basilar membrane** in the **scala media.** For any given sound frequency, there is a point along the membrane where displacement (i.e., amplitude of the traveling wave) is maximal. This **tonotopic organization** of the basilar membrane is the principal means of pitch determination. The organization is quite regular, with high-pitched sounds generating maximum vibration at the taut narrow base of the membrane near the oval window and low-frequency sounds generating maximum vibration at the loose, broad apex. The wave travels through the **scala tympani** to the **round window,** which bulges out into the middle ear and dissipates the wave (Fig. 8-15).

The Organ of Corti

The **organ of Corti** runs the length of the basilar membrane and is affected by movements of the membrane. The **stereocilia** of the **hair cells** are embedded in the **tectorial membrane,** which moves independently of the basilar membrane because of its different point of attachment. The relative motion of the basilar membrane with respect to the tectorial membrane produces a shearing action that bends the stereocilia (Fig. 8-16). This causes a change in ion permeability (potassium and calcium influx), generating a depolarization in the hair cell and causing an excitatory synaptic trans-

Fig. 8-15. The components of the cochlea, shown as it would appear if uncoiled (upper diagram) and shown in cross section (lower diagram). (From M. B. Wang. Neurophysiology. In J. Bullock, J. Boyle, and M. B. Wang. *Physiology* [2nd ed.]. Baltimore: Williams & Wilkins, 1991.)

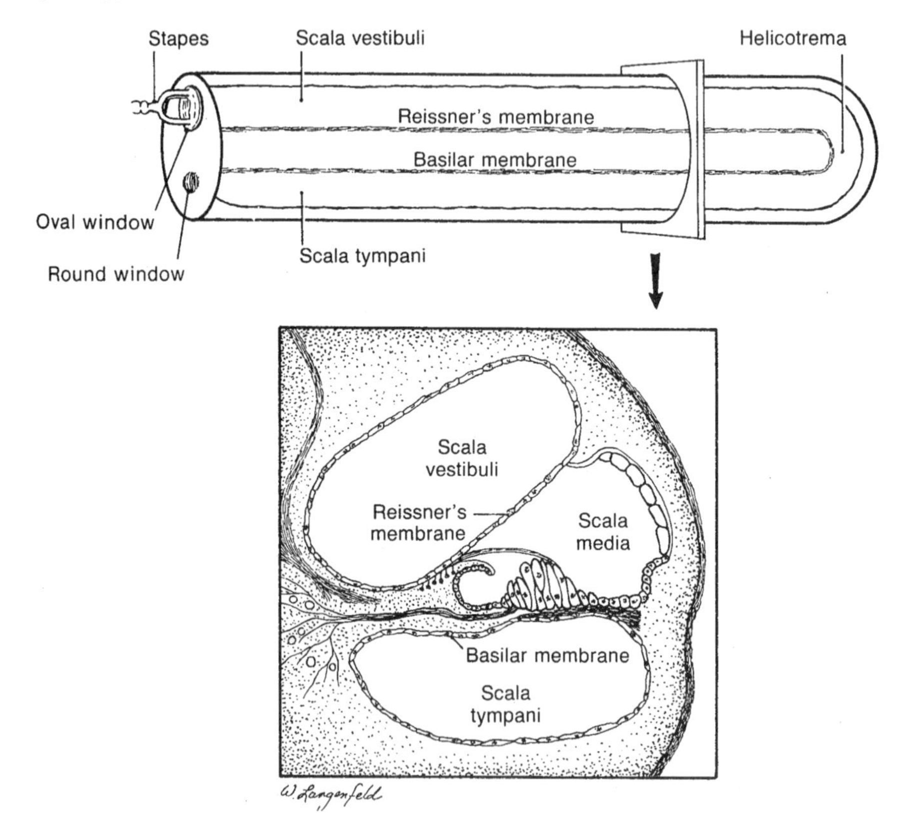

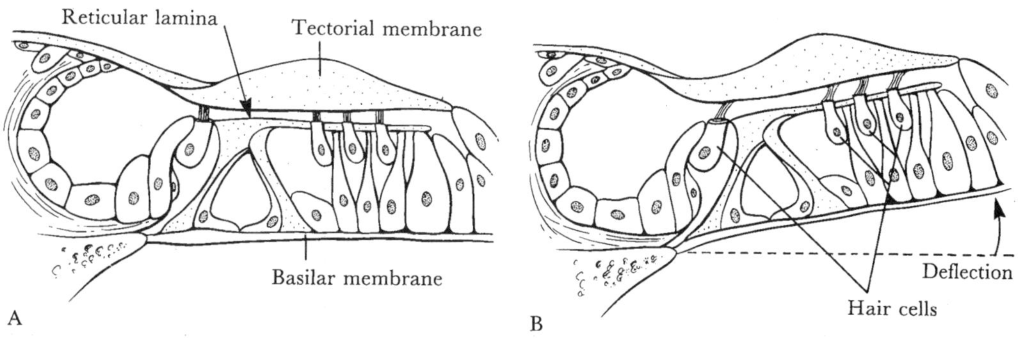

Fig. 8-16. Effect of movement of the basilar membrane on the organ of Corti. A. The organ of Corti at rest, illustrating the different points of attachment of the basilar membrane and the tectorial membrane. B. Deflection of the basilar membrane produces a shearing force across the microvilli because of the relative movement of the basilar membrane with respect to the tectorial membrane. This results in bending of the microvilli. Note, however, that the body of the hair cells remains in a fixed position relative to the basilar membrane. (Reprinted by permission of the publisher from *Principles of Neural Science* [2nd ed.], by E. K. Kandel and G. H. Schwartz. P. 374, 1985, by Elsevier Science Publishing Company, Inc.)

mitter to be released (possibly glutamate). As with other systems, stimulus intensity (loudness) is encoded by frequency of action potentials and total number of receptors stimulated.

Neural Pathways

The potential change in the hair cell leads to activity in the afferent fibers of the **cochlear division of cranial nerve VIII,** which innervate the bases of the hair cells. Ninety-five percent of the fibers go to the single row of inner hair cells, with most fibers innervating only one cell (low convergence). Each nerve fiber is most sensitive to the characteristic frequency of the hair cell that it innervates, and nerve fibers are thus **tonotopically organized.** This tonotopic arrangement continues through the higher centers of the auditory pathway: the **cochlear nuclei** in the medulla, the **superior olivary complex,** the **inferior colliculus** (via the lateral lemniscus, where many fibers cross the midline), the **medial geniculate body** in the thalamus, and the **auditory cortex** (Brodmann's area 41 in the superior temporal lobe) via the **auditory radiation.** The primary auditory cortex is concerned with recognition and analysis of sound patterns and properties, as well as sound localization. Because all centers at levels higher than the olivary complex receive bilateral input, lesions of the central auditory pathways do not give rise to mononeural deficits. More importantly, binaural input enables the auditory centers to determine the direction from which a sound emanates. Two mechanisms are used: (1) difference in the intensity of the sound between the two ears and (2) a lag in the time of arrival of the sound at the two ears (a phase delay). The hair cells receive prominent efferent innervation from the superior olivary nucleus via the **olivocochlear bundle.** These inhibitory efferents to the cochlea probably improve attention to specific sounds by helping screen out distracting "noise."

Reflexes

The ear has two small muscles that can contract reflexly with loud sounds. The **tensor tympani** (cranial nerve V) pulls the manubrium of the **malleus** medially to lessen vibration of the eardrum, and the **stapedius muscle** (cranial nerve VII) pulls the foot plate of the **stapes** out of the oval window. Contraction dampens the movement of the ossicles, which decreases the pressure of sounds reaching the inner ear, thus attenuating sensitivity of the acoustic apparatus. This action may also help to decrease the noises produced by one's own speech.

Deafness

Presbycusis is the decreasing ability to hear high-frequency sounds with age. Deafness can result from impaired sound transmission in the external or middle ear **(conduction deafness),** or from damage to the hair cells or neural pathways **(nerve deafness).** Conduction deafness can result from chronic otitis media (which causes thickening and rigidity of the tympanic membrane), destruction or rigidity of the ossicles **(otosclerosis),** or obstruction of the external ear canal. Nerve deafness can be caused by prolonged exposure to loud noise, aminoglycoside antibiotics (hair cell degeneration), or damage to the vestibulocochlear nerve or its connections (tumors, vascular accidents). There are simple tests that can be performed to see if hearing loss is due to conduction or nerve deafness (Table 8-9).

The Vestibular System

The vestibular apparatus consists of the three **semicircular canals** and the two **otolith organs** (the **utricle** and **saccule**) and is located in the membranous labyrinth. It is surrounded by perilymph and contains endolymph. The semicircular canals detect **angular (rotational) acceleration,** while the utricle and saccule detect **linear acceleration** (e.g., gravity). The hair cell is the receptor cell common to these systems. Bipolar afferent neurons from the vestibular division of the eighth cranial nerve synapse at the base and sides of each hair cell. Mechanical forces induce bending of the hair cells, which have 60 to 100 stereocilia and one tall **kinocilium** protruding from the surface. Bending of the stereocilia toward the tall kinocilium depolarizes the hair cell via K^+ influx, which causes release of a neurotransmitter and increases the rate of nerve impulses in the vestibular nerves. Bending away from the kinocilium causes decreased K^+ influx with resultant hyperpolarization. Along the axis toward the kinocilium, the hairs progressively increase in height; along the axis perpendicular to this, the hair cells are all the same height. Bending of the hair cells in the perpendicular axis causes no change in membrane potential, whereas bending of the hair cells in an intermediate direction (between the two axes) causes changes in polarization that are proportional to the direction toward or away from the kinocilium.

Table 8-9. Common tests with a tuning fork to distinguish between nerve and conduction deafness

	Weber	Rinne	Schwabach
Method	Base of vibrating tuning fork placed on vertex of skull	Base of vibrating tuning fork placed on mastoid process until subject no longer hears it; then held in air next to ear	Bone conduction of patient compared with that of normal subject
Normal	Hears equally on both sides	Hears vibration in air after bone conduction is over	
Conduction deafness (one ear)	Sound louder in diseased ear because masking effect of environmental noise is absent on diseased side	Vibrations in air not heard after bone conduction is over	Bone conduction better than normal (conduction defect excludes masking noise)
Nerve deafness (one ear)	Sound louder in normal ear	Vibration heard in air after bone conduction is over, as long as nerve deafness is partial	Bone conduction worse than normal

From W. F. Ganong. *Review of Medical Physiology* [16th ed.]. East Norwalk, CT: Appleton & Lange, 1993. P. 163.

Semicircular Canals and Cupulae

The semicircular canals (horizontal, anterior, and posterior) are arranged perpendicularly to one another and thus can detect motion in all three planes. The functional anatomy of the semicircular canals is shown in Fig. 8-17. The expanded end of each canal **(ampulla)** has a receptor structure called the **crista ampullaris,** which rises from the floor to the ceiling of the ampulla, thus closing it off and ensuring that the crista will be pushed by movement of endolymph in the canal. Each crista has hair cells embedded in a gelatinous membrane (the **cupula**) similar to the cochlear tectorial membrane in function. When the head begins to rotate, the semicircular canals turn, but the endolymph remains stationary because of its inertia. This creates a relative fluid flow in the canals, which pushes on the cupula and deforms it, bending the hair cells. With continued angular acceleration, the inertia of the endolymph is overcome and the endolymph moves at the same speed as the semicircular canal. There is no relative motion, the cupula returns to its resting position, and impulse rates return to the basal level. When rotation stops, the opposite situation arises: The semicircular canals are stationary but the endolymph is still moving. This relative motion bends

Fig. 8-17. Functional anatomy of the semicircular canals. (Modified from A. C. Guyton. *Textbook of Medical Physiology* [7th ed.]. Philadelphia: Saunders, 1985. P. 621.)

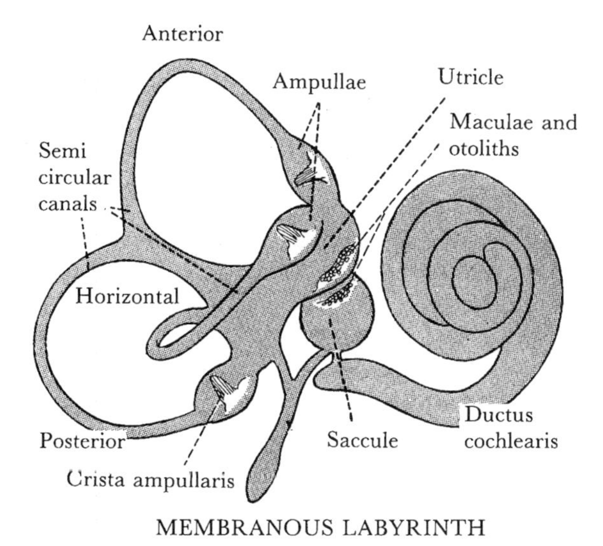

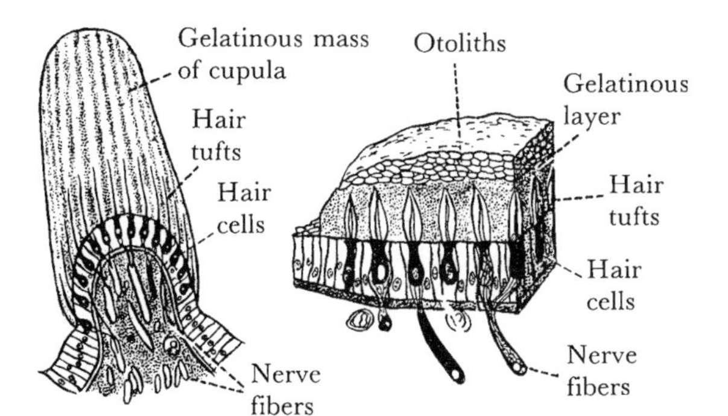

Neuro

the cupula and therefore the hairs in the opposite direction, and the impulse rate is appropriately altered. Rotation causes maximal stimulation of the semicircular canal most nearly in the plane of rotation. The canals are mirror images on opposite sides of the head, and therefore work in directly opposite fashions during rotation (i.e., left side stimulated while right side is inhibited). Linear acceleration fails to displace the cupula and does not stimulate the cristae.

Utricle, Saccule, and Otolithic Organs

On the floor of the utricle is a **macula,** or otolithic organ. Another macula is located on the wall of the saccule in a semivertical position. They respond to linear acceleration, including gravity, in a manner essentially equivalent to that of the cupula, with the utricle responding to horizontal acceleration and the saccule to vertical changes. The hair cells of the utricle and saccule are located in the macula. They are again regularly ordered with respect to the kinocilium, but the organization is planar, not linear. The functional equivalent of the cupulae are the **otoliths,** with a higher specific gravity than neighboring tissues and fluids, and which therefore exhibit inertial delay during linear acceleration. The hair cells are capped by the gelatinous **otolithic membrane** in which are embedded crystals of calcium carbonate, the otoliths. Because of their mass, they move to the lowest point in the utricle or saccule, in the opposite direction from linear motion, thereby stimulating the hair cells. Since the hair cells are arranged in a variety of directions, a single structure can respond to tilt or to linear acceleration in any of several directions. The maculae signal a (held) static position.

Neural Pathways

Afferent impulses travel in the vestibular nerves from the semicircular canals, utricles, and saccules to the **flocculonodular lobe** of the cerebellum and to the four **vestibular nuclei** in the medulla. The patterns of these nerve impulses are analyzed by the vestibular nuclei and cerebellum to provide information about the position of the head in space essential for the coordination of motor responses, eye movement, posture, and equilibrium. From the vestibular nuclei, information also travels to the oculomotor nuclei via the medial longitudinal fasciculus (for **conjugate eye movements** mediated by the **vestibuloocular reflex**), to the reticular formation in the brainstem, and to the thalamus and cortex for mediation of conscious sensation of motion necessary for orientation in space. Descending pathways to the spinal nerves (vestibulospinal tracts) exert facilitory influences on extensor tone in postural trunk and neck muscles via alpha and gamma motor neurons. Efferent paths from the vestibular nuclei and reticular formation to the maculae and cristae, via the vestibular nerve, exert excitatory influences and ameliorate effects of **motion sickness** and its associated nystagmus. Motion sickness is due to excessive stimulation of the maculae. **Ménière's disease** probably results from abnormal circulation of vestibular endolymph, and symptoms include **vertigo** (the sensation of rotation in absence of same), **tinnitus** (ringing in the ears), nausea, and vomiting. Malfunction of the vestibular system can be compensated for by visual, proprioceptive, and cutaneous (touch and pressure) receptor input.

Reflexes

Nystagmus is a reflex that maintains visual fixation on a stationary object while the body rotates. It is not related to vision and is still present in people who are blind. When rotation starts, the eyes move slowly in the direction opposite the rotation, maintaining visual fixation (vestibuloocular reflex). When the limit of eye movement is reached, the eyes jump back to a new fixation point **(fast component).** The **slow component** is initiated by impulses from the labyrinths, and the fast component is triggered by a center in the brainstem. The direction of eye movement in nystagmus is by the direction of the fast component (which is the same as the direction of rotation).

The Chemical Senses

Taste

Both taste and smell receptors are chemoreceptors that are stimulated by molecules in solutions in the fluids of the mouth and nose to produce receptor potentials that then lead to action potentials. The flavor of a food is due to a combination of taste and smell. The four primary tastes, from which all taste sensations are derived, are sour (due to hydrogen ion concentration), salt (anions of ionizable salts), sweet (organic compounds), and bitter (alkaloids). **Taste buds** are the sense organs for taste and are located in the mucosa of the tongue, palate, and pharynx. There may be special binding proteins in the mucosa of the tongue (as well as the nasal mucosa) that help to concentrate and transport substances into the chemoreceptors of the mouth. The pathway of taste sensation resembles that of somatic sensation for the tongue, traveling in the chorda tympani (anterior two-thirds of the tongue), glossopharyngeal nerve (posterior third of the tongue), and vagus (pharynx). Impulses are transmitted to the **nucleus of the tractus solitarius** in the medulla and, via the **medial lemniscus,** are relayed to the **VPM nucleus** of the thalamus. The taste projection area in the cortex is near the cutaneous sensory area for the face in the postcentral gyrus. Many drugs, especially those with sulfhydryl groups (captopril, penicillamine), can cause temporary loss or alteration of taste.

Olfaction

Humans can distinguish between 2000 and 4000 odors. The basis for olfactory discrimination is not known. Intensity discrimination for both smell and taste is relatively crude and adaptation relatively rapid. To be smelled, a substance must be volatile, and oil and water soluble. **Olfactory receptors** are neurons located in the olfactory mucous membrane. At the dendritic ends are thickenings called **olfactory rods** from which cilia protrude into the mucosa. Axons pierce the cribriform plate of the ethmoid bone to enter the **olfactory bulbs.** Impulses travel via the olfactory striae to the primary olfactory cortex near the uncus. Epileptic seizures that begin near the uncus may start with an illusion of an unpleasant smell. A unique feature of the olfactory pathway is that there is no thalamic relay in transmission of olfactory input to the cortex. Endings of trigeminal pain fibers are found in the nasal mucosa and may be responsible for reflex sneezing elicited by nasal irritants.

Neural Centers That Regulate Visceral Function

Neural regulation of the circulatory, respiratory, renal, endocrine, and gastrointestinal systems is addressed in the corresponding chapters of the text. This section discusses regulation of other visceral functions, namely, temperature, hunger, and sexual activity.

Temperature Regulation

In the body, heat is produced by muscular activity, assimilation of food, and processes that contribute to the basal metabolic rate. Heat is lost from the body by radiation, conduction, convection, and evaporation of water in the respiratory passages and on the skin. Temperature regulation is controlled by the hypothalamus. Information from the temperature-sensitive cells in the anterior hypothalamus is compared with the **hypothalamic set point** (normally 37°C), and the appropriate response is instituted. If the hypothalamic temperature is greater than 37°C, sweating and cutaneous vasodilation are activated. If the temperature is less than the set point, cutaneous vasoconstriction, shivering, and increased secretion of catecholamines are instituted. Fever is produced when the hypothalamic set point is raised above 37°C. Tissue breakdown products and bacterial toxins can raise the set point. Polymorphonuclear leukocytes and macrophages are believed to act on bacteria and bacterial breakdown products to produce a substance known as **endogenous pyrogen.** Endogenous pyrogen is believed to cause fever by inducing formation of **prostaglandin E_1,** which then acts on the hypothalamus to raise the set point. The temperature receptors then signal that the actual temperature is less than the new set point, and temperature-raising mechanisms are initiated. Despite the fever, the person subjectively feels cold because body temperature is less than the set point. When the set point is returned to normal, the situation is reversed and the fever declines.

Hunger

Regulation of food intake depends primarily on the interaction of two hypothalamic centers, a **feeding center** in the lateral nuclei and a **satiety center** in the ventromedial nuclei. Stimulation of the feeding center causes an animal to eat, while stimulation of the satiety center causes cessation of eating. Lesions of the feeding center cause **anorexia,** and lesions of the satiety center cause **hyperphagia.** The satiety center inhibits the feeding center when food intake is sufficient for the organism's nutritional status. Nutritional status is probably determined by the level of glucose utilization of cells within the satiety center rather than the level of circulating blood glucose. The amount of adipose tissue in the body is probably important for long-term regulation of feeding.

Sexual Activity

Sexual activity in humans is a very complex phenomenon. Reflexes integrated in spinal and lower brainstem centers make up the act of copulation. Parasympathetic fibers mediate lubrication and erection; sympathetic fibers mediate ejaculation and emission. The limbic system and hypothalamus help regulate more complex behavioral components as does learned behavior. The sex hormones also influence sexual behavior in complex ways.

Circadian Rhythms

The source of the circadian rhythms in the brain appears to be the **suprachiasmatic nucleus** (SN) of the hypothalamus, which receives projections from the retina. The SN exerts control over eating and drinking, sleep and wakefulness, secretion of melatonin and cortisol, and level of activity, all of which exhibit circadian rhythms.

Higher Functions

Electrical Activity of the Brain

The **electroencephalogram** (EEG) is a recording of the surface potentials of cortical cells analogous to the recording of the surface potentials of cardiac cells recorded by the electrocardiogram. It is clinically useful because normal patterns are altered over

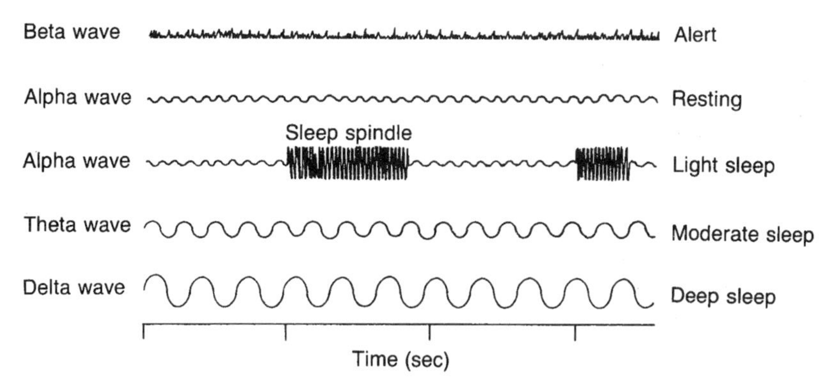

Fig. 8-18. As an individual passes from wakefulness to deep sleep, the EEG wave increases in amplitude and decreases in frequency. Sleep spindles indicate the presence of light sleep. (From M. B. Wang. Neurophysiology. In J. Bullock, J. Boyle, and M. B. Wang. *Physiology* [2nd ed.]. Baltimore: Williams & Wilkins, 1991.)

brain areas that are diseased, damaged, or injured. **Beta waves** (18–30 Hz) predominate in the awake adult. **Alpha waves** (8–12 Hz) are seen at rest with the eyes closed. **Theta** (4–7 Hz) and **delta** (less than 4 Hz) are seen in deep sleep (Fig. 8-18). **Evoked cortical potentials** are recordings of repeated summations of the EEG changes seen with a specific peripheral stimulus (i.e., auditory, visual, or somatosensory). A complex wave pattern characteristic of the nerve being stimulated is generated over the cortical area activated, and is used to trace the integrity of the sensory pathway from the periphery to its central termination.

The Reticular Activating System, Wakefulness, and Sleep

The **reticular activating system** (RAS) is a complex, diffuse network of neurons located in the **reticular formation** (RF) of the brainstem. It contains many serotonergic, noradrenergic, and adrenergic cell bodies and fibers. The RAS is responsible for producing the alert, awake state. Consciousness is a result of the interplay between three neuronal systems, all part of the RF, one causing arousal and the other two causing sleep by opposing the tonic activity of the RAS. One of the **sleep centers** is in the **midline raphe nuclei area** of the medulla (serotonergic); the other is the rapid eye movement (REM) sleep center in the **locus ceruleus area** of the pons (noradrenergic). Serotonergic agonists make one drowsy. Sleep is an active process and not merely the absence of wakefulness. Stimulation of the RAS in a sleeping animal causes the animal to awaken suddenly and results in generalized activation of the cerebral cortex. The RAS receives inputs from all sensory modalities directly or through collaterals as well as retrograde from the cerebrum. Its outputs are to all portions of the cortex either directly or through thalamic relays. Descending RF fibers inhibit transmission in sensory paths in the spinal cord. Various reticular areas are concerned with spasticity and adjustment of stretch reflexes. Because of the complex and diffuse connections within the RAS, most RAS neurons are equally activated by various afferent stimuli; they are not modality specific. In turn, activation of the RAS leads to generalized activation of the cortex. However, there is evidence for some selective cortical activation via the thalamic relays. Presumably this plays a role in attention. Permanent damage to the RAS produces coma while temporary inhibition produces sleep. General anesthetics may produce unconsciousness by depressing conduction in the RAS (via activation of $GABA_A$ [gamma-aminobutyric acid] receptors with resultant hyperpolarization of neurons). Unfortunately, sleep and anesthesia have not yet been explained neurochemically. **Brain death** occurs when the brain can no longer achieve consciousness. A strict set of criteria, used to define brain death, includes an electrically silent EEG, loss of all brainstem regulatory systems (blood pressure, respiration), and traumatic or ischemic anoxia (as opposed to toxic or metabolic causes, which may be reversible). These criteria must be present for at least 6 to 12 hours for a clinical diagnosis of brain death.

There are two types of sleep: (1) **slow-wave** or **non-rapid-eye-movement (NREM) sleep** and (2) **paradoxical** or **REM sleep.** Slow-wave sleep is deep restful sleep characterized physiologically by a decreased heart rate, blood pressure, respiratory rate, and basal metabolic rate. It is divided into four stages based on EEG patterns. Brain waves in the early stages of slow-wave sleep are high frequency and low amplitude (primarily alpha), while those in the last stages are low frequency, high amplitude, and synchronized (theta and delta waves). In contrast, REM sleep is characterized by irregular high-frequency beta waves similar to those recorded during alert wakefulness. People awakened from REM sleep often report that they were dreaming. Although some muscular twitches persist in REM sleep (e.g., rapid eye movements), muscle tone in REM sleep is decreased compared to NREM sleep. This pronounced hypotonia is the result of active spinal inhibition by centers located in the mid pons. Additionally, the threshold for sensory arousal is increased in REM sleep compared to NREM sleep. Thus, REM sleep appears to be a time of cortical activity disengaged from sensory input and motor output. Typically, REM sleep comprises 25 percent of the total sleep time, occurring approximately every 90 minutes, with about five REM periods each night. In the initial cycles, slow-wave sleep is more prominent, but as the night progresses, there is less NREM stage 3 and 4 and more REM sleep. **Narcolepsy** is a sleep disorder in which sudden onset of REM sleep overcomes a person. Barbiturate therapy decreases REM sleep. Animals deprived of REM sleep for long periods lose weight in spite of adequate caloric intake and eventually die. The amount of sleep necessary decreases throughout life, from around 16 hours per day as an infant to less than 6 hours per day as an elderly adult.

The Limbic System and Hypothalamus

The **limbic system** (amygdala, hippocampus, and septal nuclei) and **hypothalamus** are closely involved with the production of emotions, drives, and affective aspects of sensation, for example, pleasantness or unpleasantness. Punishment **(avoidance)** and reward **(approach)** centers located here are believed to play a role in learning and in providing the motivation for the performance or avoidance of certain behaviors. Stimulation of the punishment center can produce avoidance of the stimulus and a rage response, while stimulation of the reward centers produces placidity and tameness. A variety of habit-forming substances may act by increasing dopaminergic activity in the reward system (cocaine, nicotine, amphetamines). The limbic cortex appears to function as an association area for the control of behavior. Phylogenetically, it is among the oldest parts of the cortex and has only five layers, compared to the six-layer neocortex. Neocortical activity does modify emotional behavior though there are few connections from the limbic system to the neocortex. Limbic circuits have prolonged afterdischarge, which may explain why emotional responses often outlast the duration of the stimulus. Lesions of the amygdala produce **hyperphagia** and the **Klüver-Bucy syndrome** (Table 8-10); stimulation of the amygdala produces feeding behavior. Additionally, the hypothalamus plays a major role in the control of **vegetative functions,** as discussed earlier in this chapter.

Language and Speech

In humans, language functions depend more on one cerebral hemisphere (the **categorical** or **dominant hemisphere**) than the other (the **representational** or **nondom-**

Table 8-10. Selected lesions affecting higher function

Lesion site	Deficit
Wernicke's area	Fluent aphasia (also called receptive or sensory aphasia) Comprehension impaired Fluent but meaningless speech
Broca's area	Nonfluent aphasia (also called expressive or motor aphasia) Comprehension intact Very limited speech, marked difficulty forming words
Hippocampi	Anterograde amnesia (unable to form new memories)
Amygdala (temporal lobes)	Klüver-Bucy syndrome (hyperoral, hypersexuality, passivity, psychic blindness)
Frontal lobe	Increased distractibility Decreased moral and social inhibitions, increase in inappropriate behavior
Parietal lobe	Unilateral inattention and neglect
RAS	Coma

inant hemisphere). In right-handed individuals the left hemisphere is dominant in 96 percent, while in left-handed individuals it is dominant in 70 percent. Neuroanatomic studies reveal that areas specialized for language function are larger in the categorical hemisphere. If the categorical hemisphere is damaged, language functions can be developed in the remaining hemisphere up until the early teen years.

The **aphasias** that result from lesions in the categorical hemisphere (see Table 8-10) have helped researchers discover the way in which language is processed. **Wernicke's area** receives input from the auditory and visual cortex and interprets this sensory input as meaningful language. This information is then projected to **Broca's area** in the categorical hemisphere via the **arcuate fasciculus.** Broca's area processes the information and, through its output, coordinates the activity of the motor cortex for the production of meaningful sounds. The cerebellum is important in the production of fluent speech as well. The circuitry for processing written language appears to be closely related to, and in part superimposed on, this system. Wernicke's and Broca's areas illustrate the importance of association cortex in integrating and transferring information to generate complex processes.

The representational hemisphere is more concerned with **visuospatial relationships,** such as recognition of faces and identification of forms, music, or drawing. Lesions here produce **agnosias,** the inability to recognize an object by a specific sensory modality even though the sensory modality itself is intact.

The corpus callosum and the anterior and midbrain commissures provide communication between corresponding areas of the two hemispheres. Lesions of the corpus callosum illustrate the specialization of the cerebral hemispheres in terms of both right versus left and language versus visuospatial function.

Learning and Memory

More is known about the psychology of learning and memory than is known about the corresponding neurophysiology. Lesion studies indicate that learning can occur at subcortical and spinal levels (e.g., biofeedback effects on the ANS) though advanced learning is primarily a cortical function. Learning involves structural changes in the cortex through the development of new synapses (long-term learning), and through increases in synaptic efficiency (short-term learning). **Habituation** and **sensitization** phenomena are also involved. Here, strength of response is modified. For example, in the snail *Aplysia,* biochemical modification of ion channels results in a "learned" withdrawal response. Memory is a complex process consisting of encoding, storage in short-term and then long-term memory, and retrieval. The hippocampus appears to be involved in **encoding** and the temporal lobes in **retrieval.** Removal of the hippocampal formation severely disrupts recent memory function so that no new long-term memory can be stored. Memory does not appear to be stored discretely in any single part of the brain. Current knowledge indicates that protein synthesis is somehow involved in the storage of memory, and there may be changes in the proportions of nucleotides in RNA during learning.

Questions

Directions: Select the one best answer for the following questions.

1. Which of the following is **true** regarding cerebrospinal fluid?
   A. It circulates in the epidural space
   B. It is an ultrafiltrate of plasma, with a higher protein level
   C. It is produced in the arachnoid granulations
   D. It is produced in the choroid plexuses
   E. Its production is inhibited by various antibiotics
2. Regarding the blood-brain barrier, it is **true** to say that
   A. It consists of the capillary endothelium and basement membrane of cerebral vessels
   B. It excludes lipid-soluble substances
   C. It facilitates systemic uptake and release of hormones
   D. It is not affected by blood pressure
   E. Its tight junctions do not allow water or oxygen to cross
3. Regarding cerebral blood flow, it is **true** to say that it
   A. Can be decreased somewhat over time to offset an increase in the intracranial pressure
   B. Changes over the range of normal blood pressure, depending on the activity level of the brain
   C. Does not change in hypertensive patients
   D. Is not affected by medications
   E. Is very sensitive to changes in PO_2
4. Which of the following is **true** regarding cutaneous sensory receptors?
   A. They can detect temperature changes of less than 0.5°C
   B. They include phasic but not tonic receptors
   C. They produce an electrical potential when vibratory stimuli are applied to Ruffini's corpuscles
   D. They send information via Ia and Ib nerve fibers
   E. They show no adaptation to stimuli over time
5. Regarding nociceptors, it is **true** to say that they
   A. Adapt rapidly to painful stimuli so that pain can be ignored
   B. Are encapsulated nerve endings found in high concentrations in the brain
   C. Can be activated by mechanical, thermal, or chemical stimuli in the skin, viscera, and deep somatic regions of the body
   D. Can exhibit denervation hypersensitivity when sympathetic discharge is enhanced
   E. Send information via B fibers

6. The process of reciprocal innervation begins with
   A. Contraction of ipsilateral extensors
   B. Relaxation of contralateral extensors
   C. Contraction of contralateral flexors
   D. Contraction of the ipsilateral flexors
7. Which of the following statements best describes the pyramidal system?
   A. It consists primarily of sensory fibers in the corticospinal and corticobulbar tracts
   B. Its fibers end primarily on upper motor neurons, allowing for a wide signal spread
   C. It initiates and directs voluntary movement of large muscle masses
   D. It includes the basal ganglia and the brainstem
   E. It is responsible for fine motor control, as in delicate hand movements and speech
8. Regarding the cerebellum, it is **true** to say that it
   A. Has few connections with the rest of the central nervous system
   B. Modulates motor activity to help maintain equilibrium and muscle tone
   C. Receives inhibitory input from mossy and climbing fibers
   D. Receives excitatory input from Purkinje cells
9. Synaptic transmission in the autonomic ganglia is usually
   A. Adrenergic
   B. Peptidergic
   C. Cholinergic
   D. Mediated by substance P
   E. Noradrenergic
10. Which of the following statements is **true** regarding the parasympathetic nervous system?
    A. Acetylcholinesterases are inhibited by some sympathomimetics
    B. Muscarinic side effects include tachycardia and bronchial dilatation
    C. The action of acetylcholine is terminated by its breakdown into acetate and choline
    D. Receptors are divided into C1 and C2 types
    E. Preganglionic fibers travel a short distance from their origin to synapse on postganglionic fibers
11. Norepinephrine is removed from the synaptic cleft by all of the following **except**
    A. Monoamine oxidase
    B. Cholinesterases
    C. Diffusion
    D. Active reuptake into the nerve endings
    E. Catechol O-methyl transferase
12. Visual accommodation is due to
    A. Relaxation of the iris sphincter muscle
    B. Increased tension of the lens ligaments
    C. A decrease in the curvature of the lens
    D. Contraction of the ciliary muscle
    E. Pupillary dilation
13. A 40-year-old woman presents to an ophthalmologist with the complaint of decreasing vision. She has long had problems with night blindness, but she now shows a concentric narrowing of her visual fields, with central vision preserved. On ophthalmoscopic examination, dark irregular patches are seen on her retina. Which of the following statements is **not true**?
    A. A likely diagnosis is retinitis pigmentosa
    B. Her children might be expected to inherit the disease

C. Pigments accumulating in the optic nerve cause blindness
D. Her problem is not likely to be caused by cataracts
E. The primary lesion is a degeneration of pigments found in the rods and cones, and the displacement of pigment into more superficial parts of the retina

14. A 68-year-old man has noticed a decreasing ability to hear normal conversation over the last few years. A diagnosis of conduction deafness due to otosclerosis is made based on the results of the Weber and Rinne tests. Conduction deafness
   A. Is the decreasing ability to hear high-frequency sounds with age
   B. Results from impaired sound transmission in the external or middle ear
   C. Results from damage to the hair cells or neural pathways
   D. Can be seen with prolonged exposure to loud noise, or aminoglycoside antibiotics
   E. Is present if the sound is louder in the normal ear on the Weber test
15. Stimulation of which of the following areas can cause sleep?
   A. Medullary reticular formation near the raphe nuclei
   B. Deep cerebellar nuclei
   C. Parietal cortex
   D. Amygdala
   E. Hypothalamus

Directions: Each group of items in this section consists of lettered options followed by a set of numbered items. For each item, select the one lettered option that is most closely associated with it. Each lettered option may be selected once, more than once, or not at all.

A. Otolithic membrane
B. Tectorial membrane
C. Black and white vision
D. Color vision
E. Cupula

16. Organ of Corti
17. Macula
18. Crista ampullaris
19. Rods
20. Cones

Answers

1. D Cerebrospinal fluid is produced in the choroid plexuses in the ependymal lining of the ventricles, and in the capillary endothelium of the brain. It is actively secreted, and is thus not an ultrafiltrate, and is nearly protein free. Cerebrospinal fluid is absorbed into the venous sinuses via the arachnoid granulations. Production can be inhibited by corticosteroids and diuretics.
2. A The blood-brain barrier (BBB) is composed of the capillary endothelium and basement membrane of the cerebral vasculature. Lipid-soluble substances cross the BBB more easily than most water-soluble substances. The BBB prevents systemic uptake and release of hormones. Rapid changes in blood pressure can have serious consequences by disrupting the BBB. Water and oxygen diffuse through the BBB with ease.
3. A Cerebral blood flow (CBF) may be able to offset changes in intracranial pressure if these changes occur over time. The CBF remains constant over the normal blood pressure range (autoregulation), but the curve may be shifted to the

right in hypertensive patients. Many drugs change cerebral blood flow including vasodilators and vasopressors. Cerebral blood flow is much more sensitive to changes in PCO_2 than PO_2.

4. A Cutaneous sensory receptors can detect small temperature changes, when the ambient temperature is in the comfort zone. Skin receptors include phasic (pacinian and Meissner's corpuscles) and tonic (Ruffini's corpuscles and Merkel's disks) receptors. When Ruffini's corpuscles detect pressure changes in the skin, an action potential is produced. Information from skin receptors travels via type II, III, and IV fibers; Ia and Ib fibers carry information from muscle spindles and Golgi tendon organs. All sensory fibers will show some adaptation to stimuli over time.

5. C Pain receptors in the skin, viscera, and deep somatic tissues can be stimulated by mechanical, thermal, or chemical stimuli. These receptors adapt extremely slowly to painful stimuli, so that painful signals will not be ignored. Nociceptors are free nerve endings and are not found in the brain. Denervation hypersensitivity, an increased sensitivity to circulating neurotransmitter, can result when normal synaptic input is removed from a neuron. Fast pain information is transmitted by A-delta fibers, and slow pain by C fibers.

6. D Reciprocal innervation follows contraction of the ipsilateral flexors, and consists of relaxation of the ipsilateral extensors, contraction of the contralateral extensors, and relaxation of the contralateral flexors. All of these actions help reinforce the basic ipsilateral flexor contraction.

7. E The pyramidal system is primarily motor, though some descending fibers from the postcentral sensory cortex are included. Fibers end mostly on interneurons, allowing for wide signal spread and thus fine motor control. The extrapyramidal system is responsible for the movement of large muscle masses and includes the basal ganglia and brainstem.

8. B The cerebellum has many connections with the spinal cord, reticular formation, vestibular nuclei, and cortex, which help it in its role to control motor activity. Input to the cerebellum is excitatory from mossy and climbing fibers, and inhibitory from Purkinje cells.

9. C Synaptic transmission is cholinergic.

10. C Acetylcholine is hydrolyzed to acetate and choline by acetylcholinesterases. Muscarinic side effects include bradycardia and bronchial constriction. The preganglionic fibers travel a long distance to synapse on parasympathetic ganglia, where receptors are nicotinic. The receptors on effector organs are muscarinic.

11. B Cholinesterases hydrolyze cholinesterase in the parasympathetic nervous system; all other choices are partially responsible for removal of norepinephrine from the synaptic cleft.

12. D Relaxation of the iris sphincter muscle (and thus pupillary dilation) is seen in the unaccommodated state, as are increased tension of the lens (zonular) ligaments and a decrease in the curvature of the lens. Accommodation results from contraction of the ciliary muscle, with decreased tension of the lens ligaments, pupillary constriction, and medial gaze convergence.

13. C Photosensitive compounds can accumulate in the retina in retinitis pigmentosa, a congenital defect in which defective pigment is not removed normally by phagocytosis. The disease can be inherited in recessive or dominant fashion and may be associated with neurologic abnormalities. The photosensitive compounds are stored in the membrane disks in the outer segments of the rods and cones; rhodopsin is stored in rods, and three separate color-specific pigments are seen in cones. The pigments decompose on exposure to light to produce changes in electrical potential. There are no pigments at the optic disks, as there are no receptor cells in this region (and therefore a corresponding blind spot in the visual field).

14. B Conduction deafness results from impaired sound transmission in the external or middle ear and can result from chronic otitis media, destruction or rigidity of the ossicles, or obstruction of the external ear canal. Presbycusis is the decreasing ability to hear high frequencies that occurs with age. Answers C, D, and E are true of nerve deafness.
15. A The midline raphe nuclei area of the medulla and the locus ceruleus area of the pons will both cause sleep when stimulated. The deep cerebellar nuclei send excitatory input to the spinal cord and thalamus. The parietal cortical region is sensory. The amygdala and hypothalamus are involved in the production of emotion and affective experiences.
16. B Hair cells of the organ of Corti are embedded in the tectorial membrane.
17. A Hair cells of the macula are embedded in the otolithic membrane.
18. E Hair cells of the crista are embedded in the cupula.
19. C Rods are responsible for black and white vision.
20. D Cones are responsible for color vision.

Bibliography

Berne, R. M., and Levy, M. N. (eds.). *Physiology* (3rd ed.). St. Louis: Mosby–Year Book, 1993.

Bullock, J., et al. (eds.). *Physiology* (2nd ed.). Baltimore: Williams & Wilkins, 1991.

Carpenter, M. B. *Neuroanatomy* (4th ed.). Baltimore: Williams & Wilkins, 1991.

Despopoulos, A., and Silbernagl, S. (eds.). *Color Atlas of Physiology* (4th ed.). New York: Thieme, 1990.

Ganong, W. F. *Review of Medical Physiology* (16th ed.). East Norwalk, CT: Appleton & Lange, 1993.

Johnson, L. R. (ed.). *Essential Medical Physiology.* New York: Raven, 1992.

Newfield, P., and Cottrell, J. E. (eds.). *Handbook of Neuroanesthesia: Clinical and Physiologic Essentials* (2nd ed.). Boston: Little, Brown, 1991.

Pansky, B., et al. (eds.). *Review of Neuroscience* (2nd ed.). New York: McGraw-Hill, 1988.

West, J. B. (ed.). *Physiological Basis of Medical Practice* (12th ed.). Baltimore: Williams & Wilkins, 1990.

Index

Note: Page numbers followed by *f* indicate figures; those followed by *t* indicate tables.